Printed Antennas

This collection covers different printed microstrip antenna designs from rectangular to circular, broadband, dual-band, and millimeter-wave microstrip antennas to microstrip arrays. It further presents a new analysis of the rectangular and circular microstrip antenna efficiency and surface wave phenomena.

The book

- Covers the latest advances and applications of microstrip antennas
- Discusses methods and techniques used for the enhancement of the performance parameters of the microstrip antenna
- Presents low-power wide area network (LPWAN) proximity-coupled antenna for Internet of Things applications
- Highlights a new analysis of rectangular and circular microstrip antenna efficiency and surface wave phenomena
- Showcases implantable antennas, H-shaped antennas, and wideband implantable antennas for biomedical applications

Printed Antennas discusses the latest advances such as the Internet of Things for antenna applications, device-to-device communication, satellite communication, and wearable textile antenna in the field of communication. It further presents methods and techniques used for the enhancement of the performance parameters of the microstrip antenna and covers the design of conformal and miniaturized antenna structures for various applications. It will serve as an ideal reference text for senior undergraduates, graduate students, and researchers in fields including electrical engineering, electronics and communications engineering, and computer engineering.

Printed Antennas

Design and Challenges

Edited by

Praveen Kumar Malik
Arshi Naim
Ramendra Singh

CRC Press
Taylor & Francis Group
Boca Raton London New York

CRC Press is an imprint of the
Taylor & Francis Group, an **informa** business

First edition published 2023
by CRC Press
6000 Broken Sound Parkway NW, Suite 300, Boca Raton, FL 33487-2742

and by CRC Press
4 Park Square, Milton Park, Abingdon, Oxon, OX14 4RN

CRC Press is an imprint of Taylor & Francis Group, LLC

Library of Congress Cataloging-in-Publication Data
Names: Malik, Praveen Kumar, editor. | Naim, Arshi, 1976- editor. |
Singh, Ramendra (Associate Professor), editor.
Title: Printed antenna : design and challenges /
edited by Praveen Kumar Malik, Arshi Naim, and Ramendra Singh.
Description: First edition. | Boca Raton : CRC Press, 2023. |
Includes bibliographical references and index.
Identifiers: LCCN 2022027065 | ISBN 9781032365558 (hardback) |
ISBN 9781032388380 (paperback) | ISBN 9781003347057 (ebook)
Subjects: LCSH: Microstrip antennas–Design and construction.
Classification: LCC TK7871.67.M5 P74 2023 |
DDC 621.3841/35—dc23/eng/20221018
LC record available at https://lccn.loc.gov/2022027065

ISBN: 978-1-032-36555-8 (hbk)
ISBN: 978-1-032-38838-0 (pbk)
ISBN: 978-1-003-34705-7 (ebk)

DOI: 10.1201/9781003347057

Typeset in Sabon
by codeMantra

Contents

Preface

Printed antennas, which are also known as microstrip antennas, have a number of advantageous properties like mechanical durability, conformability, compactness, and low cost of manufacturing and have become an integral part of next-generation wireless communications and are extensively used to improve system capacity, data rate, reliability, etc. These antennas have wide applications in the military as well as commercial sectors. Printed antennas are often mounted on the exterior of aircraft and spacecraft, and these can be used as mobile radio communication devices. Medical applications of printed antennas include microwave imaging, medical implants, hyperthermia treatments, and wireless wellness monitoring.

As most of us are aware, printed antennas are now used almost everywhere, including on our mobile phones, satellites, etc. Moreover, they are also employed in biomedical applications, wearable sensors, vehicular communication, and IoT-enabled devices. With all these applications in mind, the book would aim at taking all these implementations into account.

This book broadly covers the Internet of Things-enabled printed antenna which is further extended toward the Wideband Wearable Antenna for IoT and Medical Applications. Further, broadband-printed antennas for wireless applications as well as multiband MIMO antennas for 5G applications are elaborated. Also, the effects of metamaterial on bio-inspired microstrip patch antenna, design, and study of compact bio-inspired-shaped smart MIMO array antenna for 5G-enabled healthcare systems are discussed at length. It also covers the IoT systems, environmental care systems, etc.

Editors

Dr. Praveen Malik is a Professor in the School of Electronics and Electrical Engineering, Lovely Professional University, Phagwara, Punjab, India. He received his PhD with a specialization in Wireless Communication and Antenna Design. He has authored or co-authored more than 50 technical research papers published in leading journals and conferences from the IEEE, Elsevier, Springer, Wiley, etc. Some of his research findings are published in top-cited journals. He has also published 10 edited/authored books with international publishers. He has guided many students leading to ME/MTech and students leading to PhD. He is an Associate Editor of different journals. His current interests include Microstrip Antenna Design, MIMO, Vehicular Communication, and IoT. He was invited as Guest Editor/Editorial Board Member of many international journals, invited as keynote speaker in many international conferences held in Asia, and invited as Program Chair, Publications Chair, Publicity Chair, and Session Chair in many international conferences. He has been granted two design patents and a few are in the pipeline.

Arshi Naim works in the Department of Information Systems, College of Computer Science, King Khalid University, which is an NCAAA-accredited university. He is a pioneer member who was introduced to the concept of the Quality and Accreditation process at the university and is currently a member of the Academic Development and Quality Committee. He has been involved in preparations of a self-study report for the program, evaluating course documents, and other accreditation-related documents. He has developed Business Information Systems Modules, such as E-Commerce, Accounting Information Systems, Information Systems Strategic Management, Marketing Strategies, etc. and has expertise in writing course specifications, course reports, and course learning outcomes. He has been working on developing study plans and curriculum development for undergraduate and postgraduate programs. He is an E-learning expert and has received many excellence awards in developing online courses from Quality Matters' US and is an official peer reviewer for online courses from KKU E-Learning

Deanship. He has received an International Academic Award from QM for the course E-Commerce conducted by King Khalid University and a Certificate of Appreciation for successful online instructor for conducting blended courses from 2014 to 2016 in E-Learning.

Dr. Ramendra Singh is currently working as an Associate Professor in the Department of Electronics and Communication Engineering with Inderprastha Engineering College, affiliated to Abdul Kalam Technical University, Uttar Pradesh, formerly known as UPTU.

He received his BE in Electronics from RGTU, Bhopal, in 2001; MTech in Digital Electronics and Systems from KNIT, Sultanpur, Lucknow, and PhD from Dr. RML Avadh University, Faizabad, in 2016. He has more than 18 years of teaching experience and 2 years of industrial experience to his credit. His research interests are in the areas of electronic system design, digital and analog system design, Internet of Things, etc. He has published a number of research papers with journals of repute as well as national and international conferences. He has guided many students leading to ME/MTech and guiding students leading to PhD.

He has been a resource person for various workshops and faculty development programs. Moreover, he has a keen interest in research-related activities leading to product design. He has published two patents and a few more are in pipeline. He has chaired several national and international conferences and is a life member of various professional societies.

List of Contributors

Tathababu Addepalli
Jawaharlal Nehru Technological University
Anantapur, India

Sumit Agarwal
Pennsylvania State University
State College, PA

Rayaluru Akshay
National Institute of Technology
Warangal, India

M. Anas
National Institute of Technology
Warangal, India

V.R. Anitha
Sri Vidyanikethan Engineering College
Tirupati, India

T. Balakumaran
Coimbatore Institute of Technology
Coimbatore, India

Sneha Bhardwaj
Lovely Professional University
Phagwara, India

Gyoo-Soo Chae
Baekseok University
Cheonan, South Korea

Pradeep Chindhi
Sant Gajanan Maharaj College of Engineering, Mahagaon
Kolhapur, India

John Colaco
Goa College of Engineering (Goa University/Govt. of Goa)
Farmagudi, India

Yash Deshmukh
National Institute of Technology
Warangal, India

Archana Deshpande
Thakur College of Engineering and Technology
Mumbai, India

Sangeeta Garg
Mewar Institute of Management
Ghaziabad (UP), India

R. Gowrishankar
Department of ECE
KIT-Kalaignarkarunanidhi Institute of Technology
Coimbatore, India

Geeta Kalkhambkar
Sant Gajanan Maharaj College of Engineering
Mahagaon, India

Pradeep Kamal
National Institute of Technology
Warangal, India

Nehru Kandasamy
National University of Singapore
Singapore

Sandeep Singh Kang
Chandigarh University
Mohali, India

Osamah Ibrahim Khalaf
Al-Nahrain Nanorenewable Energy Research Center
Baghdad, Iraq

Kourike Sai Kiran
National Institute of Technology
Warangal, India

Amarjit Kumar
National Institute of Technology
Warangal, India

Nanda Kumar, M.
Sreenidhi Institute of Science and Technology
Hyderabad, India

Roshan Kumar
Henan University
Henan, China

Shalini Kumari
Chandigarh University
Ajitgarh, India

Dac-Nhuong Le
Haiphong University
Haiphong, Vietnam

Rajesh B. Lohani
Goa College of Engineering (Goa University/Govt. of Goa)
Farmagudi, India

Shaktijeet Mahapatra
ITER, Siksha 'O' Anusandhan (Deemed to be University)
Bhubaneswar, India

Praveen Kr. Malik
Lovely Professional University
Phagwara, India

Mihir Narayan Mohanty
ITER, Siksha 'O' Anusandhan (Deemed to be University)
Bhubaneswar, India

Mehaboob Mujawar
Annamalai University
Chidambaram, India

Anil Kumar Nayak
University of Alberta
Edmonton, Canada

T.M. Neebha
Karunya Institute of Technology and Science
Coimbatore, India

Satheeshkumar Palanisamy
Coimbatore Institute of Technology
Coimbatore, India

T. Prabhu
SNS College of Technology
Coimbatore, India

Subuh Pramono
Universitas Sebelas Maret
Surakarta, Indonesia

D. Prasad
Sasi Institute of Technology and Engineering
Tadepalligudam, India

Rajani, H.P.
KLE Society's Dr. M. S. Sheshgiri College of Engineering and Technology
Belagavi, India

Mohammad Hayath Rajvee
Visvodaya Engineering College
Kavali, India

Ramesh, C.
KIT-Kalaignarkarunanidhi Institute of Technology
Coimbatore, India

Rashmi Roges
Lovely Professional University
Phagwara, India

S.D. Ruikar
Walchand College of Engineering
Sangli, India

Shrenik Suresh Sarade
Walchand College of Engineering
Sangli, India

P.R. Satarkar
Goa College of Engineering (Goa University/Govt. of Goa)
Farmagudi, India

Sarmistha Satrusallya
ITER, Siksha 'O' Anusandhan (Deemed to be University)
Bhubaneswar, India

T. Shanmuganantham
Pondicherry University
Pondicherry, India

B.K. Sharma
National Institute of Technology
Warangal, India
and
Planar Microwave Technologies Ltd.
Derby, United Kingdom

V. Sharma
Gurukul Kangri (Deemed to be University)
Haridwar, India

Shailendra P. Shastri
Thakur College of Engineering and Technology
Mumbai, India

Anshika Shrivastav
National Institute of Technology
Warangal, India

Kiran Deep Singh
Chandigarh University
Mohali, India

Madan Singh
National University of Lesotho
Roma, Lesotho

Takialddin Al Smadi
Jerash University
Jerash, Jordan

G. Srihari
Sree Vidyanikethan Engineering College
Tirupati, India

E. Suganya
Sri Eshwar College of Engineering
Coimbatore, India

Penchala Reddy Sura
Visvodaya Engineering college
Kavali, India

Tamirat Tagesse
College of Engineering
Wachemo University
Hosaena, Ethiopia

Udayakumar, E.
KIT-Kalaignarkarunanidhi Institute of Technology
Coimbatore, India

Patri Upender
National Institute of Technology
Warangal, India

A. Varshney
Gurukul Kangri (Deemed to be University)
Haridwar, India

B. Yakub
National Institute of Technology
Warangal, India

K. Yogaprasad
Rayalaseema University
Kurnool, India

Yogeshwaran, K.
KIT-Kalaignarkarunanidhi Institute of Technology
Coimbatore, India

Chapter 1

Introduction to Internet of Things-enabled printed antenna

Sneha Bhardwaj, Rashmi Roges, and Praveen Kumar Malik
Lovely Professional University

Sumit Agarwal
Pennsylvania State University

CONTENTS

1.1 INTRODUCTION

The term 'Internet of Things' was coined by Kevin Ashton in 1999. IoT aims at making the devices smarter so that manual efforts in various fields can be minimized to a great extent. IoT now finds its way into almost all fields including military, agriculture, medicine, smart and intelligent systems like smart cities, smart homes, smart devices, smart grids, and so on. The technological stack of evolution of IoT technology is summerised in Figure 1.1. IoT technology is expected to connect more than 60 billion things by 2025. A simplified process cycle of IoT includes sensors, actuators, or a network of sensors and actuators senses then collects information. The collected data are aggregated and communicated to collection points or central nodes [1,2]. The data are then analyzed as a combined function from various sources,

DOI: 10.1201/9781003347057-1

Figure 1.1 The technological stack in the evolution of IoT technology.

thereby establishing an artificial intelligence on the inferences and conclusions to together make our system a smart module.

Figure 1.2 is a pictorial representation of the process cycle of a smart IoT system. The major areas of application of IoT are shown in Figure 1.3, and Figure 1.4 represents the major challenges faced while implanting an IoT system.

A powerful and smart connecting module forms the backbone of IoT systems. The sensors have to communicate the collected data. These data have to be analyzed, studied, and manipulated, and this has to be then communicated with the end users or the service providers. Wired and wireless means of communication are widely used depending on the type of application. Various factors like the range of connectivity required, power limitations,

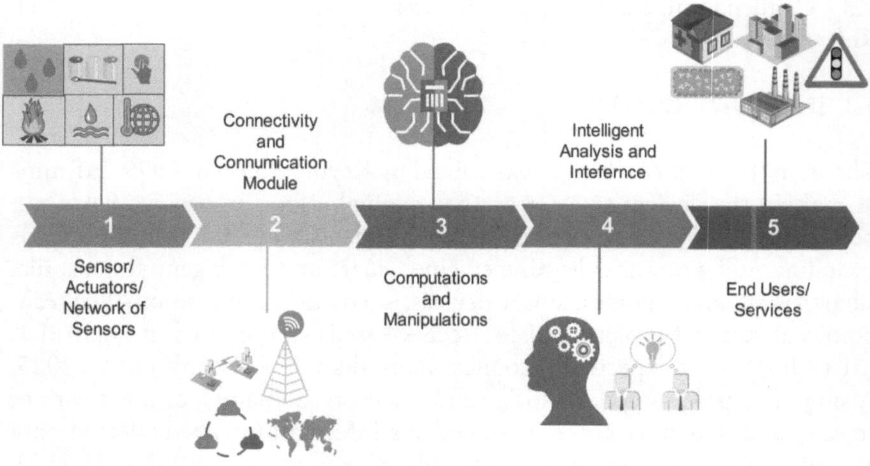

Figure 1.2 The process cycle of an IoT system.

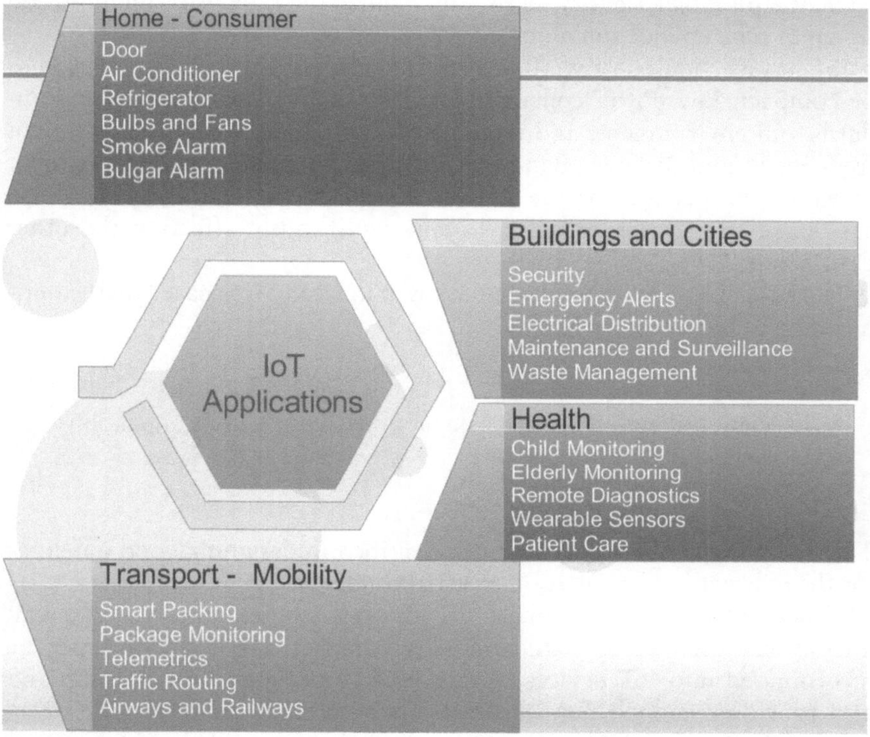

Figure 1.3 Major application fields of IoT.

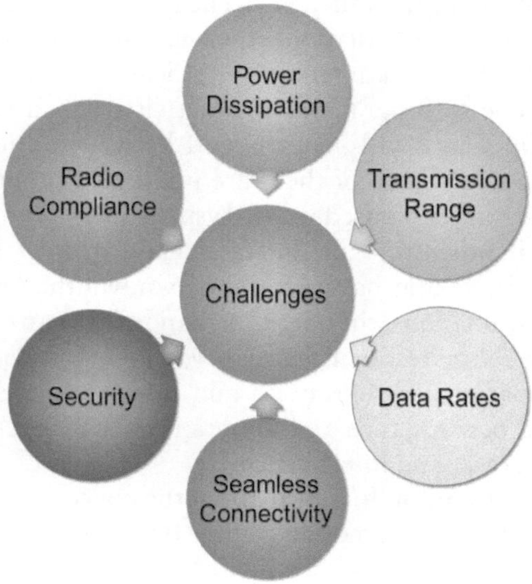

Figure 1.4 Major challenges faced in the implementation of an IoT system.

latency affordable, security levels, etc. finalize the type of communication set up in each application module [3,4].

IoT has its clutch on a wide variety of fields, but the IoT devices need to be compact, low power consuming, robust, compatible with other standards and environment, and reliable. IoT will employ different resonant frequencies and different bandwidths as per the application requirement in either the licensed or the unlicensed spectrum, and array antennas can make sure of better gain without significantly affecting the other parameters.

The base platform for implementation of IoT and 5G is based on the optimization of technologies like

- Wireless access systems
- Efficient utilization of the frequency spectrum and its allocation
- Lower power consumption
- Better antennas

The following are the research gaps identified in designing good antennas for the efficient implementation of IoT technology.

Miniaturization of Antennas: IoT devices require antennas of small dimensions and narrow bandwidth. A compact antenna can easily be incorporated into IoT devices, saving area of occupancy in the chip and also providing enough physical isolation from other modules. The interferences from the neighboring modules can be so avoided, and expected performance can be obtained from the antenna. The smaller the antenna, the better it will be [5].

Maintenance of Radiation Efficiency: The miniaturization of an antenna will greatly influence the radiation efficiency and quality factor of the antenna. While designing an antenna, the efficiency of the antenna is seen to be decreasing due to finite metal conductivity, which is an after-effect of miniaturization. There will be a tradeoff between miniaturization and radiation efficiency, and both of them are important parameters for good performance of any antenna used in IoT systems [6,7].

Ultra-Narrow Bandwidth: The antennas employed for IoT systems are usually highly sensitive and have a narrow bandwidth. As they are ultra-narrow bandwidth antennas, the operating bandwidth can easily be affected by a slight change in the resonant frequency, which, in turn, can be caused by any external parameters like noise and interference. An IoT antenna should offer good performance and maintain its attributes irrespective of its surroundings, noise, and interference.

Low Power Consumption: IoT is all about the collection, processing, and distribution of data in large amounts. These data are mostly sensed or collected by sensor networks, which itself works with energy constraints. The sensor nodes are to communicate between themselves and their hierarchical superiors through communication networks, which are mostly wireless. Antennas designed for this purpose need to be energy efficient and of the energy-harvesting type to facilitate a longer lifetime of the sensor nodes.

This can be summarized as follows:

- A highly effective, low-cost, compact, low power consumption and compatible antenna system is yet to be designed to exploit more IoT.
- Long-range low-power sensors and antennas are to be designed.
- Antenna performance parameters (gain, radiation pattern, efficiency) need to be further optimized to meet the requirements of IoT-enabled devices.
- The receiver sensitivity is another area of concern as it's affected by the coupling of other signals.
- A proper isolation technique or an antenna with good isolation characteristics would drastically improve the performance of the system.
- Compatibility and interoperability of standards is a key component as IoT is all about connecting everything irrespective of the standard and technology employed in the devices.

Building up an effective and efficient antenna system for an IoT application will require enhanced multiplexing techniques, proper frequency spectrum allocation, and scheduling and interference mitigation to work hand in hand [8,9]. The following are the challenges faced while designing an efficient IoT antenna:

- Highly compact and good performance efficiency are the key demands of IoT antennas as more and more technologies are incorporated into a single system.
- Obtaining isolation between multiple antennas when they are literally placed very close to each other inside a PCB system.
- Choosing the right antenna for the correct application from thousands of available designs.
- Optimization of transmission techniques employed in MIMO antenna systems.
- Antenna contribution toward lower power consumption and better data rates while attaining good transmission range.
- Providing seamless and secure connectivity while sticking to the radio spectrum compliance.

The realization of IoT networks will involve a heterogeneous mix of wireless technologies. The range that needs to be covered is different for different applications. Wi-Fi, Zigbee, Z-Wave, and Bluetooth are employed when the coverage is within 10–100 m, whereas LTE-M, NB-IoT, SigFox, and LoRa are opted to cover a larger area (1–10 km) (Figure 1.5).

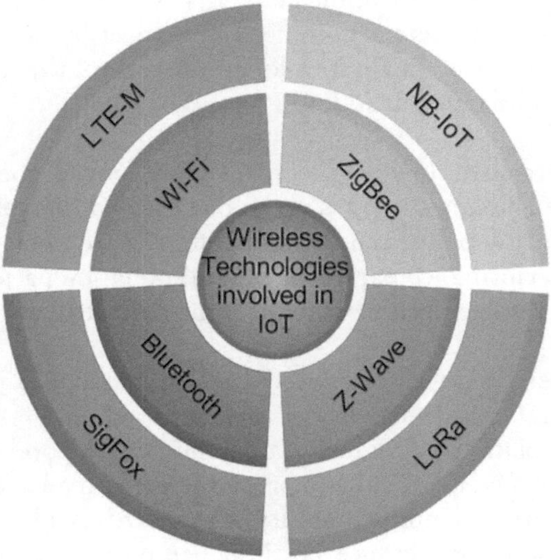

Figure 1.5 Various wireless technologies included in the implementation of an IoT system.

1.2 ANTENNA USED IN IoT

The divergent environment of the Internet of Things (IoT) is combined with the number of end devices such as radio modules, sensors, MEMS devices, batteries, energy-harvesting techniques, and antennas. Antenna plays a major and complex role in the dynamics of IoT. IoT needs to support a low-latency network that can incorporate various other networks for better connectivity. So, while choosing the antenna for any other applications, we have to look for some parameters such as shape, size, mounting option, operating frequency, coverage requirement, directivity, cost, and gain of the antenna. The shape and size of the antenna depend on where we need to put our antenna and in how much space we have to accommodate the antenna. On that basis, we have to consider the dimension of the antenna without affecting the other parameters. We also need to look at the mounting option of the antenna for the designing purpose, as for RF connector mount, we can choose a non-planar antenna, but for PCB mounting, it's better to choose a planar patch or printed antenna. Frequency of operation is always a top priority while choosing an antenna as it will be directly related to the frequency of application for which the antenna will work efficiently [10]. The IoT network smartly transmits/receives small chunks of data in an IoT network in various network topologies like a star, mesh, and/or point-to-point. Most of the time, the job of establishing such types of links is accomplished by some common and relatively less complex types of omnidirectional antennas such as wire antennas, rubber duck, patch, whip, PCB, and on-chip antenna structures. Several IoT development kits and platforms such as Arduino GSM and Qualcomm IoE make use of similar antenna structures for GPS, Bluetooth, and WiFi communication. Some applications such as Medical Body Area Networks (MBAN) and wearable electronics also require antennas to be low profile and conformal to the surface of the device. Range and directivity are also very important parameters when selecting the types of antennas for IoT applications. Directivity explains the concentration of beam radiation in a particular direction and omnidirectional antenna radiated somewhat evenly in all directions. It is related to the outdoor and indoor application of the antenna as we can say that for indoor application, we mostly require an omnidirectional antenna, and for outdoor application, a directional antenna will work better with a good coverage range. IoT applications are widely used for domestic and industrial markets for different applications such as computations, sensing as well as in creating a data-connected environment. There are various types of printed antennas used in IoT-enabled devices as per requirement.

1.2.1 Types of printed antenna

1.2.1.1 Monopole antenna

As mentioned in the literature, the monopole antenna is a microstrip line of quarter wavelength and its width is defined by using some basic convention formula. Various modifications are made to improve the characteristics of the antenna as per the application which leads to changes in its conventional design to new designs such as spiral and many more.

1.2.1.2 Dipole antenna

A dipole antenna is a basic type of conducting element whose length is the half-length of the maximum wavelength that needs to be transmitted from the antenna. To make it dipole, it is cut in the middle and fed with a coaxial cable. In the planar antenna, the dipole design has two-quarter wavelength straight microstrip lines and the antenna is fed in the mid of these two lines and the combined effective length of this antenna is $\lambda/2$.

1.2.1.3 Loop antenna

Loop antenna is designed by making loops in any shape, and it is cheap and simple in design. Its gain and radiation are also good, and it is widely used as an antenna in the communication field. The point which needs to be considered is the radius of the loop which should be less than the resonating wavelength.

1.2.1.4 Vivaldi antenna

Vivaldi antenna can be implemented by several designs as mentioned in the literature. It consists of mainly two symmetric conductors on the substrate.

Both conductors may lie on the same plane or the other plane, but they behave as a single element and help in improving the gain and efficiency of the planar antenna. As mentioned in this literature, the first element of the Vivaldi antenna is designed on top and the other one is designed on the bottom of the substrate. This antenna can be fed easily by soldering the connector to the two sides of the PCB material. The top side of the substrate acts as a conductor, and the bottom side of the substrate acts as the ground of the antenna.

1.2.1.5 Fractal antenna

The fractal antenna is derived from an iterative method. It can be of any shape like a circle, rectangle, square, leaf, etc. The antenna design can be a monopole, dipole, patch, array helical, etc. These techniques help in improving the bandwidth of the antenna, increasing the resonance of the antenna, and making the antenna resonate at multiple bands. These antennas are small in size and cheaper than the conventional antennas. The result is one fractal antenna that can replace many traditional antennas.

1.2.1.6 Planar inverted F antenna

Planar inverted F antenna (PIFA) is more in demand these days because of its low profile and radiation characteristics. Most handheld devices consist of PIFA. Its basic structure is the same as the patch antenna except for the shorting pin and its inverted F patch design. As PIFA has a low profile, it is commonly used. The shoring pin and feed location play important roles in designing the PIFA. The impedance is decreased toward the shorting pin location and increases when we move away from the pin location. This concept is used in tuning the PIFA elements.

1.2.1.7 Array antenna

It is a group of similar radiating elements arranged in a regular structure to form a single antenna that can produce radiation patterns. These patterns of antenna help in improving the directivity and gain of the antenna drastically. These types of antennas are widely used in a high-gain application at higher frequencies mostly.

1.2.2 Different types of techniques to improve the performance of antenna

Many kinds of literature mention various techniques that help to improve antenna performance. The performance of the antenna is improved by enhancing one or more characteristics of an antenna as gain, bandwidth, compact size, reducing coupling, and radiation efficiency.

- **Dielectric Resonator (DRA)**
  A dielectric resonator antenna (DRA) consists of a block of ceramic material of various shapes, the dielectric resonator, mounted on a metal surface, a ground plane. As the frequency increases, conductor losses increase, and antenna efficiency will decrease. Conversely, only losses present in the DRA are due to imperfections in the dielectrics which are very small compared to conductor losses. So, it provides high gain and efficiency of the antenna. DRA offers a low-cost, small size, and low losses antenna. So, it is very much useful at a higher frequency where the conductor losses become severe.
- **Substrate Selection**
  The substrate provides a mechanical support to the planar antenna on which the radiating element and ground are fabricated, but it also plays a very important role in the gain and efficiency of the antenna. Different substrates have different dielectric constants and loss tangents; therefore, by changing the substrate, the thermal, mechanical, and electrical properties of the antenna get changed. The substrate directly affects the size of the antenna, resonance frequency, bandwidth, gain directivity, and efficiency. For better gain, we choose a substrate with low-permittivity and low-loss tangent. A substrate with high dielectric constant tends to decrease the size of the antenna at the cost of antenna performance, as the dielectric losses increase.
- **Electromagnetic Bandgap Structure (EBG)**
  It is designed with a dielectric or metallic structure which is arranged periodically on the substrate. The dielectric properties of the substrate degrades the gain and radiation characteristics of the antenna, so the EBG structures designed on the substrate help in improving the performance of the antenna. Although the designing of the EBG structure is a complex task as its periodicity should be a half-wavelength at the center frequency. This structure behaves as a path for electromagnetic waves that helps in improving the efficiency of the antenna by decreasing the losses. EBG affects the size of the antenna, bandwidth, and losses in the antenna.
- **Slot Technique**
  This technique is done by cutting a slot from the radiating element of the antenna that tends to increase the bandwidth of the antenna and adds a new resonance frequency to the antenna. The slots behave as an antenna when a high-frequency field is present across the slots on the metallic plane, and the dimension of the slots defines the resonating frequency. The ground and patch coupled with each other by this method help in improving the gain and efficiency of the antenna. When the slots are designed on the ground plane, then the gain of the antenna increases, but when we design slots on the patch, the overall gain decreases as the radiating area of the antenna decreases. The slot's width and its distance from the radiating edges affect the

resonance of the antenna. During slot designing, we must consider the surface current and impedance matching of the patch element.

- **Corrugation**
  The corrugation process involves the etching of edges from the radiating metallic patch. By corrugation, irregularity is introduced at simple patch edges or we can say it is designed into different types of edges by introducing zigzag cuts at the edges, by rounding off the edges and by introducing rectangular regular cuts on edges. Because of these changes, the surface current follows a different path than the conventional patch antenna. These corrugation techniques help in improving the bandwidth, resonance, and size of the antenna.
- **Defected Ground Structure (DGS)**
  In this technique, the ground of the planar antenna is cut in a proper manner or slots are introduced in the ground that help in improving the efficiency and bandwidth of the antenna. DGS has been used in the field of microstrip antennas for enhancing the bandwidth and gain of microstrip antenna and to suppress the higher mode harmonics, mutual coupling between adjacent elements, and cross-polarization for improving the radiation characteristics of the microstrip antenna. The periodic defected ground structures help in the miniaturization of the antenna. DGS also helps in improving the cross-polarization of the antenna leading to improving the radiation efficiency of the antenna.
- **Multielement Radiator**
  Multielement antenna consists of multiple antennas that work as a single antenna helping in improving the gain and coverage of the antenna. Its working principle is the same as an array antenna. Multielement antenna consists of two or more techniques to enhance the performance of the antenna effectively. There are many applications where the single element doesn't fulfill the requirement, so we add multiple elements to reach the requirement.
- **Metamaterial Concept**
  Metamaterial concept includes the arrangement of periodic and aperiodic structures with a spacing between them comparable with wavelength.

  The arrangement of these structures tends to bend electromagnetic waves, which aids in the behavior of larger antennas. The overall geometry causes the material to behave as a substrate of negative permittivity and permeability, and because such a material does not exist in nature, it is referred to as a metamaterial. This technique helps in improving the gain, bandwidth, coupling loss, and overall size of the antenna.
- **Frequency Reconfigurable technique**
  This technique helps in improving the mutual coupling loss and bandwidth of the antenna. These antennas can change their frequency and

Table 1.1 The pros and cons of different performance enhancement techniques

Antenna types	*Advantages*	*Disadvantages*
Dielectric resonator antenna	Enhance the bandwidth Improves efficiency Improves gain	Design is complex
Substrate selection	Low permittivity, enhances the gain, efficiency, wide bandwidth High permittivity reduces the size of the antenna	Low-permittivity materials are costly and rarely available High-permittivity material tends to low gain
Energy band gap (EBG)	Improves impedance matching and front to back ratio Improves gain Helps in size reduction	Design is complex
Slot technique	Improves the gain, bandwidth, and efficiency of the antenna.	Difficult to design with impedance matching.
Corrugation	Improves the gain, efficiency, and bandwidth of the antenna	Impedance matching is a bit complex
Defected ground structure	Easy to implement, improve bandwidth and efficiency	Position of defects is a complex analysis
Multielement radiator	Improves the gain, bandwidth, and efficiency of the antenna.	Feeding location is difficult to design, and the size of the antenna increases.
Metamaterial concept	Improves gain, bandwidth and efficiency, compact size	Difficult to design
Frequency reconfigurable technique	Size reduction Improves gain, bandwidth and efficiency	Required electronics component and circuit

radiation properties dynamically, with the help of varactor diode, MEMS switches and pin are used. These reconfigurable antennas have less mutual coupling and show high diversity gain and efficiency (Tables 1.1 and 1.2).

1.3 CHALLENGES IN DESIGNING IoT ANTENNA

In IoT implementation on a large scale as in smart cities, we are facing a lot of challenges such as security, privacy, latency, connectivity, and many more.

1. Energy Consumption
   It is our top priority to make the device less power-consuming so it will work for a long time without interrupting the services.

Table 1.2 The pros and cons of different planar antenna technologies

Antenna types	*Advantages*	*Disadvantages*
Monopole	Simple to design and fabricate Easy integration	Less gain Require large area ground
Dipole	Simple to design and fabricate Easy integration Low cross-polarization	Less gain Less range Low BW
Loop	Easy to design Good bandwidth	Low gain
Vivaldi	Enhance gain Good bandwidth	Need more space Low gain at L band
Fractal	Compact size Wider bandwidth Impedance matching is good	Design is complex
PIFA	Low profile Enhance front to back ratio	Low gain
Array	High gain Good bandwidth	Complex design Coupling loss

2. Latency
   Latency is a time when data are sent from a connected device to a server and when they return to the same device as the response. In the case of IoT vehicles, industry high latency may lead to a collision, or some unlikely incident may happen due to delay. It makes the system difficult to get real-time data.
3. Massive Connectivity
   With such a large number of devices, connectivity with various devices with different provider systems is a challenge as our devices are always not acceptable and even the entire system is always not compatible with each device and system.
4. Security, Privacy, and Trust
   Because data are transmitted through a server that is connected to thousands of devices, security and privacy are always at a risk from intruders who can manipulate the data causing devices to crash and hackers who can access personal data and use that for their benefit. Trust is also a concern as the communication should be done with only trusted devices, not with other devices.
5. Identification and Authentication
   In IoT, the identification of each device should be done properly so that unwanted devices cannot connect with the IoT server. The authentication process also helps to save the data from unauthorized attacks.
6. Error Correction
   In IoT technology, data are transmitted over the channel for long-distance communication. They may be corrupted or lost due to fading

and interference. The existing error correction schemes of IoT networking, like hamming code, cannot help in data corruption or loss efficiently.

REFERENCES

1. *Planar Antenna: Design, Fabrication, Testing, and Application*, Nova Science Publishers Inc, New York, Oct 15, 2021, ISBN: 9781536198980.
2. *Smart Antennas: Latest Trends in Design and Application*, Malik, P., Lu, J., Madhav, B.T.P., Kalkhambkar, G., Amit, S. (Eds.), Springer, 2021. ISBN: 9783030766368, DOI: 10.1007/9783030766368.
3. *Microstrip Antenna Design for Wireless Applications*, Malik, P.K., Padmanaban, S., Holm-Nielsen, J.B., Taylor and Francis, Aug 04, 2021, ISBN: 9780367554385.
4. *Smart Antennas: Recent Trends in Design and Applications*, Malik, P.K., Kumar, P., Kumar, S., Singh, D.K., Bentham Science, Sharjah, Aug 2021, ISSN: 2717-5421 (Print), ISSN: 2717-543X (Online), ISBN: 9781681088600.
5. Malik, P.K., Chapter 4. Mathematical Modeling and Principle of Wireless Communication, *Energy Harvesting Technologies for Powering WPAN and IoT Devices for Industry 4.0 Up-Gradation*, Nova Science Publishers, Inc., Hauppauge, NY, Apr 2020, ISBN: 9781536169430.
6. Roges, R., Malik, P.K. Planar and printed antennas for Internet of Things-enabled environment: Opportunities and challenges. *Int J Commun Syst.* 2021; 34(15): e4940. https://doi.org/10.1002/dac.4940, ISSN: 1099-1131.
7. *Analysis and Design of Fractal Antenna for Efficient Communication Network in Vehicular Model, Sustainable Computing: Informatics and Systems*, Rahim, A., Malik, P.K., Elsevier, Vol 31, 2021, 100586, ISSN: 2210-5379, https://doi.org/10.1016/j.suscom.2021.100586.
8. Shaik, N., Malik, P.K. A comprehensive survey 5G wireless communication systems: Open issues, research challenges, channel estimation, multi carrier modulation and 5G applications. *Multimed Tools Appl.* 2021. https://doi.org/10.1007/s11042-021-11128–z.
9. Tiwari, P., Malik, P.K. Wide band micro-strip antenna design for higher "X" band. *Int J e-Collaboration (IJeC).* 2021; 17(4): 60–74. http://doi.org/10.4018/IJeC.2021100105, ISSN: 1548-3673.
10. Wadhwa, D.S., Malik, P.K., Khinda, J.S. High gain antenna for n260- & n261-bands and augmentation in bandwidth for mm-wave range by patch current diversions. *World J Eng.* 2021. https://doi.org/10.1108/WJE-03-2021-0133, ISSN: 1708-5284.

Chapter 2

Design and analysis of different rectangular-shaped four-element wideband multi-band MIMO antenna with enhancement of correlation coefficient

Shrenik Suresh Sarade[1*], *S. D. Ruikar*[2]

[1*]shreniks2k7@rediffmail.com

[1,2]Walchand College of Engineering, Sangli, Maharashtra, India

CONTENTS

2.1 INTRODUCTION: BACKGROUND AND DRIVING FORCES

In recent years, wireless technology is rapidly developed with increasing tremendously of its users. Therefore, there are high demands of data rate and reliability to handle the growth of users. MIMO antenna system has attracted the attention of researchers for the high data rate, channel capacity and reliability. Meanwhile, the today's wireless systems require the more antennas, which operate on multiple bands with wideband. In MIMO antenna, the multiple antennas at transmitter and receiver is present and placed close to each other. Hence, isolation and correlation

DOI: 10.1201/9781003347057-2

coefficients is become high. Therefore, enhancement of isolation and correlation parameter of MIMO antenna is critical. Different techniques are used for the improvement of isolation of the multiple-input multiple-output (MIMO) antenna. The decoupling structure is used for the enhancement of isolation [1–12]. In Refs. [13–18], a different-shaped defected ground structure (DGS) is used on the ground to change the distribution of the inductance, capacitance, and effective dielectric constant of the microstrip line to reduce mutual coupling. In Ref. [19], The electromagnetic bandgap (EBG) was designed for improvement of isolation. The mutual coupling is enhanced using the metamaterial [20] and the neutralization line [21,22]. The mutual coupling is reduced using the parasitic elements between the radiating elements [23,24] and using the traditional methods [25].

In this chapter, the compact MIMO antenna is investigated for various wireless applications such as Bluetooth, Wi-Fi, IoT, 5G band n77, Wi-MAX, WLAN, and C-band for satellite communication and X-band for radar communication. This antenna consists of four different rectangular-shaped radiating elements having dimensions designed for different frequencies. Out of that patch-1, patch-2, patch-3, and patch-4 are designed for 6, 3.45, and 2.54 GHz, respectively. The length and width of the radiating elements, ground plane, and feeding line are optimized using the transmission line method (TLM). The MIMO antenna is designed on the FR4 substrate Fwith a dielectric constant of 4.4 and a substrate height of 1.6 mm. Four rectangular radiating elements are placed on this substrate such that the spacing between elements is less than half a wavelength for the 6 GHz frequency. The MIMO antennas are designed using two methods, i.e., by traditional method and DGS method [27–29].

2.2 MIMO ANTENNA DESIGN AND ANALYSIS

The MIMO antenna consists of multiple input and output element. Therefore more information with high data rate is transfered from one place to another place. Because, data rate is linearly increases with number of radiating elements increases in the MIMO antenna. The more than two elements are present in the MIMO system, hence this elements are placed close to each other. Therefore, isolation and cross correlation are high. It is affects on the antenna performance such as data rate and channel capacity are decreases. So, proposed antenna is highlight the operation of antenna with enhancement performance parameter. In this chapter, the following MIMO antennas are designed using two different methods:

A. Four-element MIMO antenna using the traditional method
B. Four-element MIMO antenna using defected the ground structure method

2.3 MIMO ANTENNA DESIGN USING TRADITIONAL METHOD AND ANALYSIS

2.3.1 Antenna design

This antenna is designed for different frequencies, which are 2.54, 3.45, and 6 GHz. This antenna consists of four radiating elements. Patch-1 and patch-2 are designed for 6 GHz frequency. Patch-3 and patch-4 are designed for 3.54 and 2.54 GHz frequencies, respectively. The geometrical structure of a rectangular shape for 2.54, 3.45, and 6 GHz frequencies is shown in Table 2.1 using the rectangular transmission line model (TLM) [26]. The radiating elements are energized using inset feed line method. It has length (L_f) and width (W_f) as shown in Table 2.1. The FR4 substrate having 4.4 dielectric constant and 1.6 mm height is used with dimensions 75 mm ($L_{substrate}$) × 90 mm ($W_{substrate}$) × 1.6 mm in the high-frequency simulator structure (HFSS). The spacing between the elements is less than half of the wavelength with respect to 6 GHz frequency. The four-element antenna is designed using the HFSS software as shown in Figure 2.1. The dimensions of antenna is as shown in Figure 2.2 and has values are shown in Table 2.1. The back side of antenna is consists of ground plane has a 75 mm length ($L_{substrate}$) and 90 mm width ($W_{substrate}$).

Table 2.1 The geometrical structure of rectangular shape for different frequencies

Parameter	*Values of parameter*		
Frequency	2.54 GHz	3.45 GHz	6 GHz
Width of patch	Wp4 = 35.40 mm	Wp3 = 26.16 mm	Wp1 = Wp2 = 15.21 mm
Length of patch	Lp4 = 27.82 mm	Lp3 = 20.07 mm	Lp1 = Lp2 = 11.34 mm
Width (W_f) of feed line	Wf4 = 2.95 mm	Wf3 = 2.88 mm	Wf1 = Wf2 = 2.942 mm
Length (L_f) of feed line	Lf4 = 14.64 mm	Lf3 = 10.76 mm	Lf1 = Lf2 = 6.387 mm

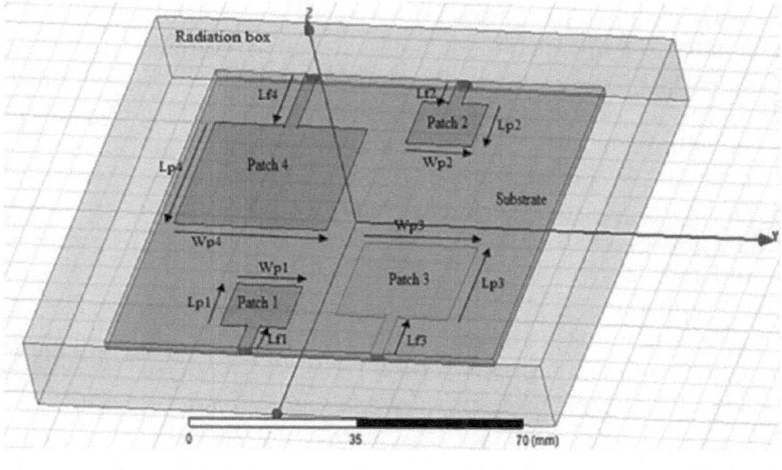

Figure 2.1 Four-element MIMO antenna.

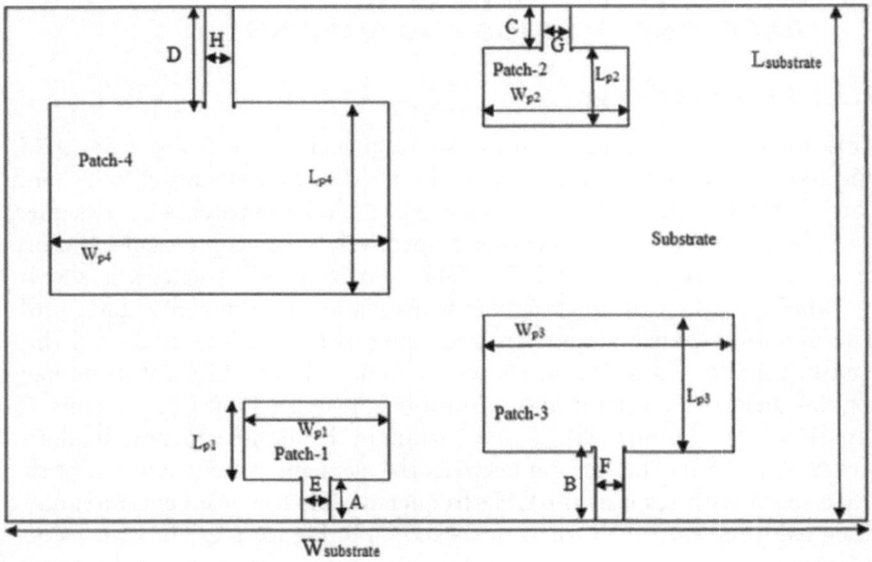

Figure 2.2 Dimensions of antenna.

2.3.2 Result analysis

The proposed antenna consists of different-shaped four radiating elements. The patch-1 is operated on the single band on 8.6 GHz frequency. The analysis of patch-1 shows that the result of return loss (S_{11}), isolation (S_{12}, $S_{13,}$ and S_{14}) and bandwidth obtained which are –17.35 dB, and less than –27.02 dB and 174 MHz (2.9%), respectively, as shown in Figure 2.3.

Patch-2 operates in three bands, including frequencies of 3.4, 5.2, and 8.8 GHz. When patch-2 is operated on 3.4 GHz, the analysis shows that the result of return loss (S_{22}), isolation (S_{21}, S_{23}, and S_{24}), and bandwidth obtained are, –13.66 dB, less than –33.50 dB, and 181 MHz (3.12%), respectively, as shown in Figure 2.4.

When patch-2 is operated on 5.2 GHz, the analysis shows that the result of return loss (S_{22}), isolation (S_{21}, $S_{23,}$ and S_{24}), and bandwidth obtained are –15.45 dB and less than –35.71 dB and 195 MHz (3.25%), respectively, as shown in Figure 2.4.

When patch-2 is operated on 8.8 GHz, the analysis shows that the result of return loss (S_{22}), isolation (S_{21}, S_{23} and S_{24}), and bandwidth obtained are –14.09 dB, less than –21.71 dB, and 204 MHz (3.40%), respectively, as shown in Figure 2.4.

Patch-3 is operated on the single band on 8.7 GHz frequency. The analysis of patch-3 shows that the result of return loss (S_{33}), isolation (S_{31}, $S_{32,}$ and S_{34}), and bandwidth obtained are, –12.35 dB, less than –36.99 dB, and 138 MHz (2.3%), respectively, as shown in Figure 2.5.

When patch-4 is operated on 6.4 GHz, the analysis shows that the result of return loss (S_{44}), isolation (S_{41}, S_{42}, and S_{43}), and bandwidth obtained

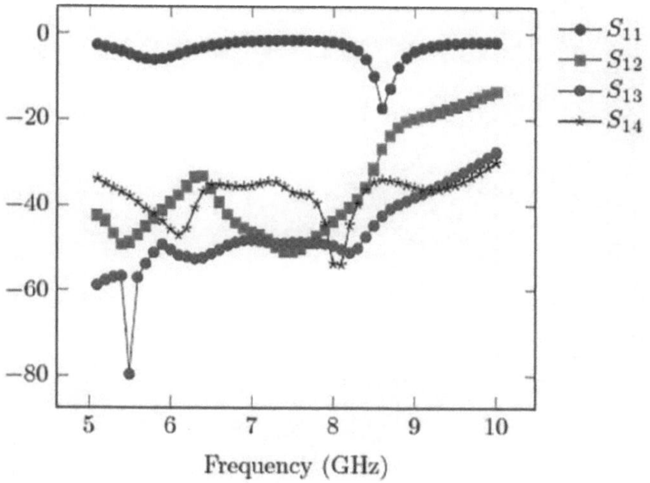

Figure 2.3 Return loss and isolation of patch-1.

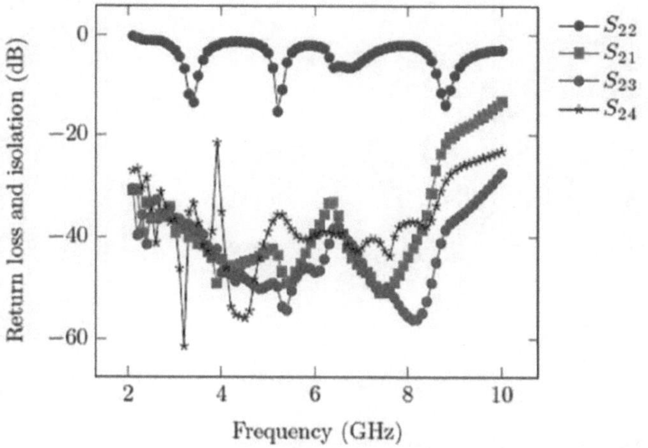

Figure 2.4 Return loss and isolation of patch-2.

are −17.14 dB, less than −30.86 dB, and 196 MHz (3.26%), respectively, as shown in Figure 2.6.

When patch-4 is operated on 7.3 GHz, the analysis shows that the result of return loss (S_{44}), isolation (S_{41}, $S_{42,}$ and S_{43}), and bandwidth obtained are −24.68 dB, less than −34.59 dB, and 115 MHz (1.916%), respectively, as shown in Figure 2.6.

When patch-4 is operated on 8.6 GHz, the analysis shows that the result of return loss (S_{44}), isolation (S_{41}, $S_{42,}$ and S_{43}), and bandwidth obtained are −17.69 dB, less than −33.92 dB, and 420 MHz (7.0%), respectively, as shown in Figure 2.6.

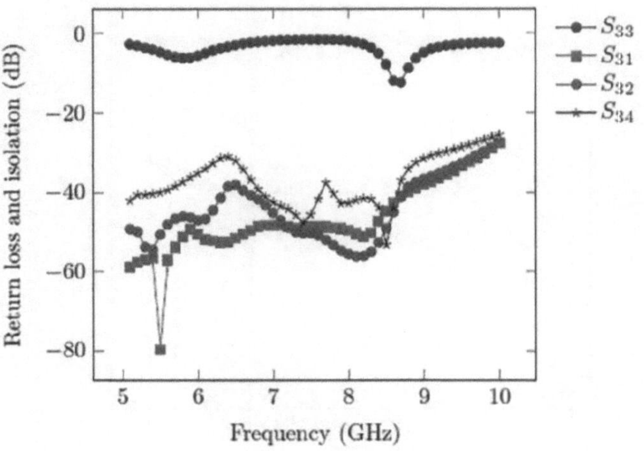

Figure 2.5 Return loss and isolation of patch-3.

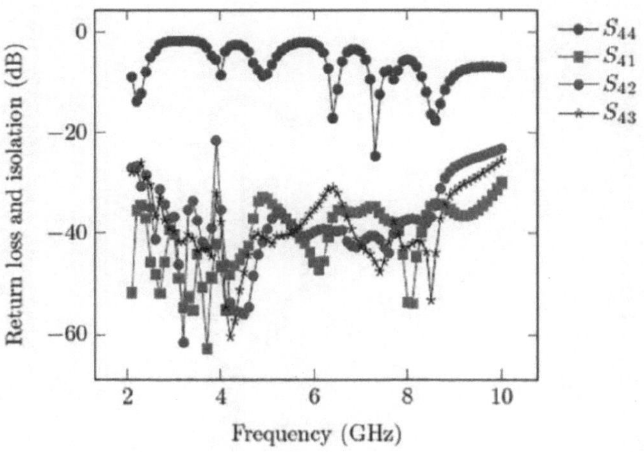

Figure 2.6 Return loss and isolation of patch-4.

The analysis of the antenna shows that the Voltage standing wave ratio (VSWR) values in dB (absolute) for patch-1, patch-2, patch-3, and patch-4 as shown in Figure 2.7 obtained which are shown in Table 2.2.

The analysis of the antenna shows that the gain as shown in Figure 2.8 and directivity as shown in Figure 2.9 are 3.84 and 5.70 dB, respectively. Figure 2.10 shows that the radiation pattern of the four-element MIMO antenna.

The total active reflection coefficient (TARC) as shown in Table 2.3, correlation coefficients (CC) as shown in Table 2.4, and envelope correlation coefficient (ECC) as shown in Table 2.4 are calculated in Ref. [26].

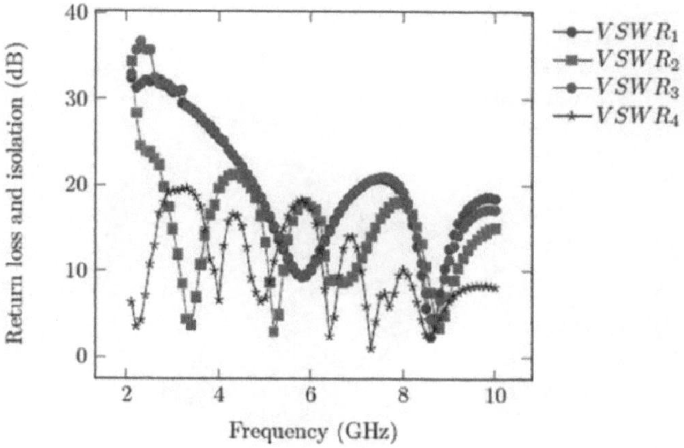

Figure 2.7 Voltage standing wave ratio (VSWR).

Table 2.2 Voltage standing wave ratio (VSWR)

Parameter	*Values in dB and absolute values with frequency*
$VSWR_1$	2.27 dB & 1.68 (8.6 GHz)
$VSWR_2$	3.6 dB & 2.29 (3.4 GHz), 2.9 dB & 1.94 (5.2 GHz) and 3.4 dB & 2.18 (8.8 GHz)
$VSWR_3$	3.8 dB & 2.39 (8.8 GHz)
$VSWR_4$	3.5 dB & 2.23 (2.2 GHz), 2.4 dB & 1.73 (6.4 GHz), 1.01 dB & 1.26 (7.3 GHz) and 2.2 dB & 1.65 (8.6 GHz)

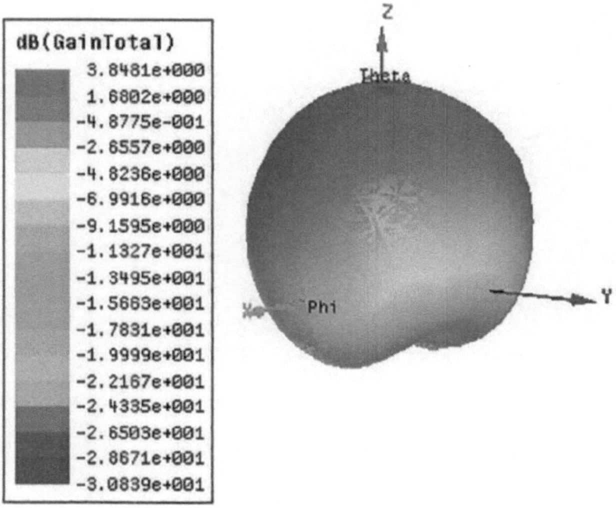

Figure 2.8 Gain (G_A) of antenna.

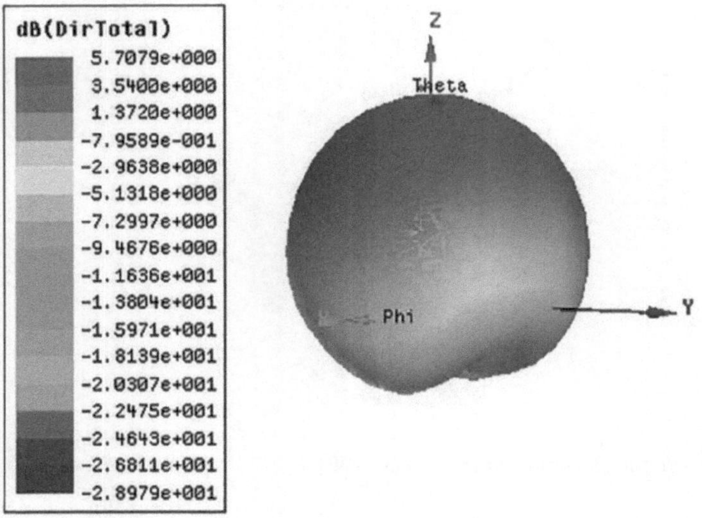

Figure 2.9 Directivity (D_A) of antenna.

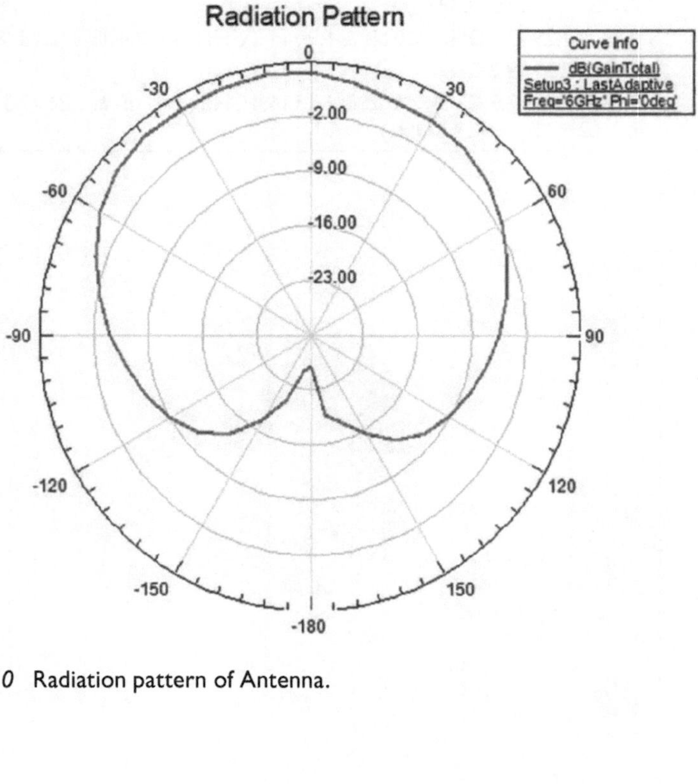

Figure 2.10 Radiation pattern of Antenna.

Table 2.3 Total active reflection coefficient (TARC)

TARC	*Absolute value & in dB w.r.t frequency*	*Between the patch*
T_{12} and T_{21}	0.033 & −14.81 (3.4 GHz), 0.024 & −16.19 (5.2 GHz), and 0.035 & −14.55 (8.8 GHz)	Patch 1 and patch 2
T_{13} and T_{31}	0.043 & −13.66	Patch 1 and patch 3
T_{14} and T_{41}	0.031 & −15.08 (2.2 GHz), 0.019 & −17.21 (6.4 GHz), 0.013 & −18.86 (7.3 GHz) and 0.017 & −17.69 (8.6 GHz)	Patch 1 and patch 4
T_{23} and T_{32}	0.041 & −13.87 (3.4 GHz), 0.045 & −13.46 (5.2 GHz) and 0.049 & −13.09 (8.8 GHz)	Patch 2 and patch 3
T_{24} and T_{42}	0.042 & −13.76 (2.2 GHz), 0.036 & −14.43 (6.4 GHz), 0.030 & −15.22 (7.3 GHz) and 0.033 & −14.81 (8.6 GHz)	Patch 2 (3.4 GHz) and patch 4
T_{24} and T_{42}	0.035 & −14.55 (2.2 GHz), 0.024 & −16.19 (6.4 GHz), 0.020 & −16.98 (7.3 GHz) and 0.023 & −16.38 (8.6 GHz)	Patch 2 (5.2 GHz) and patch 4
T_{24} and T_{42}	0.040 & −13.97 (2.2 GHz), 0.031 & −15.08 (6.4 GHz), 0.027 & −15.68 (7.3 GHz) and 0.030 & −15.22 (8.6 GHz)	Patch 2 (8.8 GHz) and patch 4
T_{34} and T_{43}	0.050 & −13.01 (2.2 GHz), 0.043 & −13.66 (6.4 GHz), 0.041 & −13.87 (7.3 GHz) and 0.042 & −13.76 (8.6 GHz)	Patch 3 and patch 4

Table 2.4 Correlation coefficient (CC) and envelope correlation coefficient (ECC)

Parameter	*Values*
CC	Less than 0.042
ECC	Less than 0.0017

2.3.3 Analysis

The four-element MIMO antenna is designed for a different frequency. The analysis of the antenna shows that it is operated on multi-band for 2.2, 3.4, 5.2, 6.4, 7.3, 8.6, 8.7, and 8.8 GHz frequencies with wideband bandwidth. The analysis of the antenna shows that the result of return loss, isolation, TARC, CC, and ECC obtained are less than –13 dB, less than –21 dB, less than 0.049 less than 0.042, and less than 0.0017. The analysis of the antenna shows 3.84 dB gain and 5.70 dB directivity with bandwidth in between 138 and 618 MHz. The absolute VSWR values are between 1.25 and 2.40.

2.4 MIMO ANTENNA DESIGN USING DEFECTED GROUND STRUCTURE (DGS) METHOD AND ANALYSIS

2.4.1 Antenna design

The four-element antenna using defected ground structure is designed by using the HFSS software as shown in Figure 2.11. The FR4 substrate is used for designing MIMO antenna. The characteristics of FR-4 substrate are 4.4 dielectric constant and 1.6 mm height. This MIMO antenna consists of four radiating elements. Of these, patch-1 and patch-2 are designed for 6 GHz frequency. Patch-3 and patch-4 are designed for 3.54 and 2.54 GHz frequencies, respectively. The geometrical structure of rectangular shape for 2.54, 3.45, and 6 GHz frequencies is calculated [26] as shown in Table 2.1. All dimensions of antenna is as shown in Figure 2.12. The spacing between the elements is less than half of the wavelength with respect to the 6 GHz frequency. The radiating elements are energized using inset feed line method. It has length (L_f) and width (W_f) as shown in Table 2.1. The antenna dimensions are 75 mm ($L_{substrate}$) × 90 mm ($W_{substrate}$) × 1.6 mm. The spacing between the radiating patches are positioned close to each other. So, isolation between the patches are high. This affects on the antenna performance and parameters, such as CC, ECC, TARC and DG. Therefore, there is a requirement to enhance the isolation between the radiating patch. Hence, DGS technique is used in the ground plane for enhancement of isolation.

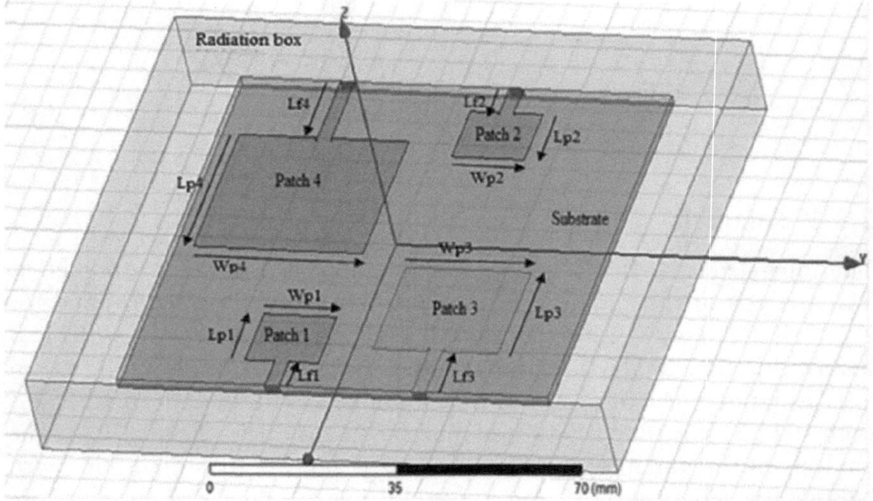

Figure 2.11 Four-element MIMO antenna using defected ground structure.

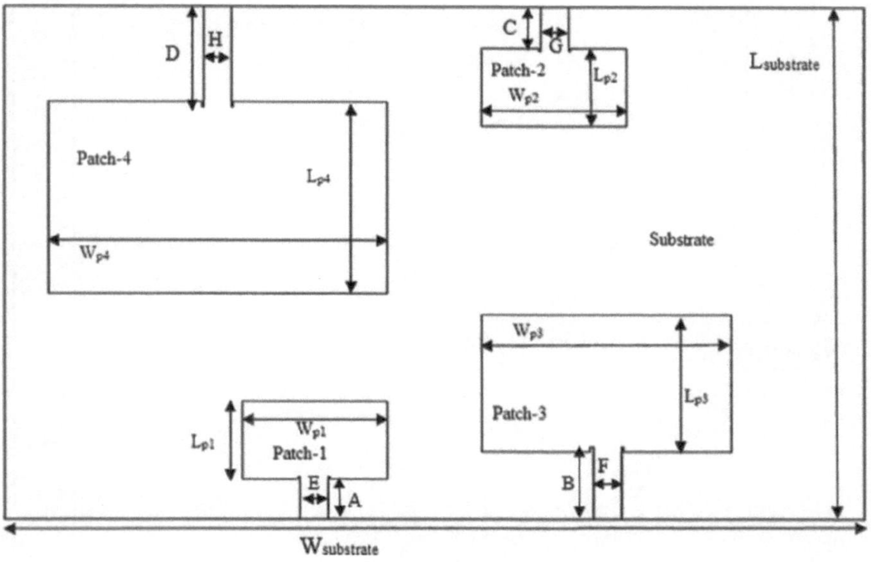

Figure 2.12 Dimensions of four-element MIMO antenna.

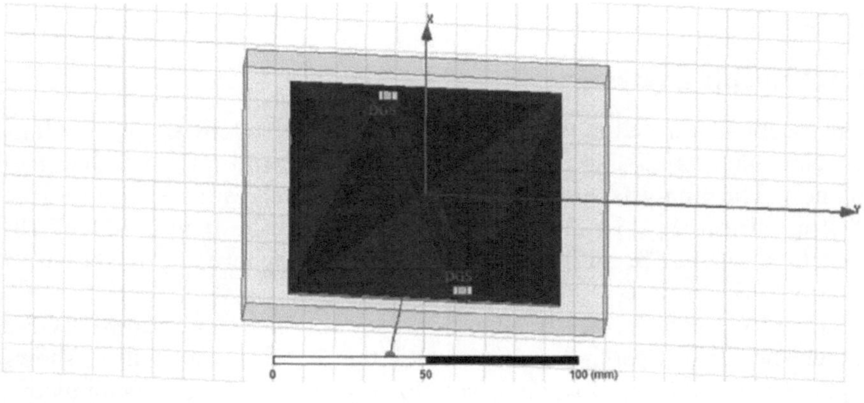

Figure 2.13 Ground plane with DGS.

The ground plane is shown in Figure 2.13. The rectangular-shaped structure having a length (X) of 2.94 mm and a width of (Y) 6.38 mm is used as a DGS in the ground plane. The dimensions of ground plane with DGS as shown in Figure 2.14.

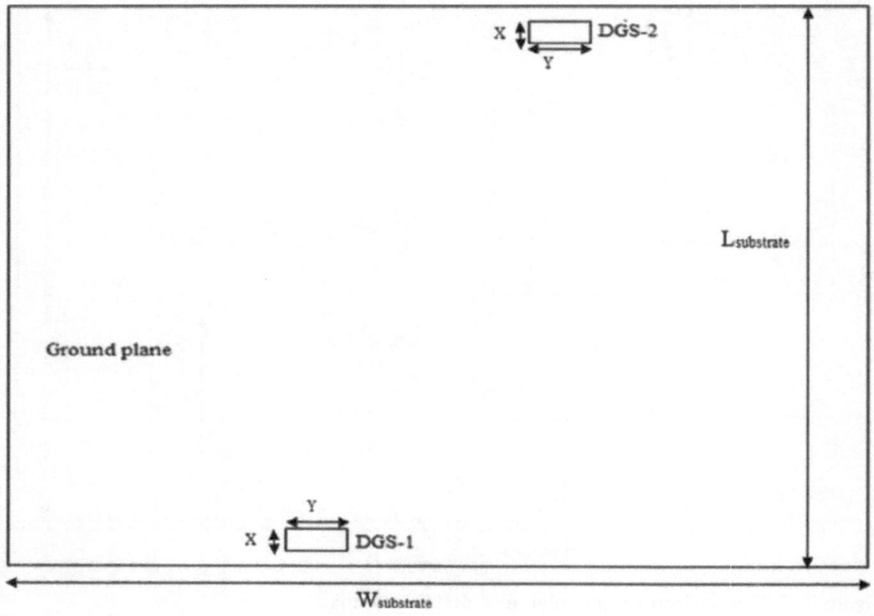

Figure 2.14 Dimensions of ground plane with DGS.

2.4.2 Result analysis

The proposed antenna is made up of four different-shaped radiating elements. Patch-1 is operated on the single band on a 5.8 GHz frequency. The analysis of patch-1 shows that the result of return loss (S_{11}), isolation (S_{12}, S_{13} and S_{14}), and bandwidth obtained is −11.07 dB, less than −39.12 dB, and 130 MHz (2.166%), respectively, as shown in Figure 2.15.

Patch-2 is operated on three bands which are of 3.3, 5.2, and 8.8 GHz frequencies. When patch-2 is operated on the 3.3 GHz, the analysis shows that the result of return loss (S_{22}), isolation (S_{21}, S_{23} and S_{24}), and bandwidth obtained is −15.52 dB, less than −24.27 dB, and 516 MHz (8.60%), respectively, as shown in Figure 2.16.

When patch-2 is operated on the 5.2 GHz, the analysis shows that the result of return loss (S_{22}), isolation (S_{21}, S_{23} and S_{24}), and bandwidth obtained is −15.56 dB, less than −35.62 dB, and 228 MHz (3.80%), respectively, as shown in Figure 2.16.

When patch-2 is operated on 8.8 GHz, the analysis shows that the result of return loss (S_{22}), isolation (S_{21}, S_{23} and S_{24}), and bandwidth obtained is −19.25 dB, less than −22.80 dB, and 224 MHz (3.733%), respectively, as shown in Figure 2.16.

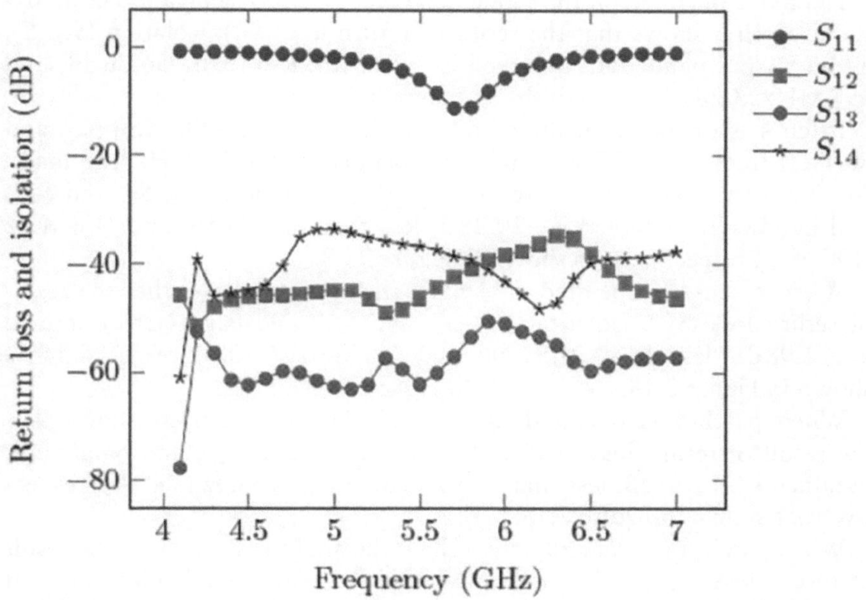

Figure 2.15 Return loss and isolation of patch-1

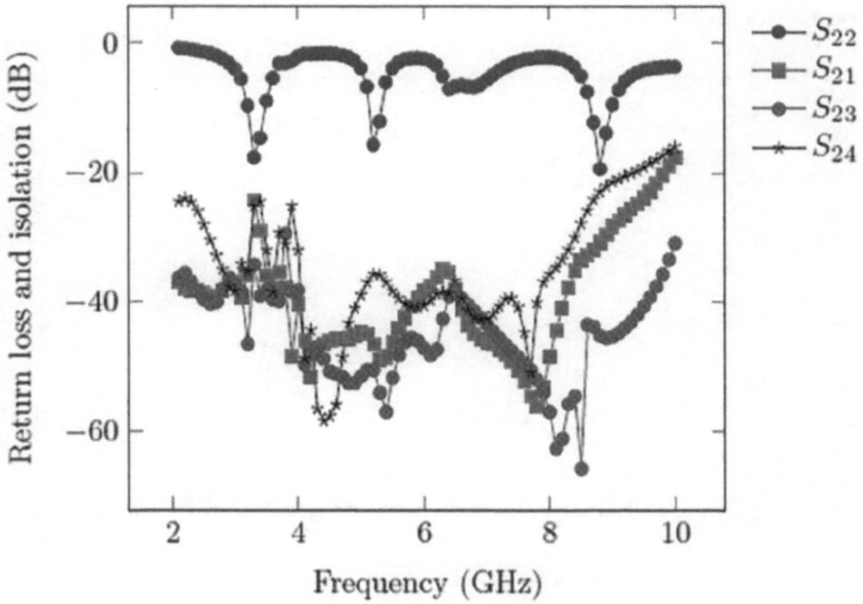

Figure 2.16 Return loss and isolation of patch-2.

Patch-3 is operated on the single band on 5.8 GHz frequency. The analysis of patch-3 shows that the result of return loss (S_{33}), isolation (S_{31}, S_{32} and S_{34}), and bandwidth obtained is –11.65 dB, less than –36.36 dB, and 197 MHz (3.28%), respectively, as shown in Figure 2.17.

Patch-4 is operated on three bands which are 2.3, 4.00, 6.5, 7.3, and 8.6 GHz frequencies. When patch-4 is operated on the 2.3 GHz, the analysis shows that the result of return loss (S_{44}), isolation (S_{41}, S_{42} and S_{43}), and bandwidth obtained is –20.28 d, less than –24.45 dB, and 646 MHz (10.76%) , respectively, as shown in Figure 2.18.

When patch-4 is operated on 4 GHz, the analysis shows that the result of return loss (S_{44}), isolation (S_{41}, S_{42} and S_{43}), and bandwidth obtained is –12.98 dB, less than –31.95 Db, and 348 MHz (5.80%), respectively, as shown in Figure 2.18.

When patch-4 is operated on the 6.5 GHz, the analysis shows that the result of return loss (S_{44}), isolation (S_{41}, S_{42} and S_{43}), and bandwidth obtained is –16.10 dB, less than –35.77 dB, and 204 MHz (3.40%), respectively, as shown in Figure 2.18.

When patch-4 is operated on 7.3 GHz, the analysis shows that the result of return loss (S_{44}), isolation (S_{41}, S_{42} and S_{43}) , and bandwidth obtained is –17.77 dB, less than –35.14 dB, and 204 MHz (3.40%), respectively, as shown in Figure 2.18.

When patch-4 is operated on 8.6 GHz, the analysis shows that the result of return loss (S_{44}), isolation (S_{41}, S_{42} and S_{43}), and bandwidth obtained is –14.66 dB, less than –25.74 dB, and 279 MHz (4.65%), respectively, as shown in Figure 2.18.

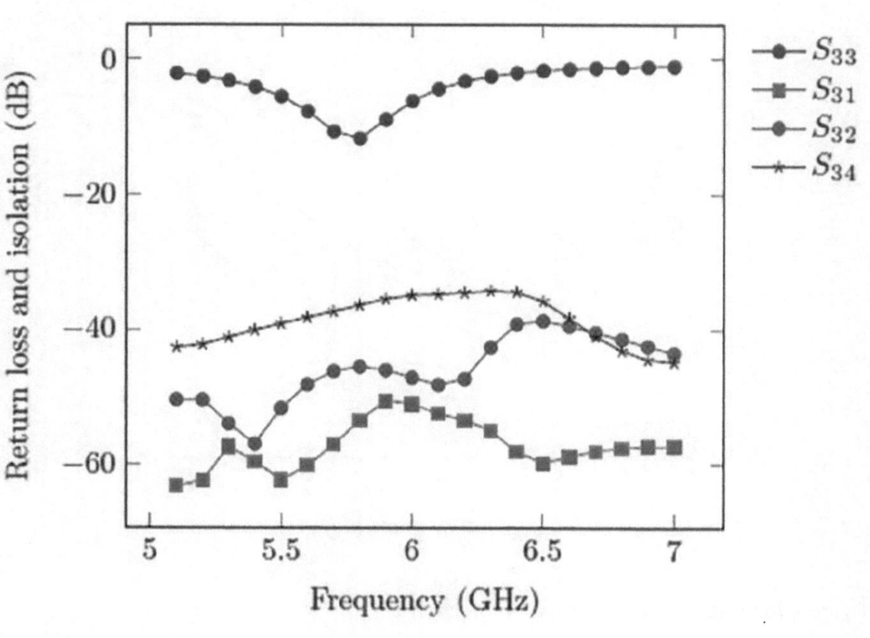

Figure 2.17 Return loss and isolation of patch-3.

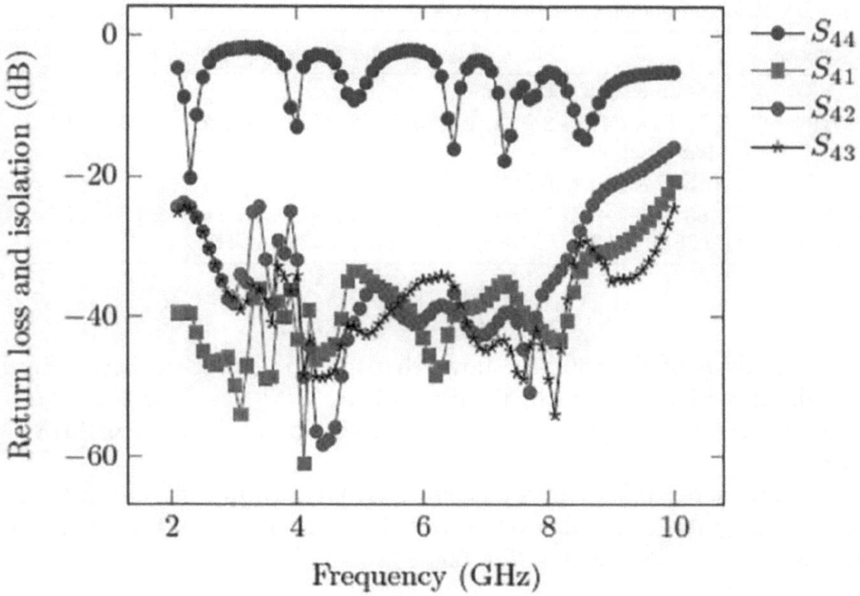

Figure 2.18 Return loss and isolation of patch-4.

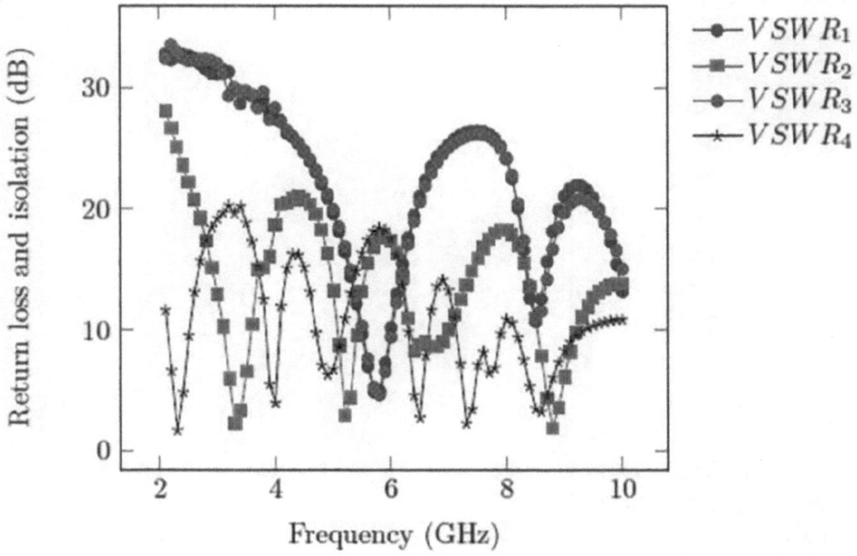

Figure 2.19 Voltage standing wave ratio (VSWR).

The analysis of the antenna shows that the VSWR values in dB (absolute) for patch-1, patch-2, patch-3, and patch-4 as shown in Figure 2.19 obtained which are shown in Table 2.5.

Table 2.5 Voltage standing wave ratio (VSWR)

Parameter	*Values in dB and absolute values with frequency*
$VSWR_1$	4.64 dB & 2.91 (5.8 GHz)
$VSWR_2$	2.32 dB & 1.70 (3.3 GHz), 2.92 dB & 1.95 (5.2 GHz) and 1.89 dB & 1.54 (8.8 GHz)
$VSWR_3$	4.64 dB & 2.91 (5.8 GHz)
$VSWR_4$	1.68 dB & 1.47 (2.3 GHz), 3.96 dB & 2.48 (4 GHz), 2.74 dB & 1.87 (6.5 GHz), 2.25 dB & 1.67 (7.3 GHz) and 3.2 dB & 2.08 (8.6 GHz)

The analysis of the antenna shows that the gain shown in Figure 2.20 and the directivity shown in Figure 2.21 are 3.62 and 5.62 dB, respectively. Figure 2.22 shows the radiation pattern of a four-element MIMO antenna.

The total active reflection coefficient (TARC) as shown in Table 2.6, CC as shown in Table 2.7, and envelope correlation coefficient (ECC) as shown in Table 2.7 are calculated in Ref. [26].

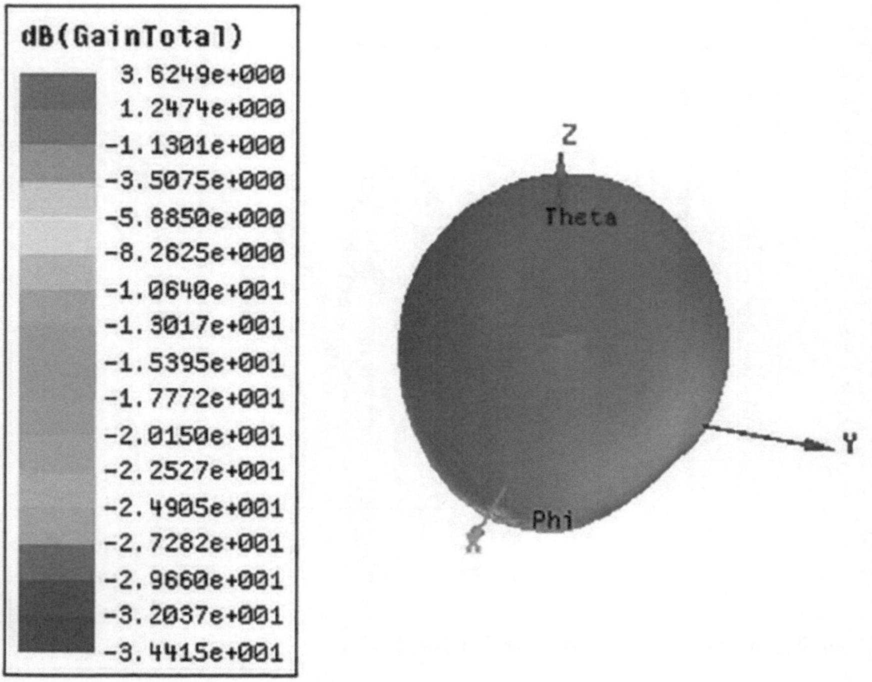

Figure 2.20 Gain of the proposed antenna.

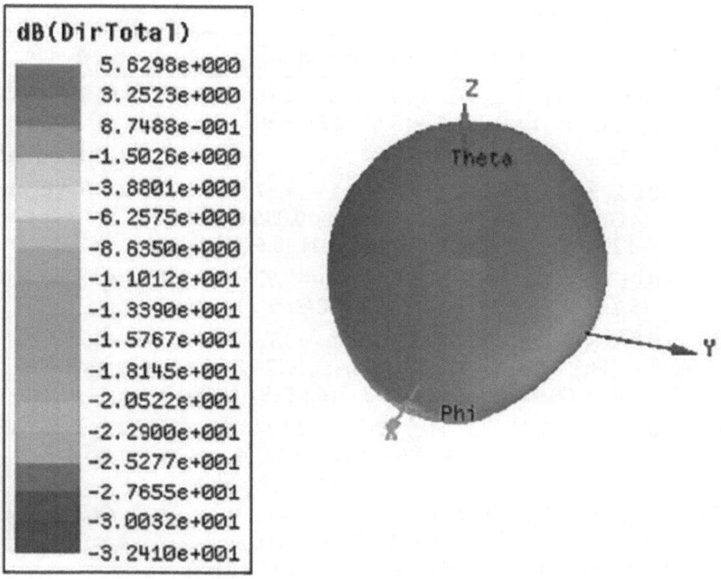

Figure 2.21 Directivity of the proposed antenna.

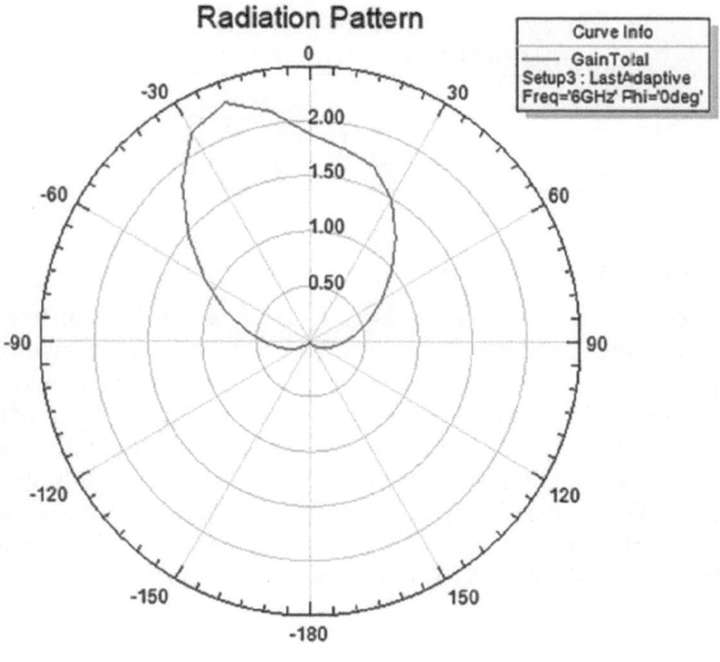

Figure 2.22 Radiation pattern of the proposed antenna.

Table 2.6 Total active reflection coefficient (TARC)

TARC	*Absolute value & in dB w.r.t frequency*	*Between the patch*
T_{12} and T_{21}	0.059 & −12.29 (3.3 GHz), 0.058 & −12.36 (5.2 GHz) and 0.056 & −12.51 (8.8 GHz)	Patch 1 and patch 2
T_{13} and T_{31}	0.073 & −11.36	Patch 1 and patch 3
T_{14} and T_{41}	0.055 & −12.59 (2.3 GHz), 0.065 & −11.87 (4 GHz), 0.058 & −12.36 (6.5 GHz), 0.056 & −12.51 (7.3 GHz) and 0.06 & −12.21 (8.6 GHz)	Patch 1 and patch 4
T_{23} and T_{32}	0.052 & −12.83 (3.3 GHz), 0.051 & −12.92 (5.2 GHz) and 0.048 & −13.18 (8.8 GHz)	Patch 2 and patch 3
T_{24} and T_{42}	0.023 & −16.38 (2.3 GHz), 0.042 & −13.76 (4 GHz), 0.028 & −15.52 (6.5 GHz), 0.024 & −16.19 (7.3 GHz) and 0.034 & −14.68 (8.6 GHz)	Patch 2 (3.3 GHz) and patch 4
T_{24} and T_{42}	0.021 & −16.77 (2.3 GHz), 0.040 & −13.97 (4 GHz), 0.025 & −16.02 (6.5 GHz), 0.022 & −16.57 (7.3 GHz) and 0.032 & −14.94 (8.6 GHz)	Patch 2 (5.2 GHz) and patch 4
T_{24} and T_{42}	0.014 & −18.53 (2.3 GHz), 0.037 & −14.31 (4 GHz), 0.020 & −16.98 (6.5 GHz), 0.016 & −17.95 (7.3 GHz) and 0.028 & −15.52 (8.6 GHz)	Patch 2 (8.8 GHz) and patch 4
T_{34} and T_{43}	0.048 & −13.18 (2.3 GHz), 0.060 & −12.21 (4 GHz), 0.051 & −12.92 (6.5 GHz), 0.049 & −13.09 (7.3 GHz) and 0.054 & −12.67 (8.6 GHz)	Patch 3 and patch 4

Table 2.7 Correlation coefficient (CC) and envelope correlation coefficient (ECC)

Parameter	*Values*
CC	Less than 0.00018
ECC	Less than 0.0000000306

2.4.3 Analysis

The four-element MIMO antenna with DGS is designed for different frequencies. The analysis of the antenna shows that it is operated on multi-band for 2.3, 3.3, 4, 5.2, 5.8, 6.5, 7.3, 8.6, and 8.8 GHz frequencies with wideband bandwidth. The analysis of the antenna shows that the result of return loss, isolation, TARC, CC, and ECC obtained is less than −11 dB, less than −22 dB, less than 0.059, less than 0.00018, and less than 0.0000000306, respectively. The analysis of the antenna shows 3.62 dB gain and 5.62 dB directivity with a bandwidth in between 130 and 646 MHz (wideband bandwidth). The absolute VSWR values are in between 1.45 and 2.95 (Figures 2.20–2.22).

2.5 CONCLUSION

The designed MIMO antenna is a multi-band antenna that provides better bandwidth. The analysis of the antenna shows that the return loss and isolation values are less than –11dB. The CC and ECC values are less than 0.001. The analysis of the antenna shows close to 10 dB diversity gain and less than –10 dB TARC. The analysis of antenna shows that, antenna with defected ground structure is better than antenna with traditional methods. The cross correlation between the radiating patches are reduces in the both designed antenna. This antenna is used for the Bluetooth/Wi-Fi/IoT (2.40–2.56 GHz), Wi-MAX (3.30–3.70 GHz), WLAN (5.15–5.30, and 5.73–5.85GHz), C-band (4–8 GHz) for satellite communication and X-band (8–12 GHz) for radar communication.

REFERENCES

1. W. Jiang, Y. Cui, B. Liu, W. Hu and Y. Xi, A Dual-Band MIMO Antenna with Enhanced Isolation for 5G Smartphone Applications, *IEEE Access*, vol. 7, pp. 112554–112563, 2019.
2. L. Cui, J. Guo, Y. Liu and C. Sim, An 8-Element Dual-Band MIMO Antenna with Decoupling Stub for 5G Smartphone Applications, *IEEE Antennas and Wireless Propagation Letters*, vol. 18, no. 10, pp. 2095–2099, Oct. 2019.
3. Z. Niu, H. Zhang, Q. Chen and T. Zhong, Isolation Enhancement in Closely Coupled Dual-Band MIMO Patch Antennas, *IEEE Antennas and Wireless Propagation Letters*, vol. 18, no. 8, pp. 1686–1690, Aug. 2019.
4. S. Nandi and A. Mohan, A Compact Dual-Band MIMO Slot Antenna for WLAN Applications, *IEEE Antennas and Wireless Propagation Letters*, vol. 16, pp. 2457–2460, 2017.
5. S. Soltani, P. Lot_ and R. D. Murch, A Dual-Band Multiport MIMO Slot Antenna for WLAN Applications, *IEEE Antennas and Wireless Propagation Letters*, vol. 16, pp. 529–532, 2017.
6. I. Nadeem and D. Choi, Study on Mutual Coupling Reduction Technique for MIMO Antennas, *IEEE Access*, vol. 7, pp. 563–586, 2019.
7. G. Li, H. Zhai, Z. Ma, C. Liang, R. Yu and S. Liu, Isolation-Improved Dual-Band MIMO Antenna Array for LTE/WiMAX Mobile Terminals, *IEEE Antennas and Wireless Propagation Letters*, vol. 13, pp. 1128–1131, 2014.
8. M.-Y. Li, Y.-L. Ban, Z.-Q. Xu, J. Guo and Z.-F. Yu, Tri-polarized 12-Antenna MIMO Array for Future 5G Smartphone Applications, *IEEE Access*, vol. 6, pp. 6160–6170, 2017.
9. Y. Li, C.-Y.-D. Sim, Y. Luo and G. Yang, Multiband 10-Antenna Array for Sub-6 GHz MIMO Applications in 5-G Smartphones, *IEEE Access*, vol. 6, pp. 28041–28053, 2018.

10. I. R. R. Barani and K.-L. Wong, Integrated Inverted-F and Open-Slot Antennas in the Metal-Framed Smartphone for 2 by 2 LTE LB and 4 by 4 LTE M/HB MIMO Operations, *IEEE Transactions on Antennas and Propagation*, vol. 66, no. 10, pp. 5004–5012, Oct. 2018.
11. Y. Liu, A. Ren, H. Liu, H. Wang and C.-Y. Sim, Eight-Port MIMO Array Using Characteristic Mode Theory for 5G Smartphone Applications, *IEEE Access*, vol. 7, pp. 45679–45692, 2019.
12. L. Cui, J. Guo, Y. Liu and C. Sim, An 8-Element Dual-Band MIMO Antenna with Decoupling Stub for 5G Smartphone Applications, *IEEE Antennas and Wireless Propagation Letters*, vol. 18, no. 10, pp. 2095–2099, Oct. 2019.
13. Z. Niu, H. Zhang, Q. Chen and T. Zhong, Isolation Enhancement in Closely Coupled Dual-Band MIMO Patch Antennas, *IEEE Antennas and Wireless Propagation Letters.*
14. H. T. Chattha, 4-Port 2-Element MIMO Antenna for 5G Portable Applications, *IEEE Access*, vol. 7, pp. 96516–96520, 2019.
15. Y. Liu, X. Yang, Y. Jia and Y. J. Guo, A Low Correlation and Mutual Coupling MIMO Antenna, *IEEE Access*, vol. 7, pp. 127384–127392, 2019.
16. Y. Li, C. Sim, Y. Luo and G. Yang, High-Isolation 3.5 GHz Eight-Antenna MIMO Array Using Balanced Open-Slot Antenna Element for 5G Smartphones, *IEEE Transactions on Antennas and Propagation*, vol. 67, no. 6, pp. 3820–3830, June 2019.
17. K.-L. Wong, C.-Y. Tsai and J.-Y. Lu, Two Asymmetrically Mirrored Gap-Coupled Loop Antennas as a Compact Building Block for Eight-Antenna MIMO Array in the Future Smartphone, *IEEE Transactions on Antennas and Propagation*, vol. 65, no. 4, pp. 1765–1778, Apr. 2017.
18. J. Choi, W. Hwang, C. You, B. Jung and W. Hong, Four-Element Reconfigurable Coupled Loop MIMO Antenna Featuring LTE Full-Band Operation for Metallic-Rimmed Smartphone, *IEEE Transactions on Antennas and Propagation*, vol. 67, no. 1, pp. 99–107, Jan. 2019.
19. S. Xiumei, Y. Liu, L. Zhao, G. Huang, X. Shi and Q. Huang, A Miniaturized Microstrip Antenna Array at 5G Millimeter Wave Band, *IEEE Antennas and Wireless Propagation Letters.*
20. P. Garg and P. Jain, Isolation Improvement of MIMO Antenna using a Novel Flower Shaped Metamaterial Absorber at 5.5 GHz WiMAX band, *IEEE Transactions on Circuits and Systems II.*
21. J. Guo, L. Cui, C. Li and B. Sun, Side-Edge Frame Printed Eight-Port Dual-Band Antenna Array for 5G Smartphone Applications, *IEEE Transactions on Antennas and Propagation*, vol. 66, no. 12, pp. 7412–7417, Dec. 2018.
22. X. Shi, M. Zhang, S. Xu, D. Liu, H. Wen and J. Wang, Dual-Band 8-Element MIMO Antenna with Short Neutral Line for 5G Mobile Handset, *Proceedings of 11th European Conference on Antennas and Propagation*, May 2017, pp. 3140–3142.
23. N. O. Parchin et al., Eight-Element Dual-Polarized MIMO Slot Antenna System for 5G Smartphone Applications, *IEEE Access*, vol. 7, pp. 15612–15622, 2019.
24. Q. Chen et al., Single Ring Slot-Based Antennas for Metal-Rimmed 4G/5G Smart- Phones, *IEEE Transactions on Antennas and Propagation*, vol. 67, no. 3, pp. 1476–1487, Mar. 2019.

25. J. Li et al., Dual-Band Eight-Antenna Array Design for MIMO Applications in 5G Mobile Terminals, *IEEE Access*, vol. 7, pp. 71636–71644, 2019.
26. S. Sarade and S. Ruikar, Developments of Multiband MIMO Antenna with Defected Ground Structure, Springer paper in ACCT-2019.
27. *Planar Antenna: Design, Fabrication, Testing, and Application*, Nova Science Publishers Inc, New York, 15 Oct 2021, ISBN: 9781536198980.
28. *Smart Antennas: Latest Trends in Design and Application*, Malik, P., Lu, J., Madhav, B.T.P., Kalkhambkar, G., Amit, S. (Eds.), Springer, ISBN: 9783030766368, DOI: 10.1007/978-3-030-76636-8.
29. *Microstrip Antenna Design for Wireless Applications*, Malik, P.K., Padmanaban, S., Holm-Nielsen, J.B., Taylor and Francis, Aug 04, 2021, ISBN: 9780367554385.

Chapter 3

Low-profile broadband printed antennas for wireless applications

Penchala Reddy Sura
PBR Visvodaya Institute of Technology and Science, Kavali, A.P., India.

Mohammad Hayath Rajvee
PBR Visvodaya Institute of Technology and Science, Kavali, A.P., India.

Tathababu Addepalli
Aditya Engineering College, Surampalem, A.P., India.

Tamirat Tagesse
Wachemo University, Ethipia.

CONTENTS

DOI: 10.1201/9781003347057-3

3.1 INTRODUCTION TO BROADBAND PRINTED ANTENNAS

The antennas working at the Third Generation (3G), LTE (Long Term Evolution), Fourth Generation (4G), Wireless Local Area Networks (WLAN), Wireless Fidelity (Wi-Fi), Bluetooth, Worldwide Interoperability for Microwave Access (WiMAX) and FSS are extremely significant in a wide range of commercial applications [1]. The advancements in the various new-fashioned technologies like 3G (1.92–1.98 and 2.11–2.17 GHz), 4G (2.3–2.4 GHz), Wi-Fi (2.4–2.485 and 5.15–5.85 GHz), Bluetooth (2.4–2.5 GHz), WLAN (2.4–2.48, 5.15–5.35 and 5.72–5.83 GHz), mid-band 5G (3.4–3.6 GHz) and WiMAX (2.5–2.70, 3.4–3.70 and 5.25–5.86 GHz) have activated further research in the design and investigation of multiband antennas [2–4]. A compact planar CPW-fed antenna that operates across various frequency bands is shown for RF energy harvesting applications. The antenna with dimensions $48 \times 42\,mm^2$ is made up of five radiating components and a stepped ground plane. When compared to a full ground plane antenna, the antenna with a stepped ground plane has a better impedance match [5].

The last two decades were committed to the growth of contemporary telecommunications. The tremendous improvements in telecommunications in recent years have resulted in significant changes, not only in terms of the number of users but also in terms of the facilities and services supplied by new technologies. Because of the increasing need for multimedia applications, technology has switched to faster data rates. The necessity for antennas with a wide bandwidth has grown as the need for faster data rates has grown. Antennas are an important component of wireless and mobile devices. As a result, broadband antennas are used to generate larger data rates.

The authors of [6] present a new planar compact antenna made up of two crossed Cornu spirals. The clothoid is built by feeding each Cornu spiral from the centre of the linear section of the curve between the two spirals. To generate circular polarisation and enhance the effective bandwidth, a sequential phase network is used to apply the sequential rotation. Signal integrity concerns have been addressed and developed to ensure that signal propagation is of high quality.

A defected star-shaped microstrip antenna was intended for wideband applications. A monopole antenna is constructed with a defective star-shaped tuning stub and a defective ground structure that is powered through a microstrip feed line. By adjusting the size of the tuning stub, the defective ground and its notch, proper tuning of resonating modes broad frequency effect has been obtained. This antenna operates between the frequencies of 1.6638 and 6.652 GHz [7].

A compact and broadband uniplanar microstrip antenna (MSA) is proposed for end-fire radiation in the sub-6 GHz 5G frequency range.

The designed antenna consists of a semi-elliptical radiating element and a U-shaped ground plane. The use of a semi-elliptical radiating element results in a broad impedance bandwidth (BW) and a minimal dimension. By strengthening the coupling between the radiating element and the ground, the U-shaped ground plane further increases the bandwidth [8].

The patch antenna (PA) and its self-complementary construction, the slot antenna (SA), are proposed and developed for directly matching the impedance of a rectifier at its 2.45 GHz resonance frequency. These antennas are constructed in three sections: meandered-line, spiral and double-folded geometries, which enable easy adjustment of their geometrical features using design equations [9].

The researchers present a novel low-profile dual-polarised antenna. Two pairs of rhombic dipoles are stimulated by two orthogonal baluns to form the antenna. The broadband feature is obtained by incorporating a metal ring beneath the rhombic dipole, which also improves the beam widths of the radiation pattern [10].

For 5G applications, a broadband high-isolation dual-polarised antenna is proposed. The suggested antenna is composed of L-shaped parts, feeding strips in the shape of a box and a box-shaped reflector. Not only does the adoption of basic L-shaped antenna components simplify production, but it also significantly enhances the isolation between the two ports. The use of box-shaped reflectors results in stable gain and radiation patterns. This antenna operates at frequencies ranging from 3.1 to 5.2 GHz [11].

A planar meta-material-based broadband high gain Resonant Cavity Antenna (RCA) operating at C-band is suggested. The RCA is modelled using a simple ray tracing technique. This antenna operates between 4 and 3.1 GHz [12].

Three distinct low-profile broadband printed antennas for a variety of wireless applications have been built and examined in this chapter. The antennas proposed are small and have a wide bandwidth. To accomplish the intended operation, various shaped radiating patches are used. The ground plane is adjusted to improve the impedance matching quality and hence the bandwidth [13–15].

3.2 CPW-FED PRINTED CIRCULAR MONOPOLE ANTENNA

3.2.1 Antenna design configuration

Figure 3.1 illustrates the design configuration of the CPW-fed printed circular monopole antenna. The antenna is composed of a circular radiating patch mounted on a low-cost FR-4 substrate with a thickness of 1.6 mm, ε_r=4.4 and tanδ=0.02. Initially, the antenna is created using the transmission line approach, and then the HFSS tool is used to do parametric analysis

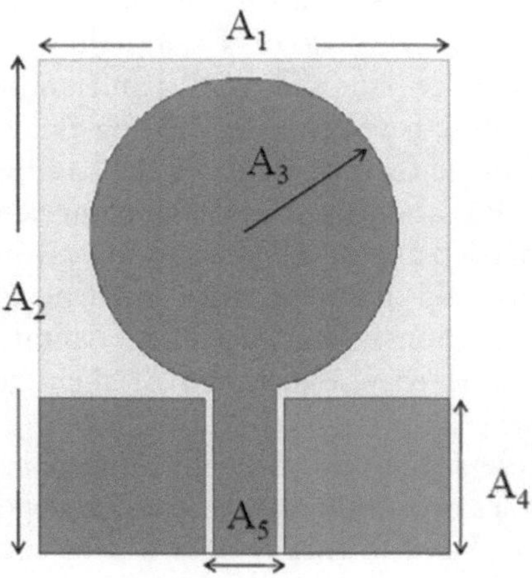

Figure 3.1 The CPW-fed printed circular monopole antenna design layout.

to obtain the optimal value. The developed printed antenna features a radiating circular patch with a radius of 12 mm on the substrate's topside and is fed by a CPW. The length, breadth and feed gap of the ground plane are 12.9, 12 and 0.6 mm, respectively. The signal line is 5 mm wide. The width and feed gap of the ground impact the antenna's effectiveness since the surface current is concentrated along the width and the feed gap plays a significant role in boosting the antenna's impedance matching.

The designed broadband printed antenna is $38 \times 32 \times 1.6\,mm^3$ in size and has a maximum gain of 5.75 dB at 9 GHz. The parametric analysis was conducted to get optimal dimensions for improved performance.

3.2.2 Result analysis

Using Ansoft HFSS software, the intended broadband printed antenna was developed and simulated. Figure 3.2 illustrates the reflection coefficient of the CPW-fed printed circular monopole antenna.

The reflection coefficient indicates the amount of power reflected into free space without being emitted. As seen in Figure 3.2, the proposed broadband antenna functions between 2.62 and 7.32 GHz with an S_{11} of less than –10 dB. The antenna performs admirably in terms of impedance.

Radiation patterns are a visual representation of three-dimensional field variations as a function of pattern and angle. Two major plane cuts are required to accurately represent the radiation pattern in two dimensions. These are referred to as the fundamental E- and H-plane patterns.

Figure 3.3 illustrates the proposed broadband printed antenna's 3D radiation patterns at two independent operating frequencies of 3.5 and 6.1 GHz. The proposed antenna has gains of 1.7 dB at 3.5 GHz and 4.43 dB at 6.1 GHz, respectively.

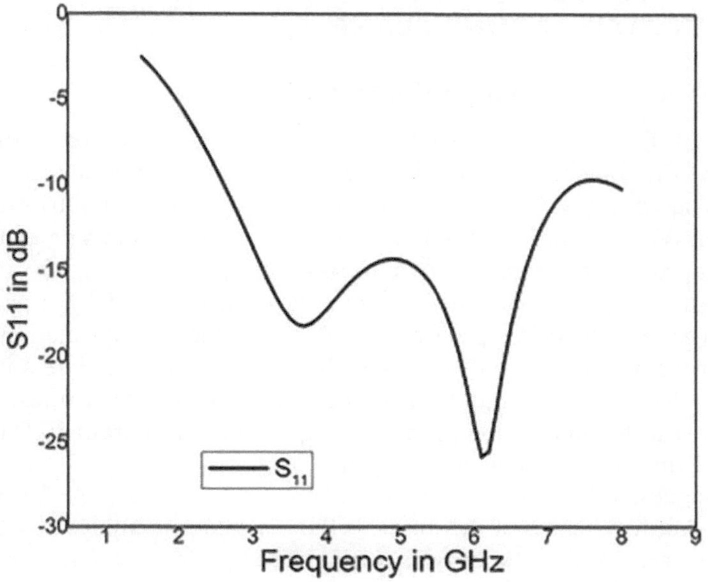

Figure 3.2 The reflection coefficient.

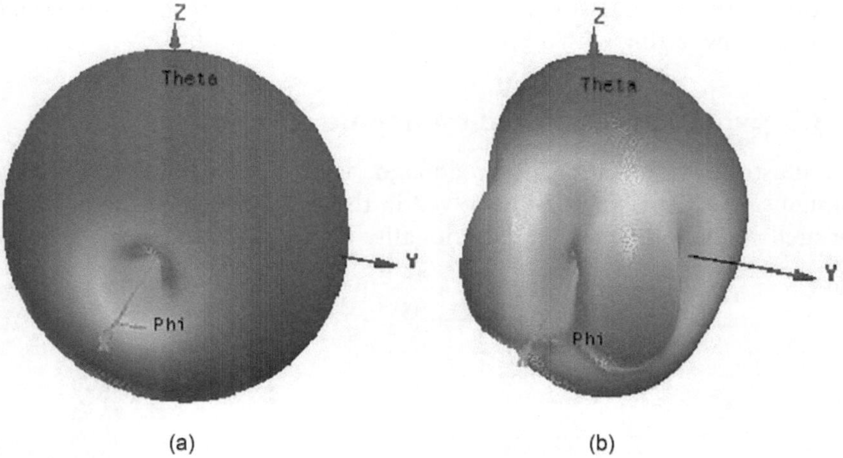

Figure 3.3 The radiation patterns.

As shown in Figure 3.3, the designed antenna has an omnidirectional radiation pattern at 6.1 GHz, and it achieves a higher gain across the operating bands with acceptable reflection loss. As a result, this antenna could be considered a better candidate for 5G, WiMAX, Wi-Fi, WLAN and C-band applications.

3.3 PRINTED MONOPOLE ANTENNA WITH DEFECTED GROUND PLANE

3.3.1 Antenna design configuration

Figure 3.1 illustrates the design arrangement for the printed monopole antenna with a defective ground plane. The planned antenna is composed of a P-shaped radiating patch mounted on a low-cost FR-4 substrate with a thickness of 1.6 mm, ε_r=4.4 and tanδ=0.02. The bottom side features a deficient ground structure in the shape of a rectangle. To begin, the antenna is created using the transmission line approach, followed by a parametric analysis in the HFSS programme to determine the optimal value. A microstrip line feed is used to power the printed antenna. The antenna's ground effect bandwidth and impedance matching are determined by the length, breadth and location of the rectangular slot. To obtain 50 Ω impedance, the microstrip line width is 3 mm.

The suggested broadband printed antenna measures $21\times23\times1.6\,mm^3$ in volume and has a gain of 3.3 dB at 4 GHz. Parametric analysis is used to determine the optimal dimensions for increased performance (Figure 3.4).

The antenna is primarily composed of a microstrip feed line with a width of 3 mm and a length of 14.5 mm that is attached to the substrate's top surface. To create the P-shaped radiating patch, three stubs are attached to the microstrip line feed. The backside is made up of DGS, which is done by loading a rectangular slot with dimensions $B_8 \times B_9\,mm^2$ to accommodate antenna downsizing with a broader IBW.

3.3.2 Evolution of printed monopole antenna

To illustrate the creation of broadband, Figure 3.5 depicts the various design stages (Ant1 to Ant4) involved in the development of the proposed printed monopole antenna. Additionally, Figure 3.6 depicts the relevant

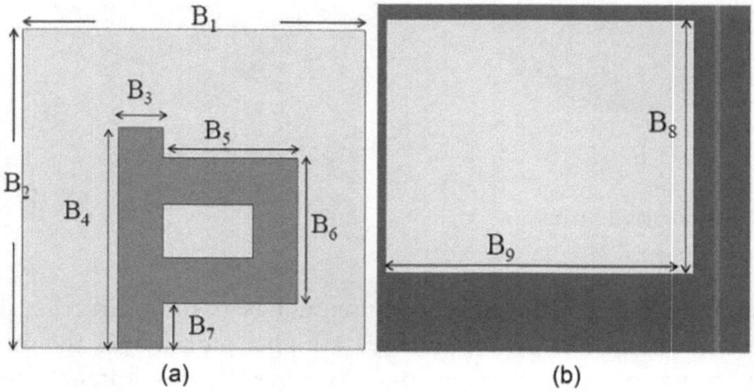

Figure 3.4 Design configuration of the printed monopole antenna.

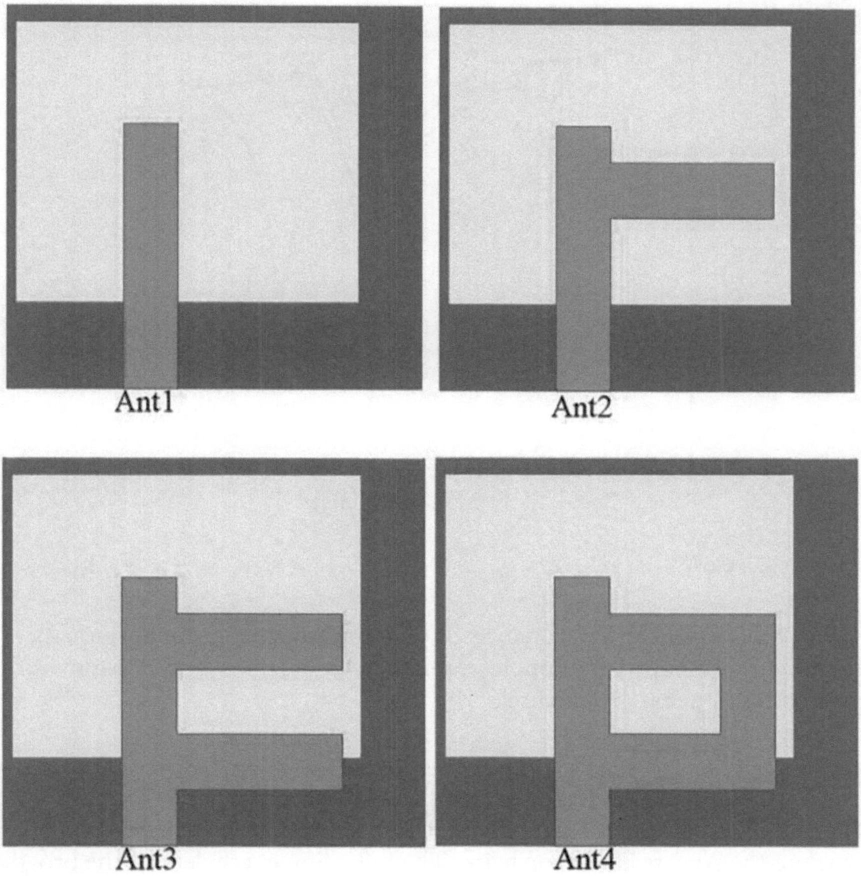

Figure 3.5 Evolution of printed monopole antenna.

input reflection coefficient (S_{11}) charts for various design phases. To begin, the antenna design starts with an asymmetrical simple printed monopole antenna with a feedline on the top surface and a defective ground on the bottom side of the substrate. This antenna radiates at 6.3 GHz (see Ant1's S_{11} response in Figure 3.6) and has a bandwidth of 1.46 GHz.

To the ground of the quarter wavelength monopole radiator, a rectangular slit with dimensions B8×B9 mm² is constructed (Ant1). The top stub, 9×3 mm², is connected to Ant1's core conductor to develop Ant2. It performs poorly at 6.3 GHz. The 9×3 mm² lower stub is put onto the central conductor of Ant2 to develop Ant3 and increase the antenna's performance. Ant3 works in the frequency range of 6–6.8 GHz. The 3.5×3 mm² side stub is joined to Ant3 to evolve Ant43 and Ant4, the proposed broadband printed antenna. The final antenna operates between 3.34 and 8.2 GHz, whereas the S_{11} operates at less than –10 dB.

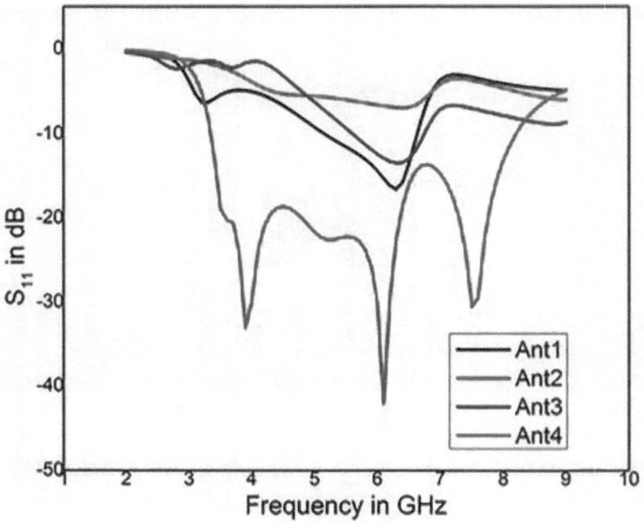

Figure 3.6 S_{11} plot.

The full wavelength resonance of the ring-shaped ground resonances merge with P-shaped monopole resonance to create a broad bandwidth with better impedance matching.

3.3.3 Result analysis

The designed broadband printed antenna has been designed and simulated by using the Ansoft HFSS software. The reflection coefficient of the broadband printed monopole antenna is illustrated in Figure 3.7.

The reflection coefficient indicates the amount of energy reflected into free space without being emitted. As seen in Figure 3.7, the planned broadband antenna functions between 3.34 and 8.2 GHz with an S_{11} of less than –10 dB. The antenna performs admirably in terms of impedance.

Radiation patterns are a visual representation of three-dimensional field fluctuations as a function of θ and ϕ. Two major plane cuts are required to accurately represent the radiation pattern in two dimensions. These are referred to as the fundamental E- and H-plane patterns.

Figure 3.8 illustrates the proposed broadband printed antenna's 3D radiation patterns at two independent operating frequencies of 4 and 6 GHz. The suggested antenna has gains of 3.3 and 3.1 dB at 4 and 6 GHz, respectively. As shown in Figure 3.8, the designed antenna has an omnidirectional radiation pattern at 3.5 GHz and a directional pattern at 6.1 GHz. Additionally, the designed antenna achieves a higher gain across the operating bands with acceptable reflection loss. As a result, this antenna could be considered a better candidate for 5G, Wi-Fi, WLAN, WiMAX and C-band applications.

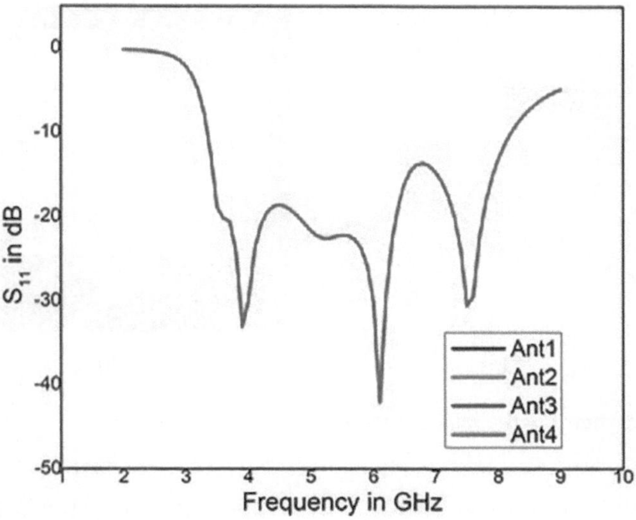

Figure 3.7 S_{11} plot of printed monopole antenna.

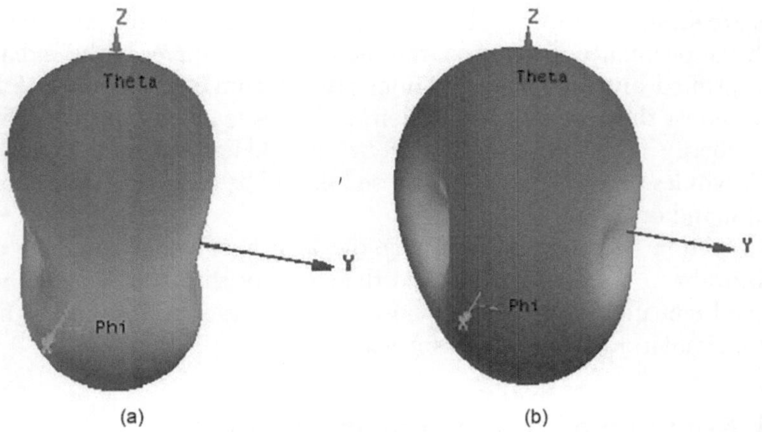

Figure 3.8 Radiation patterns of the broadband antenna.

3.4 MINIATURISED DUAL-BAND PRINTED MONOPOLE ANTENNA

3.4.1 Geometry of dual-band printed monopole antenna

Figure 3.8 depicts the design arrangement of a miniaturised dual-band printed monopole antenna for RF energy harvesting applications. The circular ring-shaped radiating patch adheres to a –1.524 mm thick Rogers

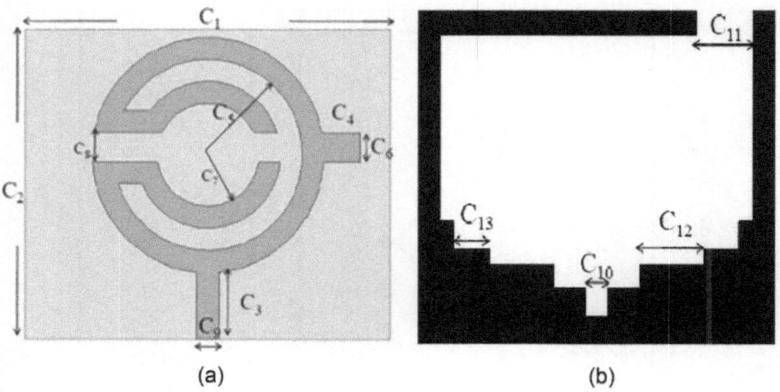

Figure 3.9 Geometry of the miniaturised dual-band antenna.

RO4003 (tm) dielectric substrate. The stepped ground plane is placed on the substrate's bottom surface. To boost the bandwidth, L- and I-shaped strips are added to the ground plane. The substrate's dielectric constant and loss tangent are 3.55 and 0.0027, respectively. The substrate's measurements are 42 mm × 50 mm. The circular's length and breadth are designed to achieve resonance in two separate bands and to increase the radiation of the printed antenna. A rectangular stub is connected to the patch to further boost the amount of radiation at the resonance frequencies. Since the RF energy harvesting circuits must be tiny and lightweight to connect to today's wireless devices, the antenna used should be low profile and capable of multiband operation (Figure 3.9).

All antenna parameters, including the length of all arms, their relative location to the circular arc, and their step in the ground plane, were evaluated repeatedly and then optimised by simulation using the 3-D finite element simulator, Ansoft HFSS software.

3.4.2 Evolution of printed monopole antenna

The major design steps (Ant1 to Ant4) involved in the construction of the proposed dual antenna are represented in Figure 3.10 to allow for an examination of the creation of dual-band signals. In addition, plots of the appropriate input reflection coefficient (S_{11}) for various design phases are shown in Figure 3.11.

Initially, the basic element used in the antenna design is a simple printed circular ring monopole antenna with microstrip line feed on the top surface and partial ground plane on the bottom side of the substrate, which radiates at 2.4 GHz (see the S_{11} response of Ant1 in Figure 3.11), represented as Ant1. Ant1 operates at a single frequency band ranging from 2.14 to 2.78 GHz with an S_{11} value less than –10 dB. In the second phase, a small

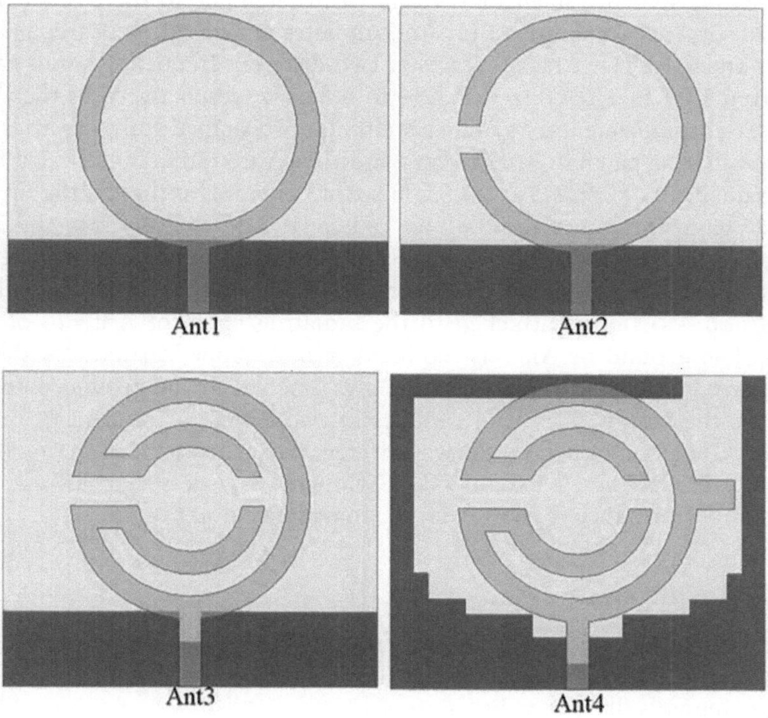

Figure 3.10 Evolution of the miniaturised dual-band antenna.

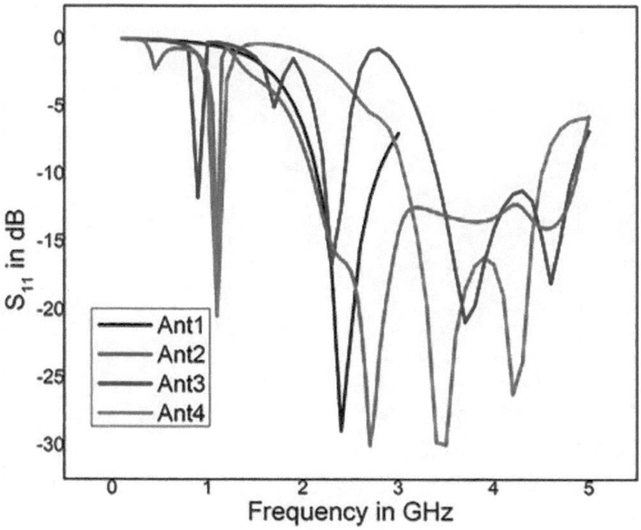

Figure 3.11 S_{11} plot.

rectangular split of dimensions 4 mm×8 mm is made to the circular ring radiator of Ant1 to obtain Ant2. Ant2 radiates at two different frequencies of 1.1 and 3.5 GHz. Ant2 operates at two different frequency bands ranging from 1.04 to 1.16 GHz and 3.34 to 4.5 GHz, respectively. In the third phase, the complementary symmetry circular arc-shaped stubs are attached to the radiating patch in Ant2 to get Ant3. The Ant3 radiates at three different frequencies of 0.9, 2.3 and 3.7 GHz. Ant3 operates at three different frequency bands ranging from 0.89 to 0.91 GHz, 2.12 to 2.43 GHz and 3.41 to 4.82 GHz, respectively, but Ant3 exhibits poor performance in terms of impedance matching. In the fourth phase, a small rectangular stub of dimension 4×6 mm is attached to the radiating patch of Ant3 to obtain Ant4. The ground of Ant3 is modified into a stepped ground as shown in Figure 3.10. L-and I-shaped strips are attached to the ground plane to increase the bandwidth. Ant4 radiates at two different frequencies of 1.1 and 2.7 GHz. Ant4 operates at two different frequency bands ranging from 1.07 to 1.12 GHz and 2.1 to 4.87 GHz while the S_{11} is less than −10 dB. The antenna performance is good from the impedance point of view.

3.4.3 Result analysis

The dual-band printed monopole antenna has been designed and simulated by using the Ansoft HFSS software. The reflection coefficient of the dual-band antenna is illustrated in Figure 3.12.

The reflection coefficient represents the amount of power reflected without radiating into free space. From Figure 3.9, it is clear that the designed

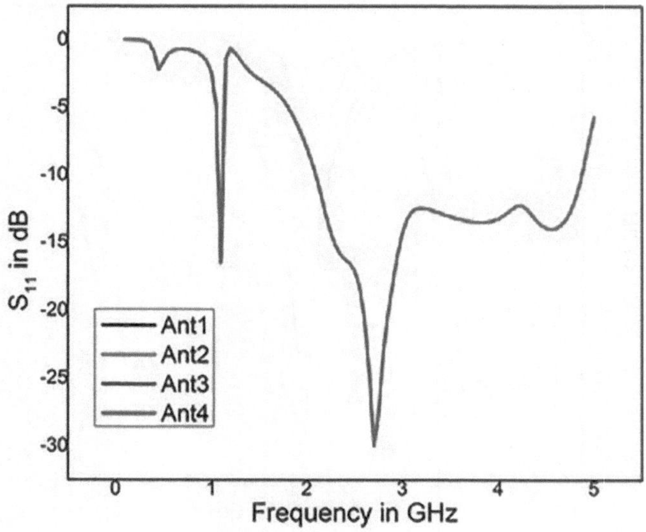

Figure 3.12 S_{11} plot of the miniaturised dual-band antenna.

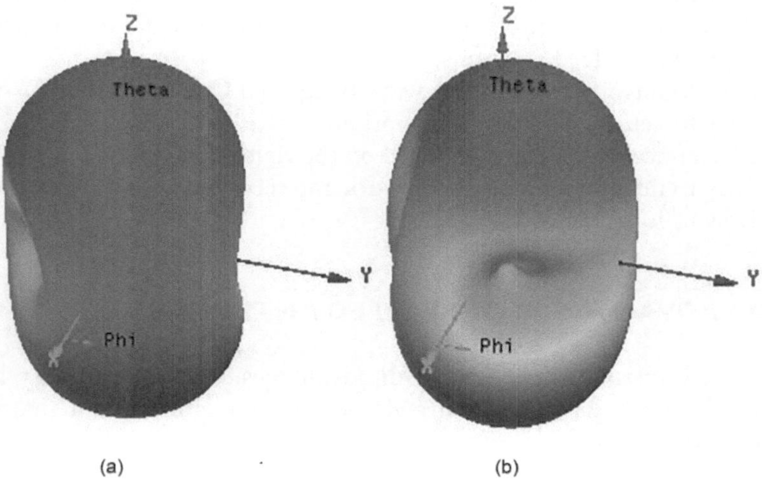

Figure 3.13 Radiation patterns of the miniaturised dual-band antenna.

dual-band antenna operates from 1.07 to 1.12 GHz and 2.1 to 4.87 GHz, while the S_{11} is less than –10 dB. The antenna performance is good from the impedance point of view.

The radiation patterns are a graphical representation of three-dimensional field fluctuations as a function of the parameters, and to define the radiation pattern in 2D, two primary plane cuts are required. These are referred to as the major E- and H-plane patterns, respectively.

Figure 3.13 depicts the 3D radiation patterns of the proposed dual-band printed antenna at two unique operating frequencies of 2.4 and 3.5 GHz, respectively, at two distinct operating frequencies of 2.4 and 3.5 GHz. When operating at 2.4 and 3.5 GHz frequencies, the suggested antenna achieves gains of 3.64 and 4.5 dB, respectively.

From Figure 3.13, it is clear that the designed antenna exhibits an omnidirectional radiation pattern at 3.5 GHz and a directional pattern at 6.1 GHz and the designed antenna attains a larger gain over the operating bands with acceptable reflection loss. As a result, this antenna could be considered as a better postulant for RF energy harvesting, 5G, Wi-Fi, WLAN and WiMAX applications.

3.5 ADVANTAGES OF PRINTED ANTENNAS

Some of the significant advantages of the printed antennas are as follows:

1. Lightweight, smaller volume and compact size.
2. Low profile, planar configuration and conformal.

3. Fabrication cost is low; hence, they can be produced in bulk.
4. Both linear and circular polarisations can be achieved simply by adjusting the feed positions.
5. Easy to integrate with Microwave Integrated Circuits (MICs).
6. Easy to achieve multiband operation.
7. Mechanically strong when fitted on the rigid surfaces.
8. Easy to integrate matching network and feed line with the antenna.
9. Easy to form large arrays.

3.6 DISADVANTAGES OF PRINTED ANTENNAS

The printed antennas have several disadvantages, even though they have several advantages. The major drawbacks of the microstrip antennas are described as follows:

1. Narrow bandwidth
2. Less efficiency
3. Lower gain
4. Unwanted radiation from junctions and feeds
5. Exhibits poor end-fire radiation
6. Power handling capability is low
7. Excitation of unwanted surface waves
8. The antenna size is large at low frequencies
9. Due to the small size, the design is complicated

3.7 APPLICATIONS OF PRINTED ANTENNAS

The printed antennas are mainly utilised in various wireless applications because of their smaller size, low cost, lightweight and low profile. Hence, they are highly compatible for the design of embedded antennas which are used in various handheld devices like cellular pagers, phones etc. The high-performance missile, spacecraft, aircraft and satellite applications also need lightweight, inexpensive and low-profile microstrip antennas. Nowadays, various commercial and government applications such as wireless communications, microwave imaging and mobile radio need microstrip antennas.

3.7.1 RF energy harvester

Energy harvesting (conversely known to as "energy scavenging") is the process of converting ambient energy in the environment into electrical energy that may subsequently be utilised to power autonomous electronic devices or circuits.

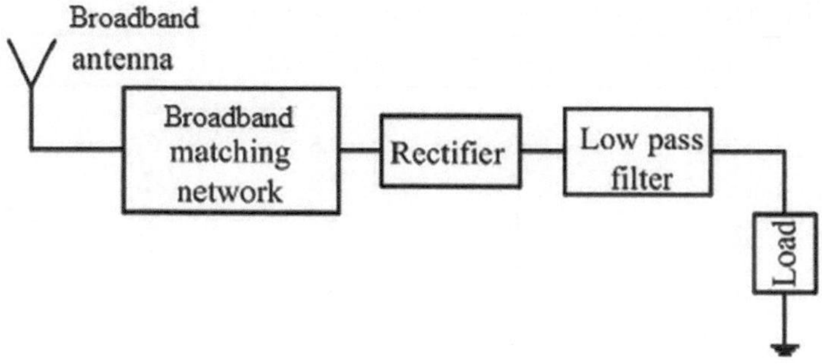

Figure 3.14 RF energy harvester.

The block diagram RF energy harvester is shown in Figure 3.14. RF energy harvesting circuits are required to be compact and lightweight to interface with current wireless devices; therefore, the antenna used must be low profile and capable of providing broadband operation.

3.7.2 5G antennas

The improvements in the field of mobile communication have had a significant impact on economic and social growth during the last few years. As a result, 5G technology has established itself as the future generation's pedestal. 5G technology is still in its infancy, yet it already offers evolutionary and revolutionary services. It is the next generation of technology that will enable ultra-high data speeds, extremely low latency, increased capacity and a high level of service quality. It's worth noting that 5G technology will create new potential for development to transcend conventional boundaries. As 5G technology also supports IoT, it enables a significant societal revolution in education, industry, healthcare and other sectors. 5G technology is intended to provide a vast IoT ecosystem in which many devices will be linked and a network can meet communication demands by balancing latency, cost and speed.

Millimetre wave bands (26, 28, 38 and 60 GHz) are 5G and support data transfer rates of up to 20 gigabits per second. Massive MIMO (Multiple Input Multiple Output – 64–256 antennas) technology enables performance "tenfold that of present 4G networks;" "Low-band 5G" and "Mid-band 5G" technologies operate at frequencies ranging from 600 MHz to 6 GHz, most notably 3.5–4.2 GHz, and enable download rates of up to 100–900 megabits per second.

The design parameters are displayed in Table 3.1.

Table 3.1 The design parameters of the proposed antennas

Parameters	*Value (mm)*	*Parameters*	*Value (mm)*	*Parameters*	*Value (mm)*
A_1	32	B_5	9	C_5	13
A_2	38	B_6	3.5	C_6	4
A_3	12	B_7	3	C_7	7
A_4	12	B_8	15.4	C_8	4
A_5	5	B_9	19	C_9	3
B_1	23	C_1	50	C_{10}	3
B_2	21	C_2	42	C_{11}	8
B_3	3	C_3	9	C_{12}	9
B_4	14.5	C_4	5.3	C_{13}	5

3.8 CONCLUSIONS

In this chapter, three different low-profile broadband printed antennas for various wireless applications have been designed and analysed. The proposed antennas are compact in size and exhibit broad bandwidth. The different shaped radiating patches are utilised to get the desired operation. The ground plane is modified to improve the quality of impedance matching and thereby increased bandwidth. The proposed printed antennas are used for RF energy harvesting, 5G, Wi-Fi, WLAN, WiMAX and C-band applications due to their wide bandwidth, compact size and better radiation patterns.

REFERENCES

1. Balanis, CA. *Antenna Theory: Analysis and Design.* 4th ed., John Wiley & Sons, Hoboken, NJ, 2016.
2. Garg R, Bhartia P, Bahl I, Ittipiboon A. *Microstrip Antenna Design Handbook*, Artech House, Boston, MA, 2001.
3. Girish Kumar KP. *Ray, Broadband Microstrip Antennas*, Artech, 2002.
4. Penchala Reddy S, Narayana Reddy S. Design of dual band bisected Psi antenna for 3G, Wi-Fi, WLAN and Wi-MAX applications. *Journal of Telecommunications and Information Technology*, 2019.
5. Sachin A, Manoj P, Pravin K. A quad-band antenna for multi-band radio frequency energy harvesting circuit. *AEU – International Journal of Electronics and Communications*, 85, 2018. doi: 10.1016/j.aeue.2017.12.035.
6. Tcheg P, Möck M, Pouhè D. A new broadband antenna of high gain: The double-cornu spiral antenna. *Progress in Electromagnetics Research* C, 118, 199–212, 2022.
7. Pandey MS, Chaudhary VS. Defected star-shaped microstrip patch antenna for broadband applications. *Progress in Electromagnetics Research* C, 118, 11–24, 2022.

8. Rajbala. Compact and broadband uniplanar microstrip antenna for endfire radiation. *Progress in Electromagnetics Research Letters*, 102, 77–85, 2022.
9. Yaseen RM, Naji DK, Shakir AM. Optimization design methodology of broadband OR multiband antenna for RF energy harvesting applications. *Progress in Electromagnetics Research B*, 93, 169–194, 2021.
10. Xu W, Fan Z. Design of a broadband low-profile dual-polarized antenna for 5G base station. *Progress in Electromagnetics Research C*, 114, 129–142, 2021.
11. Yang M, Zhou J. A broadband high-isolation dual-polarized antenna for 5G application. *Progress in Electromagnetics Research M*, 85, 39–48, 2013.
12. Satyadeep D, Sudhakar S. Design of high gain, broadband resonant cavity antenna with meta-material inspired superstrate. *AEU – International Journal of Electronics and Communications*. doi: 10.1016/j.aeue.2018.12.021.
13. Malik PK. Chapter 4. Mathematical modeling and principle of wireless communication. In: *Energy Harvesting Technologies for Powering WPAN and IoT Devices for Industry 4.0 Up-Gradation*, Nova Science Publishers, Inc., Hauppauge, NY, April 2020, ISBN: 9781536169430.
14. Roges R, Malik PK. Planar and printed antennas for internet of things-enabled environment: opportunities and challenges. *International Journal of Communication Systems*, 34(15), e4940, 2021. https://doi.org/10.1002/dac.4940, ISSN: 1099-1131.
15. Rahim A, Malik PK. Analysis and design of fractal antenna for efficient communication network in vehicular model. In: *Sustainable Computing: Informatics and Systems*, Elsevier, vol. 31, 2021, p. 100586. https://doi.org/10.1016/j.suscom.2021.100586, ISSN 2210-5379.

Chapter 4

A compact corner truncated microstrip patch antenna for radio frequency energy harvesting to low-power electronic devices and wireless sensors

Pradeep Chindhi
Sant Gajanan Maharaj College of Engineering

Rajani H. P.
Jain College of Engineering

Geeta Kalkhambkar
Sant Gajanan Maharaj College of Engineering

Nehru Kandasamy
National University of Singapore

CONTENTS

DOI: 10.1201/9781003347057-4

4.1 INTRODUCTION

Without energy, life has become impossible. Day-by-day energy demand is increasing exponentially, and at the same time, there is an increase in IoT devices. Since the cost of electricity has risen dramatically in recent years, the traditional methods of generating electricity are insufficient. On the other hand, battery deposition harms the climate and pollutes the environment. A limited battery period has stimulated industry experts and researchers to experiment with new designs and technologies to operate wireless mobile devices (WMD) for an enhanced period. Table 4.1 shows the unlicensed frequency bands that can be used for antenna design and simulation. Table 4.2 shows the power usage/requirement of various sensors which are used in the field of IoT applications.

In Ref. [1], a double patch antenna array (DPA) is presented for simultaneous wireless information and power transfer (SWIPT). Two patch antennas, one for 1.8 and the other for 2.45 GHz, were itched on the same dielectric with $W1=34.00$ and $L1=40.00$ to optimize the performance. The obtained gains are 1.5 and 1.8 dBi at 1.8 and 2.45 GHz. The inset-fed wide-band rectenna with defected ground structure is designed and analyzed for RF energy harvesting in the range of 5.336–6.194 GHz frequency. The antenna realizes a gain of 6.189 dB and directivity of 8.776 dBi with Ls (mm)=89.2 and Ws (mm)=93.2 [2]. In Ref. [3], the microstrip patch antenna with truncated edges is designed to enhance the bandwidth. A truncated arc patch antenna loaded with an innovative complementary

Table 4.1 Unlicensed frequency bands [15]

Sr. no.	*01*	*02*	*03*	*04*	*05*	*06*	*07*
Frequency bands in MHz	2400–2483	5150–5250	5250–5350	5470–5725	5725–5875	5825–5875	3300–3600

Table 4.2 Details of power usage of various sensors in IoT devices [16–18]

Sr. no.	*Sensors for IoT application*	*Power usage*	*Sr. no.*	*Sensors for IoT application*	*Power usage*
01	Smoke sensor	0.1 mW	07	Gas sensor	500–800 mW
02	Temperature sensor	0.5–5 mW	08	Infrared sensor	0.2 W
03	Optical sensor	1 mW–10 nW	09	Gyroscope sensor	0.817 W
04	Accelerometer	3 mW	10	Proximity sensor	2.4 W
05	Pressure sensor	10–15 mW	11	Photoplethysmography (PPG) sensor, average power consumption	1.47 mW
06	Image sensor	150 mW	12	Humidity sensor, average power consumption	1 mW

slotted split ring resonator (CSSRR) in the ground is discussed in Ref. [4]. The antenna accomplishes extensive bandwidth and circular polarization (CP). The determined gain of 2.476 dBi at 2.75 GHz is obtained. The impedance matching, bandwidth improvement, and frequency change with corner truncated microstrip antenna with slots is offered in Ref. [5]. The details of different rectifier topologies with impedance matching techniques for numerous frequencies of RF energy harvesting are expressed in Refs. [6–11].

The chapter is put together as follows. A truncated microstrip patch antenna has been presented in Section 4.2. Section 4.3 presents the results and discussions for the antenna; Section 4.4 wraps up the chapter [12–14].

4.2 DESIGN, SIMULATION, AND ANALYSIS

Table 4.3 shows optimized dimensions of the compact corner truncated microstrip patch antenna. The antenna dimensions are calculated and simulated on Flame Retardant-4 (FR-4) substrate of dielectric constant 4.3 with a loss tangent 0.025 and a substrate thickness of 1.6 mm. Figure 4.1a shows the proposed antenna geometry with design variables. Figure 4.1b shows the initial geometry (Case-I) to the final geometry (Case-VII) transformation.

Figure 4.2 represents simulated parametric details of |*S*11| for the initial geometry (Case-I) to final geometry (Case-VI). From Figure 4.2 and Table 4.4, it is clear that there is a frequency shift from 3.238 GHz in the initial iteration to 3.44 GHz at the final iteration. In the proposed antenna, corner truncation is accomplished to cover a new frequency band ranging from 3.3 to 3.6 GHz.

Figure 4.2 displays the simulated parametric details of *S*-parameters |*S*11| demonstrating the reflections from the antenna and frequency shift from Case-I to Case-VII. As realized from Figure 4.3, the antenna exhibits a bandwidth from 3.3437 to 3.5336 GHz which is 0.1899 GHz (189.9 MHz) Figure 4.4 depicts the impedance characteristics of the proposed antenna.

Figure 4.5 depicts the voltage standing wave ratio (VSWR) plot of the simulated antenna. It is observed that the VSWR attained at 3.45 GHz is 1.

Table 4.3 Antenna design variable

Design variables	*Wg*	*Lg*	*Wp*	*Lp*	*Lt*	*Lt1*	*Fl*	*Wf*	*lg*
Dimension in mm	36	30	27	21	8.955	5.6376	8.700	3.00	1.200

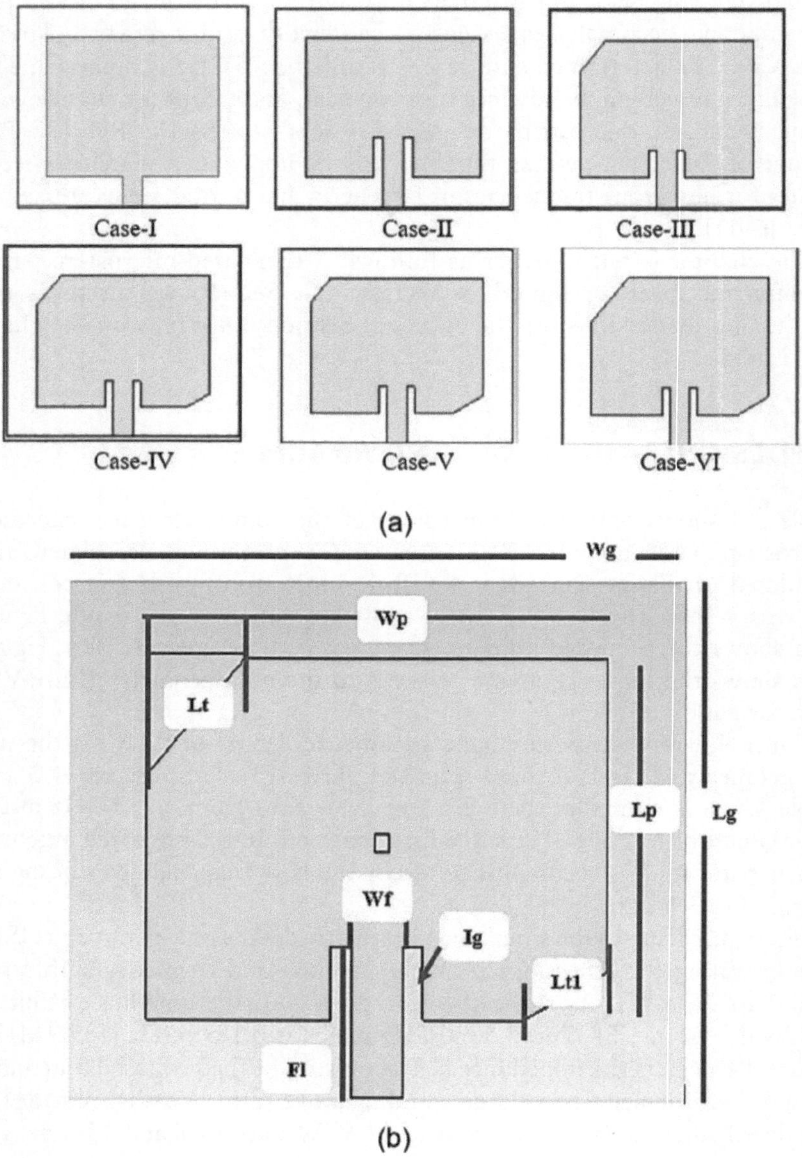

Figure 4.1 (a) Geometry with design variables (b) geometrical iteration.

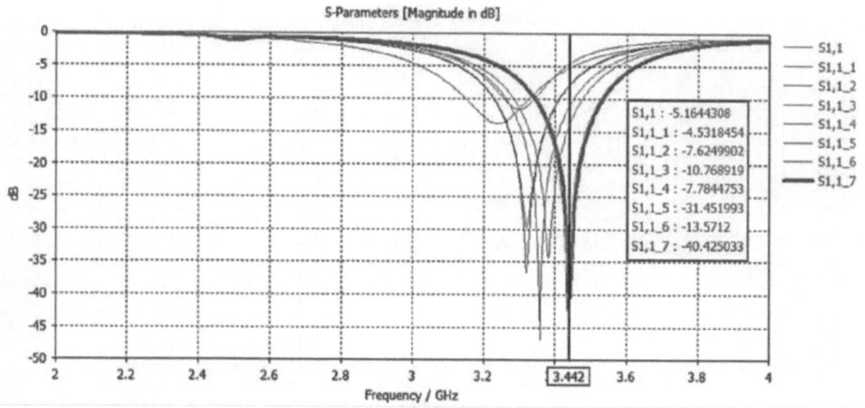

Figure 4.2 Simulated parametric details of S-parameters |S11|.

Table 4.4 Parametric |S11| details

Cases	*Case-I*	*Case-II*	*Case-III*	*Case-IV*	*Case-V*	....	*Case-VII*
fr in GHz	3.238	3.298	3.32	3.358	3.434	....	**3.442**
|S11| in dB	−13.74	−11.54	−29.67	−46.80	−42.48	....	**−40.42**

fr, resonant frequency; |S11|, S-parameters.

Bold values is in dB

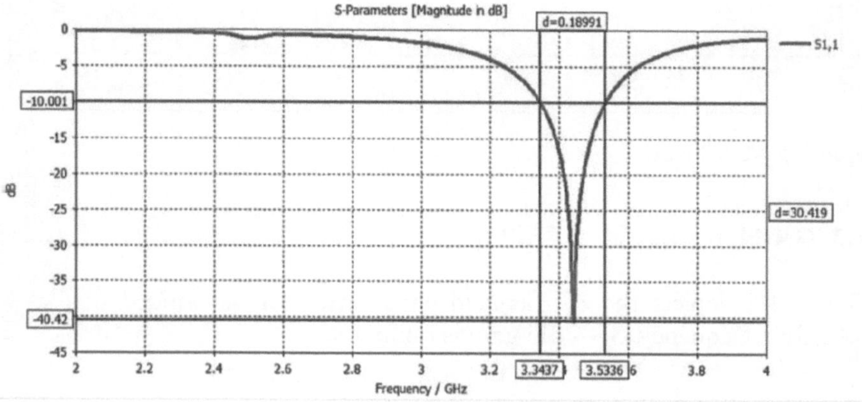

Figure 4.3 Simulated S-parameter and bandwidth.

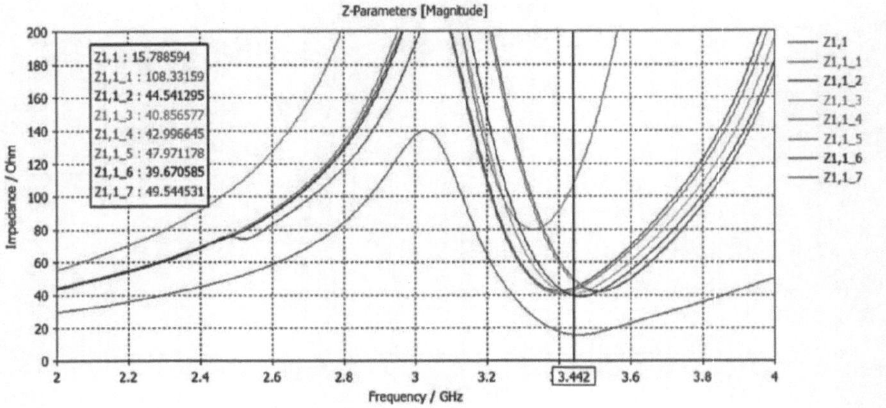

Figure 4.4 Simulated parametric Z1,1.

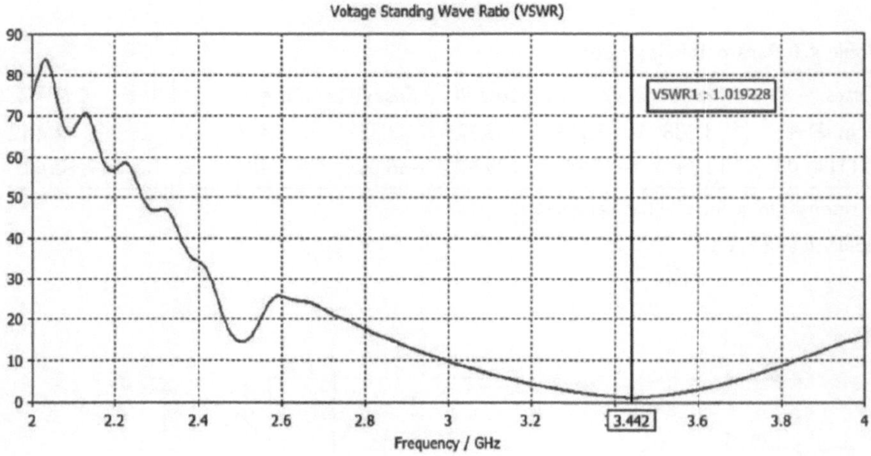

Figure 4.5 VSWR plot.

4.3 GAIN

Figure 4.6 depicts the 3D far-field gain pattern. It is noticed that at a 3.45 GHz frequency, 3.44 dBi gain is obtained.

4.4 DIRECTIVITY

Figure 4.7 depicts the directivity of the planned antenna. The attained directivity is 5.5 dBi.

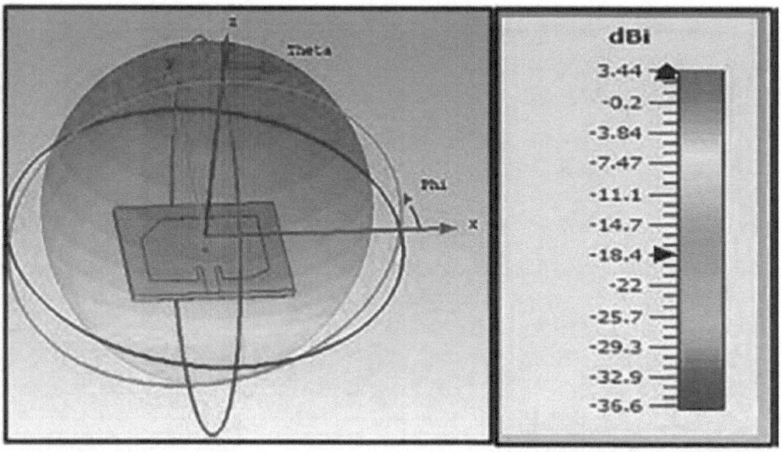

Figure 4.6 3D gain plot.

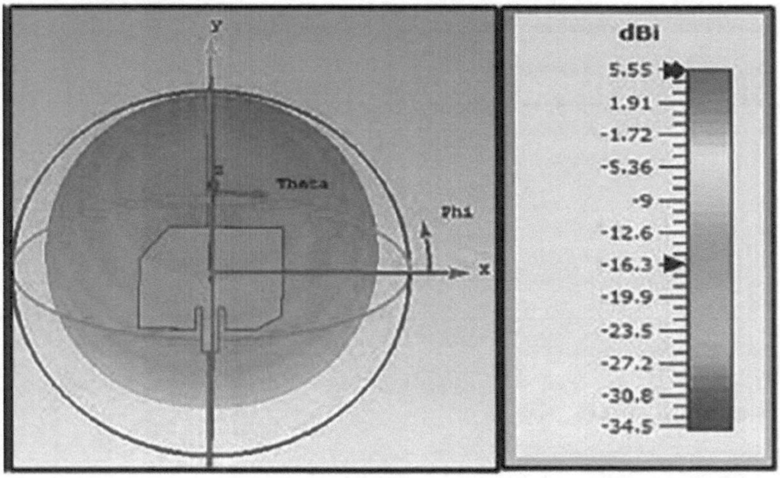

Figure 4.7 3D directivity plot.

4.5 SURFACE CURRENT DISTRIBUTION-FINAL ITERATION

The surface current distribution (SCD) is depicted in Figure 4.8. The red percentage confirms the maximum current circulation over the radiating area.

The surface current distribution (SCD) at 3.45 GHz is depicted in Figure 4.9.

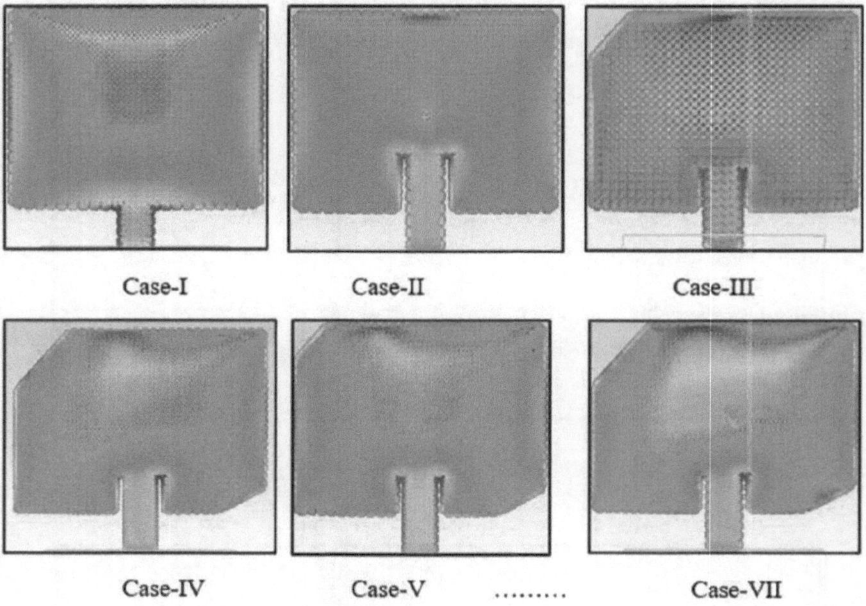

Figure 4.8 Surface current distribution for Case-I to Case-VII.

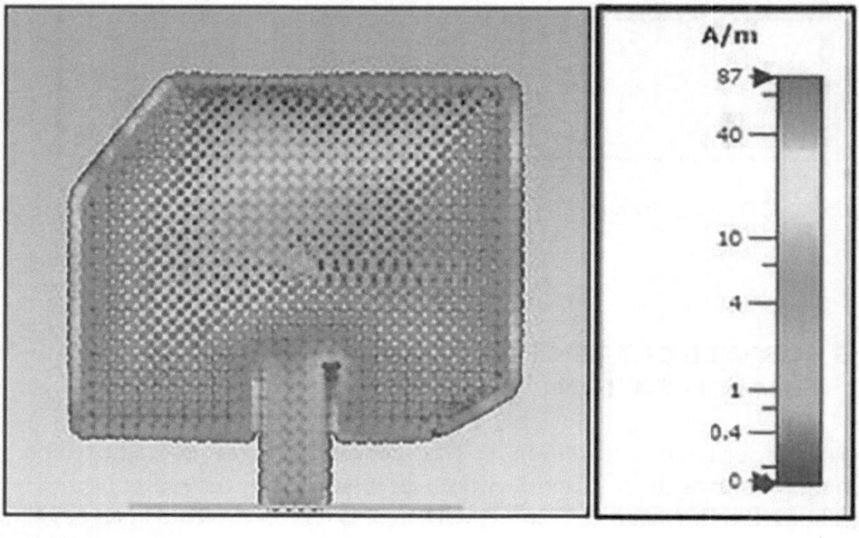

Figure 4.9 Surface current distribution at 3.45 GHz.

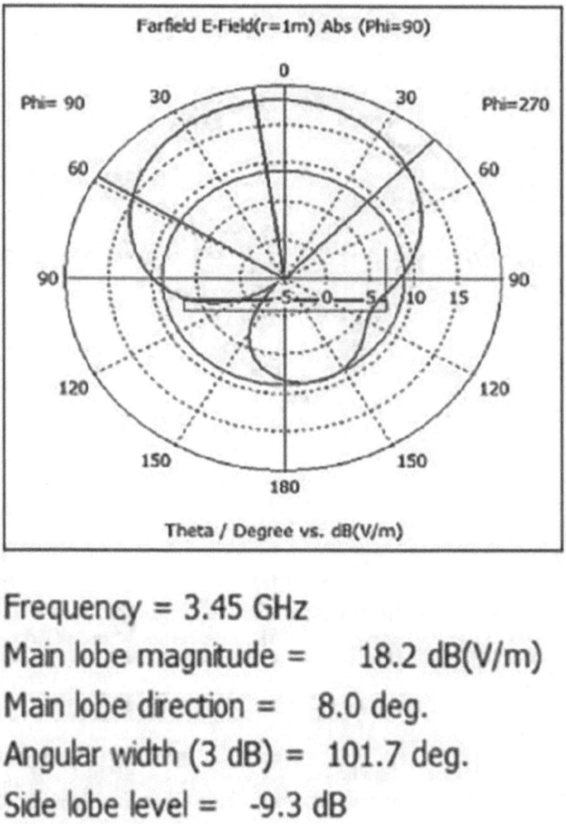

Figure 4.10 Far-field E-field, radiation patterns at 3.45 GHz. E-plane (in grey) and H-plane (in black).

4.6 RADIATION PATTERN

E- and H-plane far-field radiation patterns are shown in Figures 4.10 and 4.11, respectively. It is detected that the compact corner truncated microstrip patch antenna resonates with a maximum gain of 3.44 dBi at 3.45 GHz. The details of the main lobe magnitude and sidelobe level are shown in Figures 4.10 and 4.11.

The radiation and total efficiency of the proposed antenna are depicted in Figure 4.12. In the proposed compact corner truncated microstrip patch antenna, the maximum radiation and total efficiency are 0.6162 (61.62%) and 0.6159 (61.59%), respectively. Insert slots (of any type) in the proposed antenna increase radiation and total efficiency while reducing copper loss. While inserting slots, make sure that the performance of the antenna should not change. A minor change in the radiation and total efficiency is observed due to the insertion of a square slot in the proposed antenna.

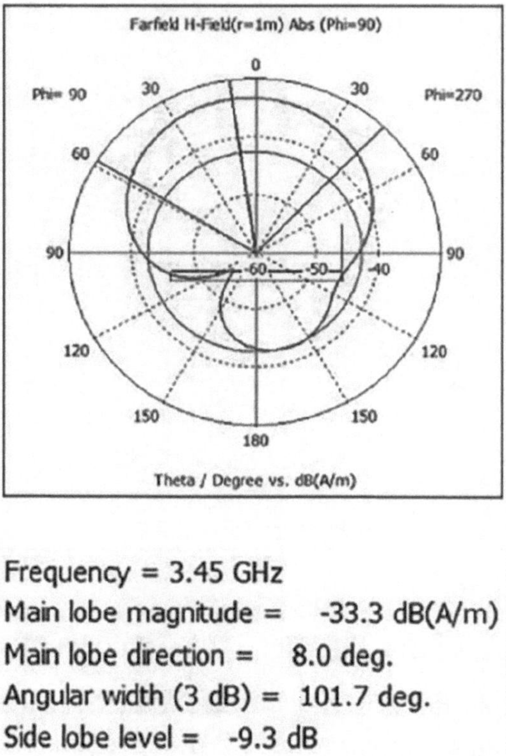

Figure 4.11 Far-field H-field, radiation patterns at 3.45 GHz. E-plane (in grey) and H-plane (in black).

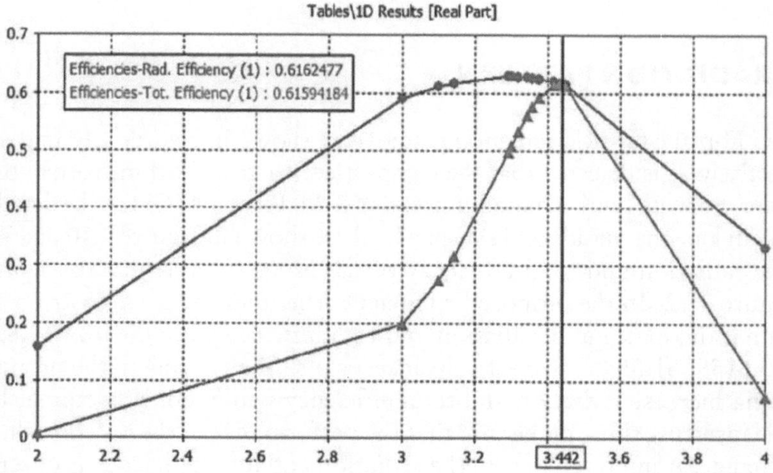

Figure 4.12 Radiation and total efficiency at 3.442 GHz.

4.7 CONCLUSION

A compact corner truncated microstrip patch antenna for RF energy harvesting has been proposed. Frequency shift, impedance matching, and gain improvement have been realized due to the truncation of two opposite edges. The optimum gain and directivity of the proposed antenna are 3.44 and 5.55 dBi. The proposed antenna finds application in RF energy harvesting for powering wireless IoT sensors.

4.8 FUTURE SCOPE

The gain and directivity of the proposed compact corner truncated microstrip patch antenna can be improved by integrating an array of antennas. The performance analysis of the antenna with different RF rectifier topologies needs to be evaluated. After performance analysis, a rectenna system can be implemented for powering wireless IoT sensors.

REFERENCES

1. Silva V S, Cambero E V, Pereira E T, Casella I R S, Capovilla C E, Double Patch Antenna Array for Communication and Out-of-band RF Energy Harvesting, *Journal of Microwaves, Optoelectronics and Electromagnetic Applications*, 19, No. 3, September 2020, http://dx.doi.org/10.1590/2179-10742020v19i378.
2. Rajawat A, Singhal P K, Design and Analysis of Inset Fed Wide-Band Rectenna with Defected Ground Structure, *Journal of Circuits, Systems, and Computers*, 29, No. 3, 15 pages, 2020, World Scientific Publishing Company, doi: 10.1142/S0218126620500474.
3. Kaur A, Malik P K, Microstrip Patch Antenna with Truncated Edges for Bandwidth Improvement for Wireless Applications, 2021. Original Paper Chapter, https://www.springerprofessional.de/en/microstrip-patch-antenna-with truncated-edges-for-bandwidth-impr/18759036.
4. Rao S M, Basarkod P I, A Novel Complementary Slotted Split Ring Resonator Loaded Truncated Arc Patch Antenna with Enhanced Performance, *Progress in Electromagnetics Research C*, 101, 203–218, 2020, doi: 10.2528/PIERC20031003.
5. Salleh S, Zakariya M A, Lee R M A, A Comparison Study of Rectifier Designs for 2.45 GHz EM Energy Harvesting, *Energy and Power Engineering*, 13, 81–89, 2021, https://www.scirp.org/journal/epe, ISSN Online: 1947-3818, ISSN Print: 1949-243X.
6. Baudha S, Kumar V D, Corner Truncated Broadband Patch Antenna with Circular Slots, *Microwave and Optical Technology Letters*, 57, No. 4, April 2015, doi: 10.1002/mop.
7. AL-Azawy M, Sari F, Analysis of Dickson Voltage Multiplier for RF Energy Harvesting, *2019 1st Global Power, Energy and Communication Conference (IEEE GPECOM2019)*, June 12–15, 2019, Cappadocia, Turkey, 978-1-5386–8086-5/19/$31.00 ©2019 IEEE.

8. Christiana Erinosho T, Adeniyi Adekola S, Akinwale Amusa K, Design of Practical Rectennas for RF Energy Harvesting, *2019 PhotonIcs & Electromagnetics Research Symposium*, Spring (PIERS I SPRING), 17–20 June, Rome, Italy.
9. Bahhar C, Aidi M, Mejri F, Aguili T, Design and Optimization of High-efficiency Rectenna for RF Energy Harvesting, *2019 PhotonIcs & Electromagnetics Research Symposium*, Spring (PIERS I SPRING), 17–20 June, Rome, Italy.
10. Rajawat A, Singhal P K, Design and implementation of a Dual-Band Rectifier Antenna for Efficient RF Energy Harvesting in Wireless Sensor Networks, *Journal of Circuits, Systems, and Computers*, doi: 10.1142/S0218126619500348.
11. Mahmoud M S, Mohamad A H, A Study of Efficient Power Consumption Wireless Communication Techniques/Modules for Internet of Things (IoT) Applications, *Advances in Internet of Things*, 6, 19–29, 2016, http://www.scirp.org/journal/ait, http://dx.doi.org/10.4236/ait.2016.62002.
12. Shaik N, Malik P K, A Comprehensive Survey 5G Wireless Communication Systems: Open Issues, Research Challenges, Channel Estimation, Multi Carrier Modulation and 5G Applications, *Multimedia Tools and Applications*, 2021, https://doi.org/10.1007/s11042-021-11128–z.
13. Tiwari P, Malik P K, Wide Band Micro-Strip Antenna Design for Higher "X" Band, *International Journal of e-Collaboration (IJeC)*, 17, No. 4, 60–74, 2021, http://doi.org/10.4018/IJeC.2021100105, ISSN: 1548-3673.
14. Wadhwa D S, Malik P K, Khinda J S, High Gain Antenna for n260- & n261-Bands and Augmentation in Bandwidth for mm-Wave Range by Patch Current Diversions, *World Journal of Engineering*, 2021, https://doi.org/10.1108/WJE-03-2021-0133, ISSN: 1708-5284.
15. https://www.livemint.com/industry/telecom/telecom-ministry-will-identify-spectrum-bands-to-roll-out-5g-network-in-india-1558844925443.html.
16. Pandey R, Shankhwar A K, Singh A, An Improved Conversion Efficiency of 1.975 to 4.744 GHz Rectenna for Wireless Sensor Applications, *Progress in Electromagnetics Research C*, 109, 217–225, 2021.
17. Vibhav K S, Syed A I, Beg M T, Energy-efficient Communication Methods in Wireless Sensor Networks: A Critical Review, *International Journal of Computer Applications*, 39, No. 17, February 2012, doi: 10.5120/4915-7484.
18. Yang Y, Yeo J, Priya S, Harvesting Energy from the Counterbalancing (Weaving) Movement in Bicycle Riding, *Sensors*, 12, 10248–10258, 2012, doi: 10.3390/s120810248.

Chapter 5

Microstrip interconnect design and modeling using reverse approach to obtain an efficient wideband MS line-to-RWG hybrid transition

A. Varshney and V. Sharma
Gurukul Kangri

Roshan Kumar
Henan University

CONTENTS

5.1 ABSTRACT

This chapter demonstrates the application of the reverse transition model to the design of a simple, low-cost, via-less transition model interconnect with the highest transmission, minimum reflections in conjunction with wideband –10 dB fractional bandwidth of more than 72%. The modeling analysis and method steps to generate RLC electrical equivalent circuits of the proposed microstrip interconnect are also presented. The RLC electrical equivalent T- and Π-circuits for reverse transition have been also realized in this chapter. The proposed modeling method is moreover pertinent to other microwave bands and millimeter-wave transitions and components as well.

DOI: 10.1201/9781003347057-5

5.2 INTRODUCTION

The reverse transition word had been primarily used by Earl R. Murphy in 1985 [1]. A reverse transition is an alternative approach to exciting transverse electric dominant mode TE_{10} by extending a portion of microstrip line main conductor into the rectangular waveguide. This microstrip interconnect plays a peculiar role in microwave and millimeter-wave (wireless and RADAR) applications. This chapter demonstrates a simplified approach to generating an inverse Ka-band two-port transition between the microstrip (MS) line and rectangular waveguide (RWG). The RT has been designed at a center frequency of 30 GHz. This MS line-to-RWG is exactly the same as an ordinary transition. The only difference is the insertion of the MS line portion into the RWG. In RT, microstrip lines' main conductor strip face is toward the back-short of the waveguide, whereas in the ordinary transition, the microstrip is facing toward the RWG port [1].

There are several kinds of inline end inserted and side inserted (broad wall or wide wall) MS line-to-waveguide transitions available [2]. These generally available techniques and methodologies have not been suitable for applications where wide bandwidth, low insertion losses (IL), and good return losses (RL) are necessary. These electromagnetic couplers are called ordinary transitions. On the other hand, there are possibilities of other kinds of MS line-to-RWG transitions, if the face of the MS line probe inserted into the waveguide is opposite to the waveguide port or we can simply say it is facing the back-short of the waveguide. These are usually known as reverse transitions [1].

In this chapter, a simplified highly accurate electrical T-equivalent RLC circuit and Π-equivalent RLC circuit of a Ka-band microstrip (MS) line-to-rectangular waveguide (RWG) reverse transition have been generated and then synthesized with a minimum number of passive lumped components. The circuit analysis gives a noteworthy compromise between the MS line-to-RWG transition 3D model and its equivalent circuits [10–12]. The exact calculated values of each electrical R, L, and C have been presented in a tabular form for both T- and Π-equivalent RLC circuits of RT design at the operating frequency. These equivalent circuits are essential for future smartphones and other laboratory electronic equipment and baby toys.

5.3 REVERSE TRANSITION MODEL

A simplified approach to generating RLC electrical equivalent circuit model of MS line interconnect using the two-port networks analysis conversion relationships [13] and the conversion algorithm steps have been explained in the subsequent sections.

5.3.1 Design equations

i. *Microstrip Line Calculations* [13]

The effective dielectric constant:

$$\varepsilon_{\text{eff}} = \frac{\varepsilon_r + 1}{2} + \frac{\varepsilon_r - 1}{2}\left(1 + 12\frac{h}{w}\right)^{-1/2} \tag{5.1}$$

where ε_{eff} = Effective dielectric constant or permitivity of microstrip line, ε_r = dielectric constant of substrate, w=width of microstrip, and h=height of the substrate.

Characteristics impedance:

$$Z_0 = \frac{60}{\sqrt{\varepsilon_{\text{eff}}}}\ln\left(8\frac{h}{w} + \frac{w}{4h}\right) \quad \text{for } \frac{w}{h} \le 1 \tag{5.2a}$$

$$Z_0 = \frac{120\pi}{\left(\sqrt{\varepsilon_{\text{eff}}}\right)\left[\frac{w}{h} + 1.393 + 0.667\ln\left(\frac{w}{h} + 1.444\right)\right]} \quad \text{for } \frac{w}{h} \ge 1 \tag{5.2b}$$

Microstrip width is calculated with the relation:

$$Z_0 = \frac{\eta}{\sqrt{\varepsilon_r}}\frac{h}{w} \tag{5.3}$$

where w=microstrip width, f_0=designed frequency width of substrate and ground; W_{sub}.$=W_{\text{gnd}}$.=9–10 times the width of the microstrip:

$$\text{i.e.}\quad W_{\text{sub.}} = W_{\text{gnd.}} = 9\text{ w or } 10\text{ w} \tag{5.4}$$

Alternatively, this can also be evaluated by

$$W_{\text{sub.}} = \frac{1}{2f_0\sqrt{\mu_0\varepsilon_0}}\sqrt{\frac{2}{\varepsilon_r + 1}} \tag{5.5}$$

ii. *Rectangular Waveguide Calculations* [13]

$$\text{Wavelength, } \lambda_0 = \frac{c}{f_0},\ c = \text{light velocity} \tag{5.6}$$

Dominant modes, TE_{10} and TM_{11}

Cut-off frequency,

$$f_c = \frac{c}{2\sqrt{\mu\varepsilon}}\sqrt{\left(\frac{m}{a}\right)^2 + \left(\frac{n}{b}\right)^2} \tag{5.7}$$

where $m=1$ and $n=0$ for TE_{10} mode, a=breadth of waveguide, and b=width of waveguide.

Guided wavelength:

$$\lambda_g = \frac{\lambda_0}{\sqrt{1-\left(\frac{f_c}{f_0}\right)^2}} \tag{5.8}$$

Waveguide impedance:

$$Z_g = \frac{\eta}{\sqrt{1-\left(\frac{f_c}{f_0}\right)^2}} \quad \text{for TE mode} \tag{5.9a}$$

$$Z_g = \eta\sqrt{1-\left(\frac{f_c}{f_0}\right)^2} \quad \text{for TM mode} \tag{5.9b}$$

where $\eta = \sqrt{\frac{\mu}{\varepsilon}}$.

Waveguide back-short distance:

$$D = \frac{\lambda_g}{4} \tag{5.10}$$

$$\text{Waveguide length, } L = N * \frac{\lambda_g}{2} \tag{5.11}$$

where N=any integer value other than 0 .

iii. Conversion Equations

From the HFSS simulation parameter optimization, we get S-parameters as S_{11}, S_{12}, S_{21}, and S_{22}. Characteristic impedances of MS line is $Z_0 = 50\ \Omega$ and that for WR-28, RWG TE_{10} dominant mode $Z_{TE} = 530\ \Omega$.

Impedance Z-parameters in terms of S-parameters [13] are given by the following conversion relationships:

$$Z_{11} = Z_0 \frac{(1+S_{11})(1-S_{22})+S_{12}S_{21}}{(1-S_{11})(1-S_{22})-S_{12}S_{21}} \tag{5.12}$$

$$Z_{12} = Z_0 \frac{2S_{12}}{(1-S_{11})(1-S_{22})-S_{12}S_{21}} \tag{5.13}$$

$$Z_{21} = Z_0 \frac{2S_{21}}{(1-S_{11})(1-S_{22})-S_{12}S_{21}} \tag{5.14}$$

$$Z_{22} = Z_0 \frac{(1-S_{11})(1+S_{22})+S_{12}S_{21}}{(1-S_{11})(1-S_{22})-S_{12}S_{21}} \tag{5.15}$$

Admittance *Y*-parameters in terms of *S*-parameters [13] are given by the following conversion relationships:

$$Y_{11} = Y_0 \frac{(1-S_{11})(1+S_{22})+S_{12}S_{21}}{(1+S_{11})(1+S_{22})-S_{12}S_{21}} \tag{5.16}$$

$$Y_{12} = Y_0 \frac{-2S_{12}}{(1+S_{11})(1+S_{22})-S_{12}S_{21}} \tag{5.17}$$

$$Y_{21} = Y_0 \frac{-2S_{21}}{(1+S_{11})(1+S_{22})-S_{12}S_{21}} \tag{5.18}$$

$$Y_{22} = Y_0 \frac{(1+S_{11})(1-S_{22})+S_{12}S_{21}}{(1+S_{11})(1+S_{22})-S_{12}S_{21}} \tag{5.19}$$

where $Y_0 = \frac{1}{Z_0}$ (5.20)

All Z are in $(R+jX)$ form and Y are in $(G+jB)$ form.

S-parameter to *Z*-parameter conversion formulas for multi-port network are given in Ref. [14]:

$$[Z] = ([I]-[S])^{-1} \cdot ([I]+[S]) Z_0 \tag{5.21}$$

where $[Z]$, $[I]$, and $[S]$ have their usual meanings.

5.3.2 Modeling methodology to generate electrical equivalent model

Modeling methodology steps have been presented in this section. The following general procedure steps have been used to generate RLC electrical equivalent circuits of any microstrip line interconnects/antennas/microwave components:

i. First, design the ordinary MS line transition in the desired frequency band by choosing suitable parameters, design software (HFSS or CST), substrate, and design procedure steps [2,15].
ii. Next, record the simulated results in terms of their return loss and insertion loss, fractional bandwidth, and VSWR.
iii. Once satisfied with the parameters of step II, observe and record the complex magnitudes as well as phases of all *S*-parameters (S_{11}, S_{12}, S_{21}, and S_{22}) at the designed frequency.
iv Now apply two-port circuit theory analysis to convert all the *S*-parameters (S_{11}, S_{12}, S_{21}, and S_{22}) in to *Z*-parameters (i.e., Z_{11}, Z_{12}, Z_{21}, and Z_{22}).
v. Then determine the T-equivalent complex impedance values from *Z*-parameters as shown in Figure 5.1a.
vi. Now comparing each T-equivalent electrical impedance with $(R+jX)$, it provides the values of *R*, *L*, and *C* at the designed frequency.
vii. Draw the RLC electrical equivalent two-port T-network for the designed 3D model of MS line interconnects.
viii. Now apply the two-port circuit theory to convert all the *S*-parameters (S_{11}, S_{12}, S_{21}, and S_{22}) into admittance *Y*-parameters (i.e., Y_{11}, Y_{12}, Y_{21}, and Y_{22}). Then determine the Π-equivalent network from *Y*-parameters.
ix. Afterward, comparing all Π-equivalent electrical impedance with $(G+jB)$, it provides the values of conductance, *G* and susceptance, and *B* values at the designed frequency.
x. Determine the values of *R*, *L*, and *C* from the values of *G* and *B*.
xi. Draw the RLC electrical equivalent two-port Π-network for the designed reverse MS line-to-RWG transition model as shown in Figure 5.1b.

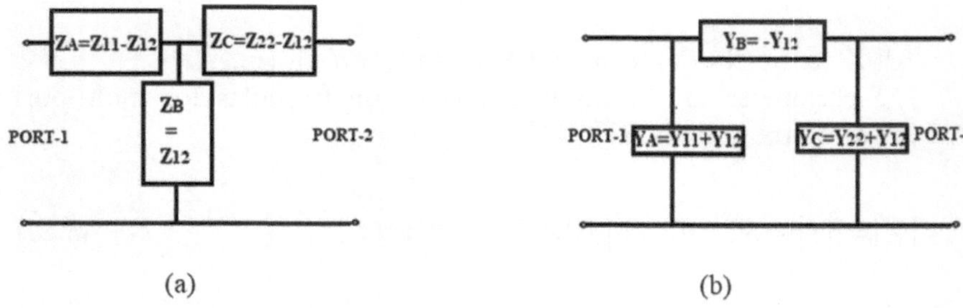

Figure 5.1 General equivalent modeling circuits for any two-port network. (a) Electrical T-equivalent circuit and (b) electrical Π-equivalent circuit.

5.3.3 Equivalent circuits of a two-port network

From the impedance and admittance parameters of Section 5.3.1 conversion formulas, electrical T-equivalent and electrical Π-equivalent circuits [5,13] can easily be obtained as shown in Figure 5.1a and b.

5.3.4 Ka-band MS-to-RWG reverse transition

A Ka-band reverse MS line-to-RWG transition at the designed center frequency of 30 GHz is shown in Figure 5.2a and b [2,15–17]. A reverse

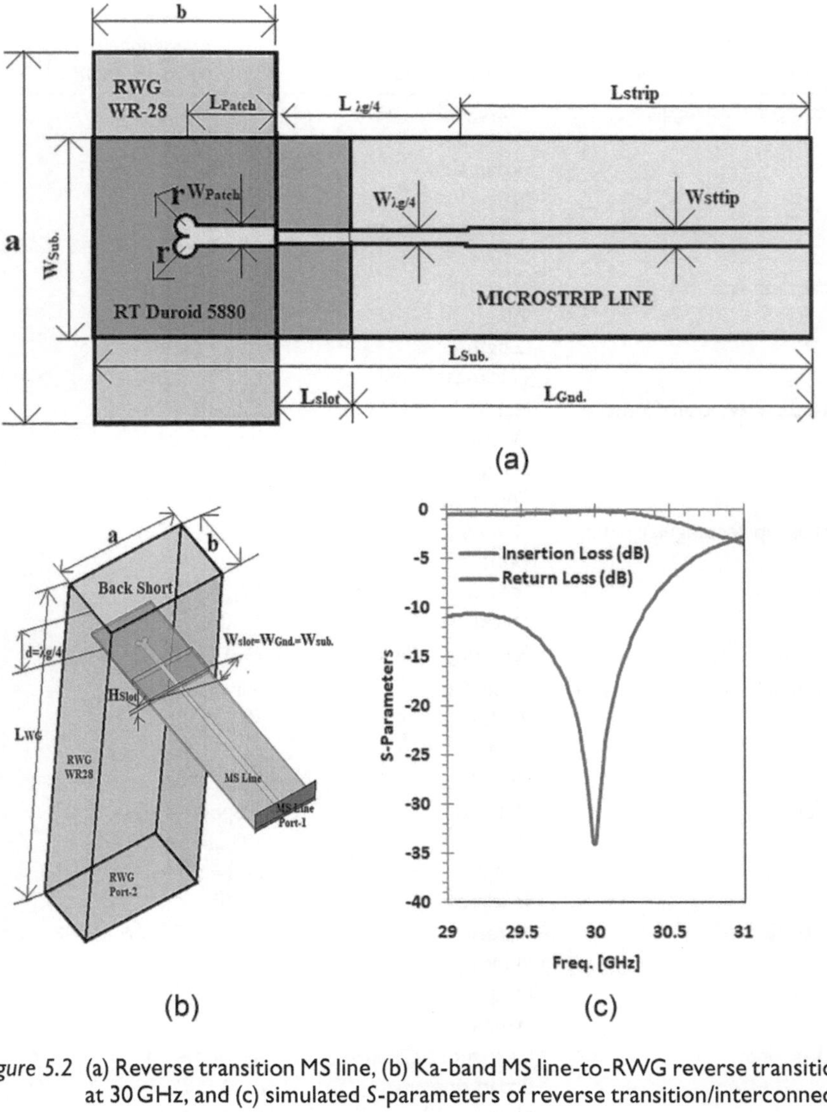

Figure 5.2 (a) Reverse transition MS line, (b) Ka-band MS line-to-RWG reverse transition at 30 GHz, and (c) simulated S-parameters of reverse transition/interconnect.

transition between MS line and WR28 is formed by placing a slot of suitable height and width equal to the substrate width of MS line in the broad wall of WR28. A microstrip line (probe facing toward the RWG back-short) is then inserted into this slot with a ω-shaped head rectangular probe of sufficient width and length to achieve maximum transmission and minimum reflections with the help of quarter wave impedance transformers between the 50 Ω strip line and probe section to match the 530 Ω impedance of RWG WR28. The optimized dimensions of design parameters are presented in Table 5.1. All optimized dimensions are in mm. The simulated

Table 5.1 Optimized dimensions of reverse transition model

Interconnects part name	*Interconnects parameters*	*Parameter designation and values (mm)*
Rectangular waveguide	Waveguide type	WR-28
	Dielectric	Air filled
	Broad side	a=7.112
	Narrow side	b=3.556
	Waveguide length	$L_{WG} \approx 3\frac{\lambda g}{4} = 21.4$
Microstrip line	Material	Copper (Cu)
	Width	W_{strip}=0.37
	Length	L_{Strip}=7.244
	Thickness	t=.05
Quarter wave transformer	Material	Copper (Cu)
	Width	$W_{\lambda g/4}$=0.24
	Length	$L_{\lambda g/4}$=3.4
	Thickness	t=0.05
Microstrip rectangular patch	Material	Copper (Cu)
	Width	W_{Patch}=0.40
	Length	L_{Patch}=1.556
	Thickness	t=.05
Two-patch circles	Each ring radius	r=0.16
E-plane 90° broad side a, slot	Dielectric	Air filled
	Width	W_{Slot}=3.8
	Length	L_{Slot}=1.48
	Height	H_{Slot}=0.32
Ground	Material	Copper (Cu)
	Width	$W_{Gnd.}$= 3.8
	Length	$L_{Gnd.}$= 9.164
	Thickness	t =.05
Substrate	Material	Roger RT Duroid 5880 (tm)
	Width	$W_{Sub.}$= 3.8
	Length	$L_{Sub.}$= 14.2
	Height	h=0.127
Back-short	Back-short distance	$d=\lambda g/4$=3.4
	Broad side	a=7.112
	Narrow side	b=3.556

resultant return loss (RL) and insertion loss (IL) of design in terms of their *S*-parameters are presented in Figure 5.2c.

5.3.5 MS-to-RWG transition/interconnect analysis

S-parameters from the interconnect simulated results of Figure 5.3c at a resonance frequency of 30 GHz are given by

$$[S]=\begin{bmatrix} S11 & S12 \\ S21 & S22 \end{bmatrix}=\begin{bmatrix} 0.02304\angle-177^\circ & 0.9885\angle-142^\circ \\ 0.9885\angle-142^\circ & 0.0086\angle 59.8^\circ \end{bmatrix}$$

The calculated *Z*-parameters and *Y*-parameters using conversion relationships of Section III (A), respectively, are $[Z]=\begin{bmatrix} Z11 & Z12 \\ Z21 & Z22 \end{bmatrix}=$

$$[Z]=\begin{bmatrix} Z11 & Z12 \\ Z21 & Z22 \end{bmatrix}=\begin{bmatrix} 61.639\angle 88.5^\circ & 259.73\angle-91.3^\circ \\ 259.73\angle-91.3^\circ & 676.34\angle 87.8^\circ \end{bmatrix}$$

$$[Y]=\begin{bmatrix} Y11 & Y12 \\ Y21 & Y22 \end{bmatrix}=\begin{bmatrix} 0.0262\angle 88.8^\circ & 0.0101\angle 89.6^\circ \\ 0.0101\angle 89.6^\circ & 0.0024\angle 89.5^\circ \end{bmatrix}$$

where the characteristics impedance of the MS line is $Z_0=50\ \Omega$ (used to calculate Z11 and Z12) and that for RWG is $Z_0=530\ \Omega$ (used to calculate Z21 and Z22).

After comparing each *Z*-parameter with $(R+jX)$, the equivalent passive elements *R*, *L*, and *C* can be easily calculated. Similarly, each *Y*-parameter is compared with $(G+jB)$ and yields the values of lumped parameters *G*, *C*, and *L*.

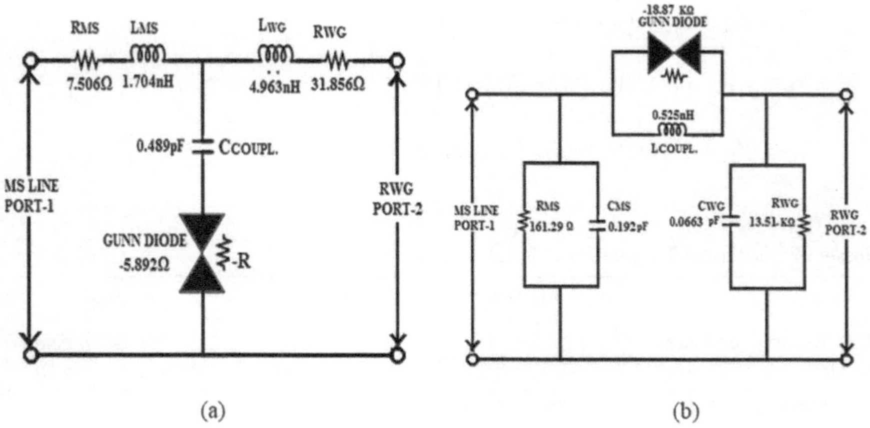

Figure 5.3 Ka-band reverse MS line interconnect. (a) Electrical T-equivalent circuit and (b) electrical π-equivalent circuit.

i. For the T-equivalent electrical network

The calculated values of T-equivalent impedances Z_A, $Z_{B,}$ and Z_C are

$$Z_A = Z_{11} - Z_{12} = (7.506 + j321.28)\Omega$$

$$Z_B = Z_{12} = (-5.892 - j259.66)\Omega$$

$$Z_C = Z_{22} - Z_{12} = (31.856 + j935.50)\Omega$$

Noteworthy that the negative value of R is represented by the Gunn diode.

The values of L and C are calculated using the very basic relationships of reactance

$$+jX = +j2\pi f_0 L$$

and

$$-jX = -j\frac{1}{2\pi f_0 C}$$

Respectively using the designed center frequency, i.e., $f_0 = 30\,\text{GHz}$. Comparing real and imaginary parts with calculated values of Z_A, Z_B, and $Z_{C,}$ respectively, respectively, Which gives the values of the passive elements (R, L, and C) of T-equivalent circuit as depicted in Table 5.2. Each combination of Z_A, Z_B, and Z_C is represented as a series combination of passive elements.

ii. For Π-equivalent electrical network:

The calculated values of Π-equivalent admittances Y_A, Y_B, and Y_C are

$$Y_A = Y_{11} + Y_{12} = (0.00062 + j0.0363)$$

$$Y_B = -Y_{12} = (-0.000053 - j0.0101)$$

$$Y_C = Y_{22} + Y_{12} = (0.000074 + j0.0125)$$

Table 5.2 Electrical T-equivalent circuit elements

	MS line		*RWG*		*Coupling*	
T-equivalent impedances (in Ω)	R_{MS} (Ω)	L_{MS} (nH)	R_{WG} (Ω)	L_{WG} (nH)	$C_{coup.}$ (pF)	*Gunn diode (−R in Ω)*
Z_A=7.506+j321.28	7.506	1.704	...	...	...	...
$Z_{B=}$−5.892−j259.66	...	...	...	...	0.489	5.892
Z_C=31.856+j935.50	...	...	31.856	4.963	...	...

Table 5.3 Electrical Π-equivalent circuit elements

	MS line		*RWG*		*Coupling*	
Π-equivalent impedances (in Ω)	R_{MS} (Ω)	C_{MS} (pF)	R_{WG} (Ω)	C_{WG} (pF)	$L_{coup.}$ (nH)	*Gunn diode (–R in Ω)*
Y_A=0.00062+*j*0.0363	161.29	0.192	-	-	-	-
$Y_{B=}$–0.000053–*j*0.0101	-	-	-	-	0.525	18.87K
Y_C=0.000074+*j*0.0125	-	-	13.51K	0.0663	-	-

Noteworthy, the value of resistance R is determined through the value of conductance G and the values of L and C are calculated using the very basic relationships of susceptance

$$R = \frac{1}{G}$$

$$-jB = -j\frac{1}{2\pi f_0 L}$$

and

$$+jB = +j2\pi f_0 C$$

Respectively using the designed center frequency, i.e., $f_0 = 30\,\text{GHz}$. Comparing real and imaginary parts with calculated values of Z_A, Z_B, and $Z_{C,}$ respectively, gives the values passive elements (R, L, and C) of Π-equivalent circuits are arranged in Table 5.3. Each combination of Y_A, Y_B, and Y_C is represented as a parallel combination of passive elements.

5.4 RESULTS AND DISCUSSIONS

The reverse transition model solution is passive at all frequencies within the specified tolerance. The simulated RL and IL are present in Figure 5.3c. RL is greatly improved to –34 dB and the IL achieved is around 0.1 dB along with 72%, –10 dB fractional bandwidth, and VSWR is very close to 1. This represents the maximum energy transfer between the microstrip line-to-rectangular waveguide ports. The resultant electrical T-equivalent circuit and Π-equivalent circuit of the MS line-to-RWG transition/interconnect have been drawn in Figure 5.3a and b, respectively. These equivalent circuits of MS line interconnect are simple and have a very less number of electrical passive components. Since the negative value of a resistance is only possible with the Gunn diode, a negative resistance is represented with the Gunn diode in the equivalent circuits. The magnitude and field vector

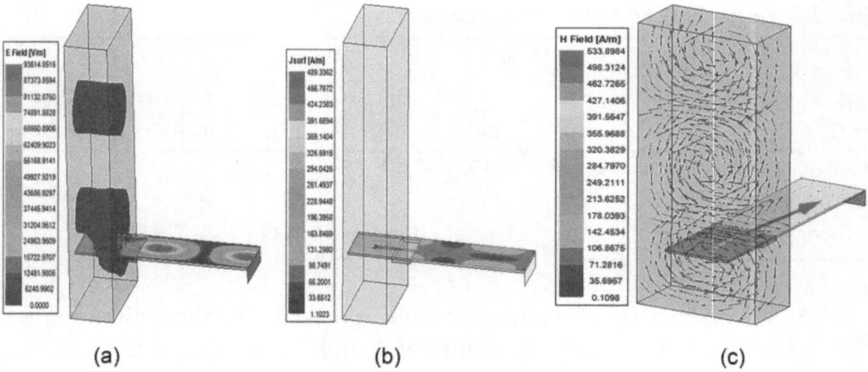

Figure 5.4 Field distributions inside reverse transition model. (a) Electric field vector and magnitude (*E*), (b) surface current density (*J*), and (c) magnetic field distribution (*H*).

Table 5.4 Optimized dimensions of reverse transition model

Ref.	*Band*	*RL (dB)*	*IL(dB)*	*Fractional BW*	*VSWR*
[14]	Q-	−23	−2.0	23%	1.23
This work	Ka-	−34	−0.100	72%	1.0173

of electric field distribution (E), surface current density (J), and distribution of magnetic field vector (H) are shown in Figure 5.4a–c.

The parameter performance comparisons of similar existing transition results with the proposed design work are presented in Table 5.4. The insertion loss (IL) and return loss (RL) are impressively improved along with VSWR.

5.5 CONCLUSIONS

The chapter shows the application of the reverse transition model to design high-performance microstrip line interconnects. Interconnects between the microstrip line and rectangular waveguide have been designed and developed using the reverse transition model for Ka-band application. Excellent performance in terms of insertion loss (−0.100 dB), return loss (−41.33 dB), and fractional bandwidth (72%) has been achieved. The simplified RLC equivalent circuit approach of microstrip line interconnects shows a good agreement between the designed frequency of MS line-to-RWG transition and the resonating frequency of the equivalent two-port T- and Π-networks. This article gives the general idea of how to develop an RLC equivalent circuit from the knowledge of S-parameters of any two-port microwave 3D model. This method could also be generalized to any multi-port microwave components to obtain its electrical passive components equivalent circuits. These circuits can be easily fabricated with very low cost and high density of integration because these occupy very less space. The novel simplified approach is not limited to the MS line interconnects. Possibly, in the future, this technique may also be applicable to obtain electrical equivalent models of other microwave components such as slot line-to-RWG transition, RWG to circular waveguide transition, microwave filter, resonator, directional coupler, circulator, isolator, phase shifter, attenuator, etc. This approach may also be applicable for the multi-port network [18,19] to obtain a microwave equivalent model from the knowledge of electrical circuits by just following the reverse procedure steps explained in Section III.B. Electrical equivalent of microstrip line interconnects are commonly used in modern mobile devices such as smartphones.

REFERENCES

1. E. R. Murphy, Microstrip to waveguide transition, US Patent 4,453,142, 1984.
2. A. Varshney, V. Sharma, A comparative study of microwave rectangular waveguide-to-microstrip line transition for millimeter wave, wireless communications and radar applications, *Microwave Review*, vol. 26, no. 2, pp. 21–36, Dec. 2020.
3. K. Chang, Chapter 2: Review of waves and transmission lines, In: *RF and Microwave Wireless Systems*, 1st edition, pp. 54–55, reprint 2015, Wiley Student edition, New Delhi.
4. R. A., Pucel, Technology and design considerations of monolithic microwave integrated circuits, In: D. K. Ferry, Ed., *Gallium Arsenide Technology*, pp. 189–248, 1985, H.W. Sams Co., Indianapolis, IN.
5. R. E. Collin, Chapter 4: Circuit theory of waveguiding systems, In: *Foundation of Microwave Engineering*, 2nd edition, pp. 220–259, reprint 2014, Wiley Student edition, New Delhi.
6. A.Gorbunova, Y.Kuznetsov, Equivalent circuit synthesis for microstrip structures design and optimisation, *2011 XXXth URSI General Assembly and Scientific Symposium*, Istanbul, pp. 1–4, 2011, doi: 10.1109/URSIGASS.2011.6050630.
7. M. X. Yu, A novel microstrip-to-microstrip vertical via transition in X-bandmultilayer packages, *International Journal of Antennas and Propagation*, vol. 2016, 8 pages, 2016, Article ID 9562854, doi: 10.1155/2016/9562854.
8. W. Wiatr, D. K. Walker, D. F Williams, Coplanar-waveguide-to-microstrip transition model, *2000 IEEE MTT-S International Microwave Symposium Digest* (Cat. No.00CH37017), n.d., doi: 10.1109/mwsym.2000.862328.
9. M. Ansarizadeh, A. Ghorbani, R. A. Abd-Alhameed, An approach to equivalent circuit modeling of rectangular microstrip antennas, *Progress in Electromagnetics Research B*, vol. 8, pp. 77–86, 2008, doi: 10.2528/pierb08050403.
10. A. Kaur, P. K. Malik, Multiband elliptical patch fractal and defected ground structures microstrip patch antenna for wireless applications, *Progress in Electromagnetics Research B*, vol. 91, pp. 157–173, 2021, doi: 10.2528/PIERB20102704, ISSN: 1937-6472.

11. N. Shaik, P. K. Malik, A retrospection of channel estimation techniques for 5G wireless communications: opportunities and challenges, *International Journal of Advanced Science and Technology*, vol. 29, no. 05, pp. 8469–8479, 2020, ISSN: 2005-4238.
12. P. K. Malik, M. Singh, Multiple bandwidth design of micro strip antenna for future wireless communication, *International Journal of Recent Technology and Engineering*, vol. 8, no. 2, pp. 5135–5138, July 2019, doi: 10.35940/ijrte.B2871.078219, ISSN: 2277-3878.
13. D. M. Pozar, Chapter 4: Microwave network analysis, In: *Microwave Engineering*, 4th edition, pp. 165–220, reprint 2016, Wiley Student edition, New Delhi.
14. Z. Y. Malik, M. I. Nawaz, M. Kashif, Mmic/Mic compatible planar microstrip to waveguide transition at Ku-band for radar applications, *Proceedings of International Bhurban Conference on Applied Sciences & Technology*, Islamabad, Pakistan, 11–14 January, 2010, pp. 51–53.
15. A. K. Varshney, A microwave rectangular waveguide-to-micro-strip line transitions @30 GHz, *International Journal of Emerging Technology and Advanced Engineering*, vol. 3, no. 8, pp. 563–568, Aug. 2013.
16. Y-C. Leong, S. Weinreb, Full band waveguide-to-microstrip probe transitions, *1999 IEEE MTT-S International Microwave Symposium Digest* (Cat. No.99CH36282), Anaheim, CA vol. 4, pp. 1435–1438, 1999, doi: 10.1109/MWSYM.
17. Y-C. Shih, T-N. Ton, L. Q. Bul, Waveguide to microstrip transition for millimeter wave applications, *IEEE MTT-S Digest*, pp. 473–475, 1988.
18. L. Bobaru, M. Stanculescu, S. Deleanu, M. Iordache, D. Niculae, V. Bucata, Using S parameters in two-port circuit analysis, *2019 15th International Conference on Engineering of Modern Electric Systems (EMES)*, Oradea, Romania, pp. 81–84, 2019, doi: 10.1109/EMES.2019.8795202.
19. T. Reveyrand, Multiport conversions between S, Z, Y, h, ABCD, and T parameters, *2018 International Workshop on Integrated Nonlinear Microwave and Millimetre-wave Circuits (INMMIC)*, Brive La Gaillarde, pp. 1–3, 2018, doi: 10.1109/INMMIC.2018.8430023.

Chapter 6

SRR-loaded octagonal Sierpinski-based carpet-shaped antenna for multiband application

K. Yogaprasad
Rayalaseema University

Nanda Kumar M.
Sreenidhi Institute of Science and Technology

V. R. Anitha
Sri Vidyanikethan Engineering College

Anil Kumar Nayak
University of Alberta

CONTENTS

6.1 INTRODUCTION

Multiple resonant frequencies are used to operate in wireless applications, which are mainly deployed to cover the microwave frequencies of the proposed design. These are L, S, C, X, and K_u bands occupying a frequency range of 1–2, 2–4, 4–8, 8–12, and 12–18 GHz. The biggest task is to design an antenna to operate in multiband applications with two or more bands [1]. It is a great advantage to substitute multiple antennas with one antenna, which reduces design costs and space constraints.

Microstrip-based antennas are used due to their simple design. They contain a patch positioned on top of the substrate. The bottom of the substrate

DOI: 10.1201/9781003347057-6

is coated with a conductor which acts as the ground and is very useful for wireless applications [2–5] because of its lower weight and low cost. In the last few decades, the growth of microstrip-based antennas has rapidly increased and has overcome many restrictions as per user applications. The general shapes used are circular, triangular, and rectangular and are changed day by day as per the user's perspective.

In the beginning, different methods were introduced to get dual resonance and these were slot, shorting pin, array, different pole dimensions, etc. [6–9]. Size miniaturization is the main task in today's scenario, and the best solution is to choose fractal-based antennas. A fractal antenna is a type of antenna that uses a fractal which means a design that gives self-similarity that is used for maximizing the length [7]. A fractal can be constructed in a much smaller size because it is used to make operational frequency independent of its scale and some of the geometries such as Minkowski fractals, Koch curves, Sierpinski triangles, etc. [2,7]

The materials available non-naturally are meta-materials that change the permittivity/permeability for exact frequency generation, and SRR and CSRR (complementary SRR) are central evidence to prove the meta-material properties. The best properties of the meta-materials are size miniaturization and multiband extraction [10,11].

In this research paper, an SRR loaded with an octagonal patch, Sierpinski carpet based on the fed defected ground by a microstrip, is used for multiband applications. The proposed design which includes microstrip design equations, Sierpinski fractal design, and design flow are discussed in Sections 6.2 and 6.3, and describe the negative extraction of SRR. Section 6.4 explains the results and Section 6.5 ends with a conclusion followed by a reference section [12–14].

6.2 ANTENNA GEOMETRY

The proposed antenna designs are shown in Figure 6.1, and the first antenna (@1) has an octagonal patch added with a split ring resonator. The bottom of the antenna introduced with the octagonal slot is indicated as antenna2 (@2), and finally, the Sierpinski-based fractal with the first iteration is introduced at the bottom of the structure as represented in antenna3 (@3). The proposed design intends to improve the resonant frequencies and implement them for multiband applications. The design integrated with the FR-4 material has a dielectric constant of 4.4, a thickness of 1.6 mm, and a copper thickness of 0.035 mm.

The parametric representation of the design has been shown in Figure 6.2, and Table 6.1 represents the optimum parameters used in this design. The electromagnetic tool is used to examine the recital of the antenna such as S_{11}, gain, VSWR, radiation patterns, and surface current. The fabrication

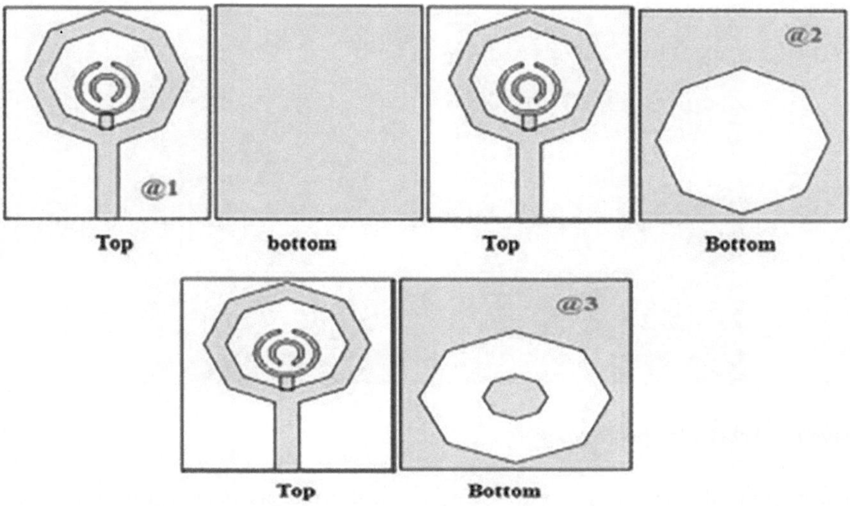

Figure 6.1 Evolution of the antenna projected.

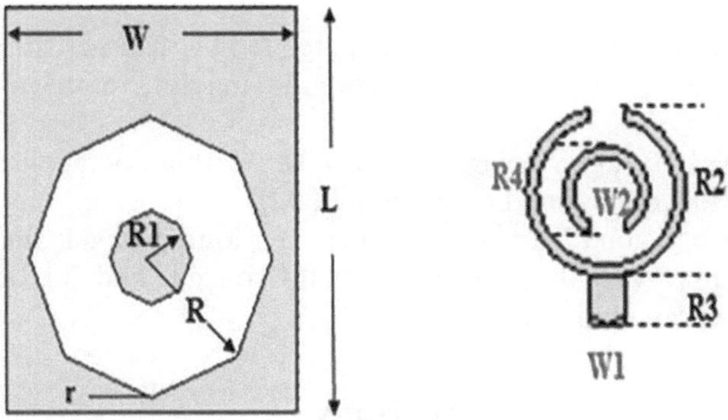

Figure 6.2 Parameter representation of the proposed design.

Table 6.1 Optimized parameters of Sierpinski-based carpet antenna

Parameters	*Values (mm)*	*Parameters*	*Values (mm)*
R2	8	*W*	24.8
R3	1.5	*L*	30
r	1.02	*R1*	10
W1	1.5	*R*	3.33
W2	1.5	*R4*	6

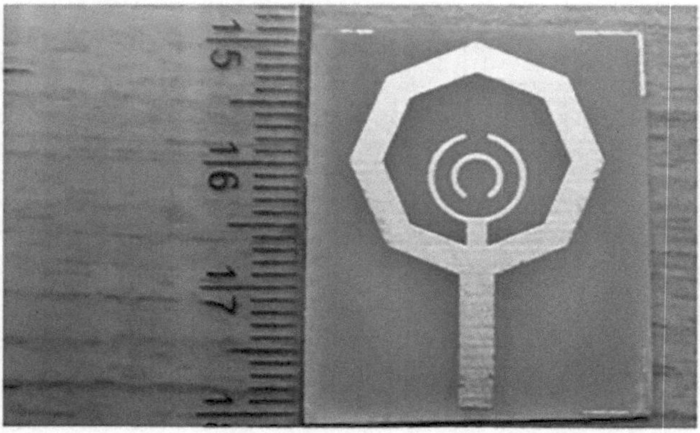

Figure 6.3 Fabricated prototype.

prototype is shown in Figure 6.3, and the antenna has been validated with a vector network analyzer (VNA) as well as an anechoic chamber. The length and width of the microstrip feed are 10.8 and 3.2 mm, respectively, and the octagonal patch's inner and outer diameters are 19.5 and 14 mm, respectively.

The steps to follow and implement the proposed antenna are as follows:

Step 1: Introduce the octagonal ring patch fed by the microstrip.

Step 2: Split ring resonator is added with the ring slot, notated as antenna1 (@1).

Step 3: The bottom of the antenna is etched with an octagonal slot with a diameter of R, represented as antenna2 (@2).

Step 4: An octagonal slot is added between antennas @1 and @2. The octagonal patch with the Sierpinski fractal concept and R1 outer patch diameter is represented as antenna3.

6.2.1 Microstrip design equations

The width (W) and height (h) of the microstrip are [15,16]

$$\frac{W}{h} = \begin{cases} \dfrac{2}{\pi}\left\{ \begin{array}{c} a-\ln(2a-1)-1+\cdots \\ \dfrac{\varepsilon_r-1}{2\varepsilon_r}\left[\ln(a-0.61)-\dfrac{0.1}{\varepsilon_r}\right] \end{array} \right\} & \dfrac{W}{h}>2 \\ \dfrac{8e^a}{e^a-2}, & W/h<2 \end{cases} \tag{6.1}$$

where

$$a = \frac{377\pi}{2Z_o\sqrt[2]{\varepsilon_r}} \tag{6.2}$$

$$b = \frac{Z_o}{60\sqrt[2]{\varepsilon_r}}\sqrt[2]{\frac{\sqrt[2]{\varepsilon_r}+1}{2}+\frac{\varepsilon_r-1}{\varepsilon_r+1}\left(\frac{0.11}{\varepsilon_r}+0.23\right)} \tag{6.3}$$

The length of the microstrip is [7,15,16]

$$L_m = n^*\lambda_g; \qquad n = 1,3,5,7\ldots \tag{6.4}$$

6.2.2 Sierpinski carpet iteration

A Sierpinski sieve is used to create the Sierpinski carpet fractal, where triangles are replaced with octagons. Cells [1] and [0] can be described in a matrix format. Equation 6.5 [17] describes the first iteration:

$$\left\{0 \to \begin{bmatrix} 0 & 0 & 0 \\ 0 & 0 & 0 \\ 0 & 0 & 0 \end{bmatrix}, 1 \to \begin{bmatrix} 1 & 1 & 1 \\ 1 & 0 & 1 \\ 1 & 1 & 1 \end{bmatrix}\right\} \tag{6.5}$$

6.3 NEGATIVE PERMEABILITY EXTRACTION IN SRR

Smith et al. invented the concept of negative permeability and permittivity with the help of *S*-parameters, and further, Nicolson Ross Wier was the first scientist to evaluate the permeability of the CSRR antenna [18]. The waveguide method is considered to evaluate the negative permeability in the SRR structure and a graphical representation of the evaluation process is demonstrated in Figure 6.4.

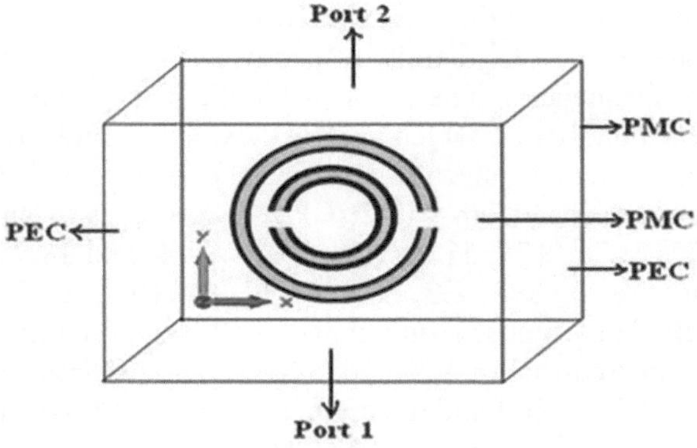

Figure 6.4 SRR validation.

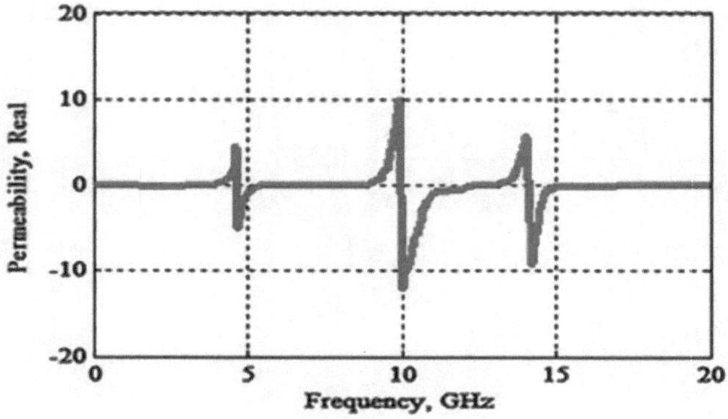

Figure 6.5 Results.

In this waveguide setup, the ports are chosen along the *x*-axis to pass the EM field and the *y*-axis to perfect electric conductor (PEC) fields. The bottom and top of the waveguide setup are assigned by a perfect magnetic conductor (PMC) along the *z*-axis. SRR's multiband Koch fractal antenna design features an exposed S_{11} notch; therefore, the CSRR's purpose to produce permeability at this frequency is experiential. Figure 6.5 depicts the actual permeability components after they have been removed. Consequently, the best proof of meta-material property's realism comes from effective material constraints.

6.4 DISCUSSION OF RESULTS

Figure 6.6 represents the reflection coefficient of all structures over the frequency of 4–20 GHz. Antenna 1 (@1) has resonated three frequencies and is useful for dual-band applications. These are the X and Ku bands, respectively. The graph is highlighted in green color. Antenna2 (@2) has resonated five frequencies that are indicated with rose color and are useful for tri-band, which are the C/X/K_u bands, respectively. Antenna3 (@3) produces seven resonant frequencies. The graph is highlighted in yellow color, and it is used for the quad-band, which are the C/X/K_u/K bands, respectively.

The proposed antenna reflection coefficient is represented in Figure 6.7 over the frequency range from 4 to 20 GHz. It resonates seven frequencies which are 4.77, 7.076, 9.75, 11.997, 14.516, 17.114, and 18.953 GHz, and their *S*11 values are –37.898, –32.22, –15.427, –24.402, –22.901, –22.678, and –16.348 dB. The proposed antenna is used for quad-band applications: C band (resonant frequencies 4.77, 7.076 GHz), X-band (resonant frequencies 9.75, 11.997 GHz), Ku-band (resonant frequencies 14.516, 17.114 GHz), and K-band (resonant frequency 18.953 GHz).

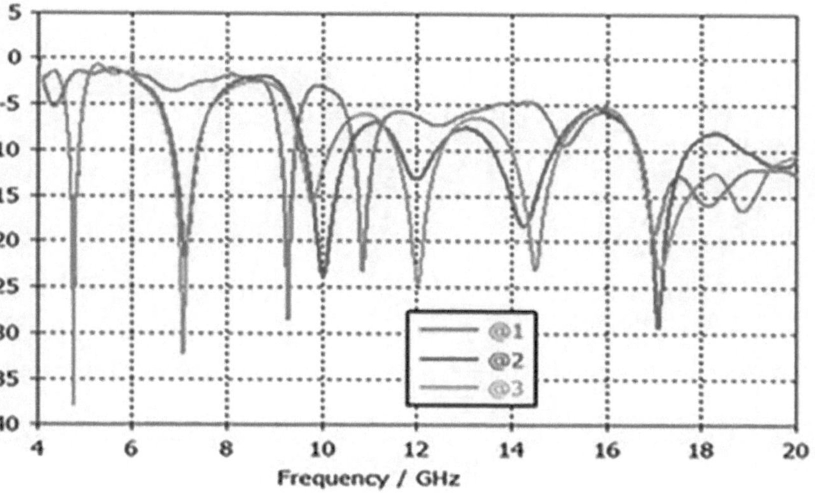

Figure 6.6 S_{11} over a frequency for the proposed antenna.

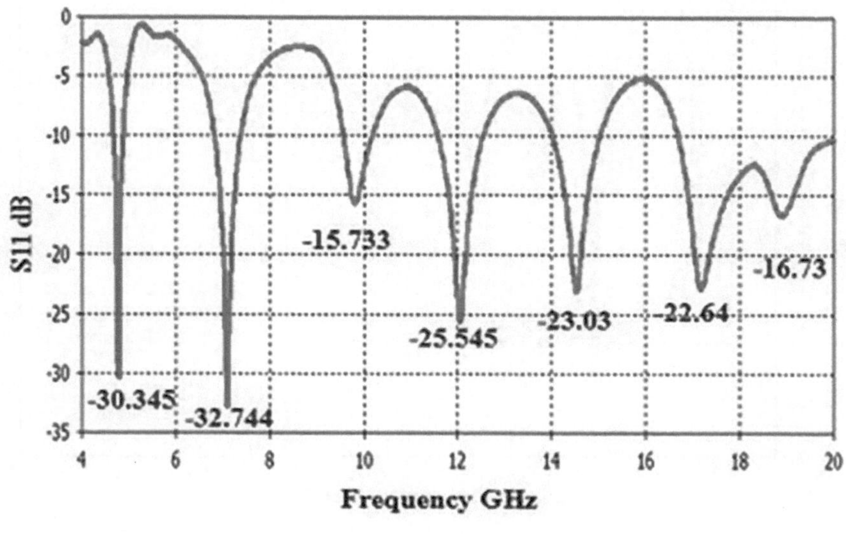

Figure 6.7 Reflection coefficient of a proposed antenna.

Figure 6.8 indicates the VSWR response of the antenna over the frequency range of 4–20 GHz. The VSWR results with respect to 2:1 are matched with the reflection coefficient of the antenna of the reference S_{11} = −10 dB. It has

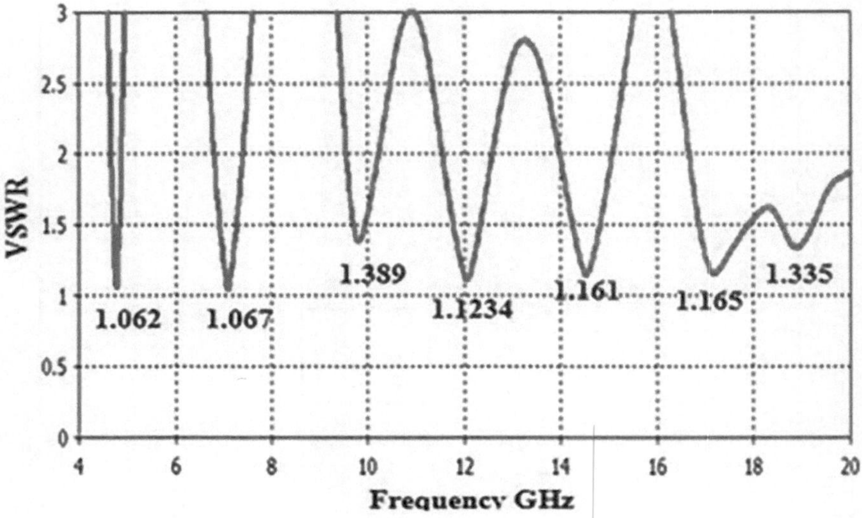

Figure 6.8 VSWR of a proposed design.

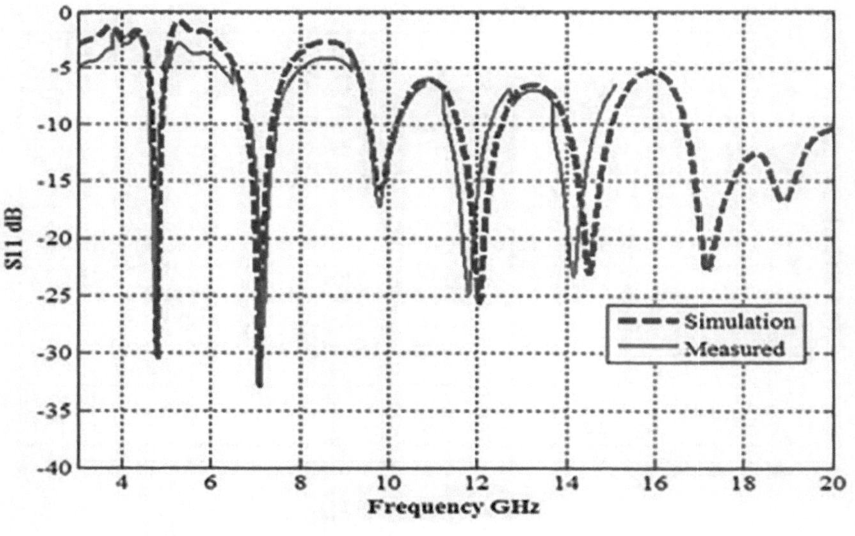

Figure 6.9 Reflection coefficient for simulation and measured results.

seven resonant frequencies and their VSWR values are 1.062 at 4.77GHz, 1.067 at 7.076 GHz, 1.389 at 9.75 GHz, 1.1234 at 11.997 GHz, 1.161 at 14.516 GHz, 1.1654 at 17.114 GHz, and 1.335 at 18.953 GHz.

An excellent correlation between the measured and simulation results can be seen in Figure 6.9, which compares the S_{11} results. The measurement results are obtained up to 15 GHz due to not having a facility. The

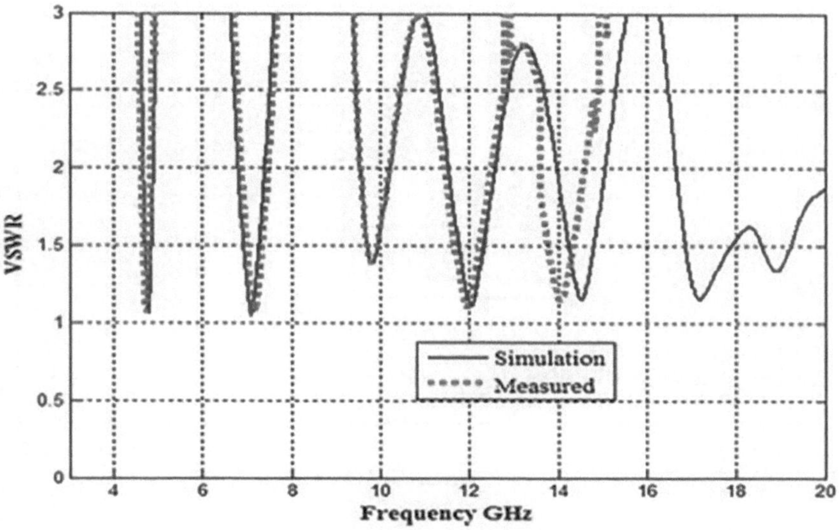

Figure 6.10 VSWR for simulation and measured results.

Table 6.2 Simulation and measurement results comparison of SRR-loaded octagonal Sierpinski

S. no.	*Resonant frequency GHz*		*Reflection coefficient dB*	*VSWR*
1	SR	4.78	−30.34	1.062
	MR	4.75	−25.02	1.1134
2	SR	7.09	−32.77	1.047
	MR	7.2	−28.05	1.09
3	SR	9.82	−15.73	1.391
	MR	9.81	−16.89	1.385
4	SR	12.03	−25.547	1.113
	MR	11.86	−24.95	1.114
5	SR	14.5	−23.03	1.155
	MR	14.15	−23	1.156
6	SR	17.1	−22.64	1.17
	MR	-	-	-
7	SR	18.93	−16.73	1.345
	MR	-	-	-

different colors are used to represent the measured and simulated results. As shown in Figure 6.10, the VSWR findings are compared to the simulation and measurement data with a close match between the simulation and measured results. Table 6.2 compares the resonance frequencies, reflection coefficient, and VSWR of the simulated and observed values.

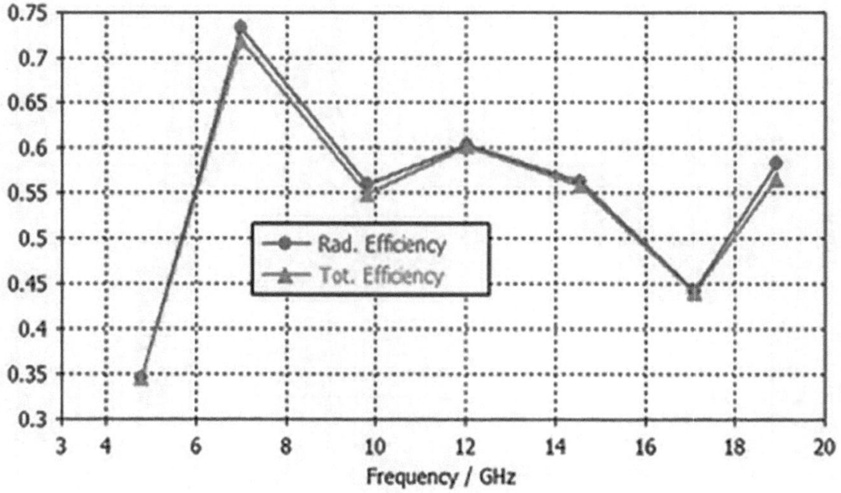

Figure 6.11 Efficiencies of the proposed antenna.

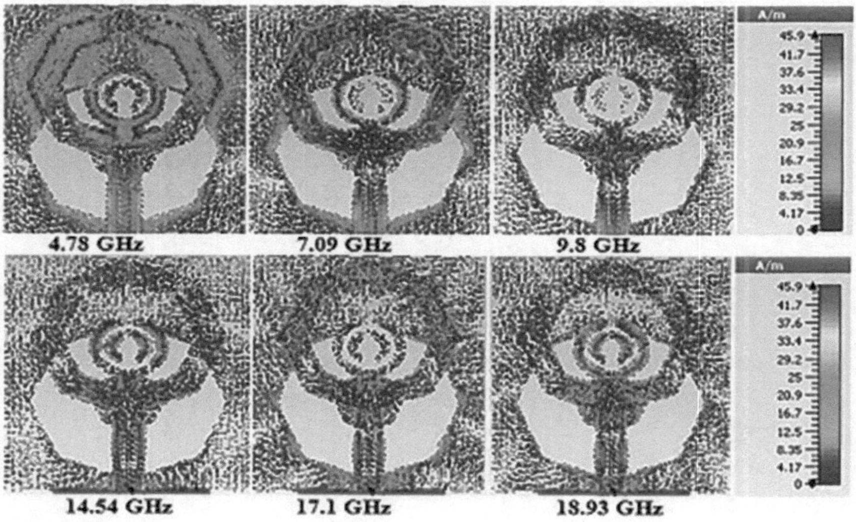

Figure 6.12 Surface current at resonant frequencies.

Figure 6.11 represents the efficiencies of the proposed antenna, where the red and green colors indicate the radiation and total efficiency of the antenna, respectively. The radiation and total efficiencies are 35%, 35% at 4.78 GHz, 74%, 73% at 7 GHz, 56%, 55% at 9.82 GHz, 60%, 59.95% at 12 GHz, 56.5%, 55.2% at 14.53 GHz, 44%, 44% at 17.14 GHz, and 58%, 56% at 18.93 GHz. The surface currents of the proposed design at resonant

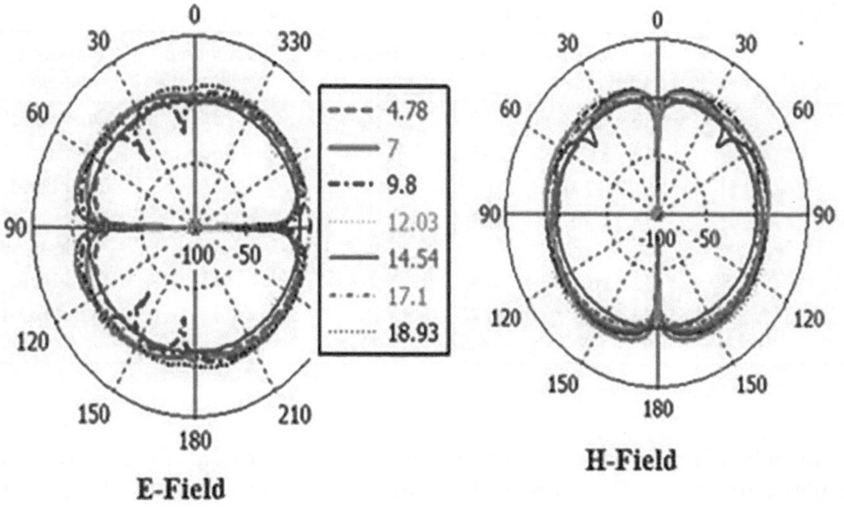

Figure 6.13 Radiation patterns.

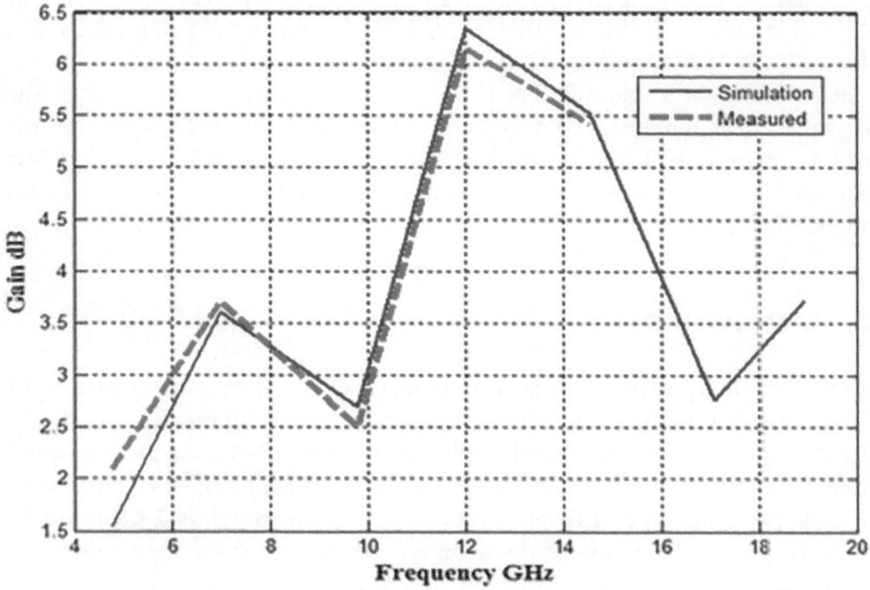

Figure 6.14 Gain over frequency GHz.

frequencies are represented in Figure 6.12 and observed at every frequency. At the time of feeding and the beginning of the patch, the current flow is at its highest.

Figure 6.13 shows the radiation patterns of the proposed antenna at resonant frequencies and different colors are used to represent the resonant

Table 6.3 Comparison of the proposed and existing literature

S. no.	*Ref.*	*Total antenna size (mm²)*	*Resonance frequency (GHz)*	*Total area used (mm²)*	*Purpose of antenna*
1.	[19]	30×45	3.4/5.8	*1350*	Dual-band
2.	[10]	28×31	3.35/5.4/7.25	*868*	Tri-band
3.	[11]	31.7×27	2.654/4.245	*855.9*	Dual-band
4.	[20]	24.3×30.8	3.85/4.0	*748.5*	Dual wide-band
5.	[21]	32×36	1.14/11.8	*1152*	Dual-band
6.	Proposed Work	24.8×30	4.77/7.07/9.75/11.99/ 14.52/17.11/18.95	*744*	Quad-band

frequency. The E-plane shows the bidirectional radiation pattern and H-plane indicates the bidirectional radiation pattern at every frequency. The gain over frequency in terms of the measured and simulated results is represented in Figure 6.14 and the measured gain values are almost close, matched with simulation results. The gain values are measured up to 15 GHz. The simulated, measured gain values are 1.614, 2.1 dBi at 4.77 GHz, 3.6, 3.67 dBi at 7.076 GHz, 2.75, 2.5 dBi at 9.75 GHz, 6.35, 6.2 dBi at 11.997 GHz, and 5.502, 5.475 dBi at 14.516 GHz.

It is shown in Table 6.3 that the existing literature is compared to the suggested design in terms of dimensions, resonant frequencies, applications, and antenna area. It has been observed that the proposed design has less size and resonated more frequencies and was implemented for multiband applications.

6.5 CONCLUSION

In this chapter, a microstrip feed octagonal patch loaded with a split ring resonator with an octagonal Sierpinski fractal shape is used in the defected ground for multiband applications. The FR-4 substrate material used to integrate an antenna has a dielectric constant of 4.4. The thickness of the antenna is 1.6 mm, and the size is projected as 24.8 mm×30 mm×1.6 mm. The results of the negative extraction of the split ring resonator are reported. The proposed antenna operated in seven frequencies which are 4.77, 7.07 GHz (S-band), 9.75, 11.99 GHz (X-band), 14.52, 17.11 GHz (Ku-band), and 18.95 GHz (K-band) with respect to a –10 dB reference and also matched the results with VSWR(1:2), and finally, the comparison of the simulation and measured results is validated and also matched in terms of S_{11}, VSWR, radiation patterns, gain, etc.

REFERENCES

1. B. Bashar, Q. Elias, Design of broadband circular patch microstrip antenna for Ku-band satellite communication applications, *International Journal of Microwave Optical Technology*, 11, 362–368, 2016.
2. D. K. Naji, Design of compact dual-band and tri-band microstrip patch antennas, *International Journal Electromagnetic Applications*, 8, 26–34, 2018.
3. V. V. Reddy, N. V. S. N. Sharma, Compact circularly polarized asymmetrical fractal boundary micro-strip ant for wireless applications, *IEEE Antennas Wireless Propagation Letters*, 13, 118–120, 2014.
4. Y. Sung, Bandwidth enhancement of a microstrip line-fed printed wide-slot antenna with a parasitic center patch, *IEEE Transactions on Antenna Propagation*, 60, 1712–1716, 2016.
5. M. Dinesh, M. N. Kumar, K. Balachandra, Micro-strip feed reconfigurable antenna for wideband applications, *Journal: Lecturer Notes in Electrical Engineering*, 665–671, 2018.
6. M. S. Sedghi, M. N. Moghadasi, F. B. Zarraabi, A dual band fractal slit antenna loaded by jerusalem crosses for wireless plus WiMAX communications, *Progress in Electromagnetic Research Letters*, 61, 19–24, 2016.
7. M. Nanda Kumar, T. Shanmuganantham, Division shaped SIW slot antenna for millimeter wirelesss/automotive radar applications, *Journal of Computer and Electrical Engineering*, 71, 667–675, 2018.
8. T. Ali, A. W. Mohammad Saadh, R. C. Biradar, A. Andujar, J. Anguera, A miniaturized slotted ground structure UWB antenna for multiband applications, *Microwave and Optical Technology Letters*, 60, 2060–2068, 2018.
9. D. K. Naji, Design of compact dual-band and tri-band microstrip patch antennas, *International Journal of Electromagnetics and Applications*, 8, 26–34, 2018.
10. T. Ali, M. M. Khaleeq, S. Pathan, R. C. Biradar, A multiband antenna loaded with metamaterial and slots for GPS/WLAN/WiMAX applications, *Microwave Optical Technology Letters*, 60, 79–85, 2017.
11. L-M. Si, W. Zhu, H-J. Sun, A compact, planar, and CPW-fed metamaterial inspired dual band antenna, *IEEE Antennas and Wireless Propagation Letters*, 12, 2013.
12. A. Rahim, P. K. Mallik, V. A. Sankar Ponnapalli, Fractal antenna design for overtaking on highways in 5G vehicular communication ad-hoc networks environment, *International Journal of Engineering and Advanced Technology (IJEAT)*, 9(1S6), 157–160, December 2019, doi: 10.35940/ijeat.A1031.1291S619, ISSN: 2249-8958.
13. P. K. Malik, H. Parthasarthy, M. P. Tripathi, Axisymmetric excited integral equation using moment method for plane circular disk, *International Journal of Scientific and Engineering Research*, 3(3), 1–3, March 2012, ISSN: 2229-5518.
14. P. K. Malik, H. Parthasarthy, M. P. Tripathi, Analysis and design of pocklingotn's equation for any arbitrary surface for radiation, *International Journal of Scientific and Engineering Research*, 7(9), 208–213, September 2016, ISSN: 2229-5518.

15. M. Nanda Kumar, T. Shanmuganantham, Broad-band H-spaced head shaped slot with siw based antenna for 60GHz wireless communication applications, *Microwave and Optical Technology letters (MOTL)*, 61(8), 1911–1916, 2019.
16. M. Nanda Kumar, T. Shanmuganantham, Broad band I shaped SIW slot antenna for V-band applications, *Applied Computational Electromagnetic Society (ACES)*, 34(11), November 2019.
17. M. N. Kumar, K. Yogaprasad, V. R. Anitha, A quad band sierpenski based fractal antenna fed by CPW, *Microwave and Optical Technology Letters*, 2, 893–898, 2018.
18. T. Ali, A. W. Mohammad Saadha, R. C. Biradar, A fractal quad-band antenna loaded with L-shaped slot and metamaterial for wireless applications, *International Journal of Microwave and Wireless Technologies*, 2018.
19. T. Mandal, S. Das, Coplanar waveguide fed 9-point star shape monopole antennas for worldwide interoperability for microwave access and wireless local area network applications, *Journal of Engineering*, (4), 155–160, 2014.
20. P. Dawar, N. S. Raghava, A. De, Ultra wide band, multi-resonance antenna using swastika metamaterial, *International Journal of Microwave and Optical Technology*, 11, 413–420, 2016.
21. R. Mishra, A. Pandey, Asymmetric crescent shaped dual band monopole antenna for UWB and GSM 1800/1900 applications, *International Journal of Microwave and Optical Technology*, 11, 356–361, 2016.

Chapter 7

Antenna in RFID smart systems

Shailendra P. Shastri and Archana Deshpande
Thakur College of Engineering and Technology

CONTENTS

7.1 INTRODUCTION

Smart RF systems are generally very compact to be portable. The world has been witnessing advancement in the area of integrated technology which has led to the development of very tiny chips, and today, we can have any computing machine which is even smaller than our palm. These devices are equally powerful to carry all the required functions. Many of these devices are used to make RF systems for monitoring movement and record-keeping of goods, multi-person tracking systems for malls, smart parking systems, and many more. The success of such a smart system completely depends on communication. The communication is not only on board among different modules, but it is also off-board where the antenna plays a very important role. A primary concern is given to accommodate electronic boards in any electronic system, whereas the antenna casing has got limited space to care for the entire look of the product. These smart systems are generally compact in size to make them very handy and portable, but these features put a lot of challenges for the RF engineers involved in the designing of antennas. Although the antenna casing has got limited space, its orientation and location impose constraint on the design of the antenna to have a suitable direction of radiation either from an edge or from the face of the case. There are various antennas available in the literature but a selection of antenna is not only dependent on antenna parameters but its suitability to get accommodated in the finalized casing which is designed to attract customers.

DOI: 10.1201/9781003347057-7

7.2 RELATED LITERATURE

There is an abundance of literature available which covers the design of compact antennas. In Ref. [1], a broadband planar monopole antenna using two parasitic square open-loops and a defective ground plane is investigated. The two loops are printed on both sides of the substrate to obtain broadband circular polarized (CP) radiation. The defected ground structures having horizontal and vertical slots are used to obtain additional CP resonances for further broadening of the axial ratio, AR. In Ref. [2], a compact conformal antenna for wearable applications is designed using a finite-sized metallic sheet backed with an anisotropic metasurface. The metasurface is employed not only to act as the ground plane for isolation but also acts as the main radiator. The metallic sheet-backed metasurface helps in reducing the maximum 1 g-averaged SAR value. A gain of about 6.2 dBi, FBR of 23 dB and excellent SAR makes the proposed antenna highly suitable for medical sensing and monitoring applications. A monopole antenna modified into a T-shaped structure and supported with a set of split-ring resonators (SRR) can have an electrically smaller size. The miniaturization is attained after loading SRRs in proximity to the T-shaped radiator. The proposed electrically small monopole has a gain of 1.76 dBi and an efficiency of 78.5% [3]. Another way to have a wide-band axial ratio (AR) monopole antenna is to introduce a stepped-shaped ground plane. This stepped-shaped ground plane is created by cutting notches. The stepped-shaped ground plane excites all the electric components required to generate CP radiation of 36.5% bandwidth [4]. The CP radiation shows good characteristics for both RHCP and LHCP. Besides, this antenna has a simple structure, a small size of dimensions $16 \times 22 \times 1\,mm^3$, and is suitable for various wireless communication systems. The AR bandwidth can be further increased to 70% by feeding the shifted monopole at the edge of the substrate, shortening the conventional ground plane, and adding a short vertical extension [5]. A wide AR can alternately be achieved with a coplanar waveguide (CPW)-fed monopole antenna incorporated with an orthogonal slit in the ground plane. Any impedance mismatch in the antenna can be circumvented with the help of a stub running parallel to the monopole [6]. A compact planar monopole antenna for wearable applications can also be designed using EBG not only to isolate the antenna from the human body but also to contribute toward enhanced radiation efficiency and gain. The EBG material is supported with a backing sheet to reduce the SAR levels inside the body. The antenna designed can provide a gain of 6.88 dBi with 70% efficiency [7]. This chapter gives the background information and literature review associated with the design of compact antennas [8–10].

7.3 ANTENNA DESIGN METHODOLOGY

The selection of the antenna depends on certain parameters which the antenna should fulfill such as range of communication and area coverage by the antenna. There are various antennas available in the literature but a selection of antenna is not only dependent on antenna parameters but its suitability to get accommodated in the finalized casing which is designed to attract customers. There are certain applications where a microstrip patch antenna can be a better candidate but it acquires larger space. A reduction in antenna size will influence its efficiency and radiation characteristics. Another and very noticeable antenna is a monopole. It is simple to design and occupies very a small space. Unfortunately, the radiation property of the monopole is not directional, therefore, it cannot be a good candidate for scanning and monitoring applications. But the omnidirectional nature of the monopole antenna makes it suitable to craft it to have face side or edge side radiation. Therefore, a monopole antenna is taken into consideration here to demonstrate the challenges available to an antenna designer and its mitigation techniques.

A simple monopole antenna occupies a maximum length of $\lambda_o/4$ in space, where λ_o is the free-space wavelength. A printed monopole appears as shown in Figure 7.1, placed along the Y-axis in the X-Y plane. Figure 7.1 includes the radiating portion of the antenna along with the feeding structure comprising a microstrip line which elongates the overall dimension of the antenna.

There are different shapes of monopoles available in literature but we will begin with a simple monopole antenna and later will show various modification techniques to make it suitable for real applications.

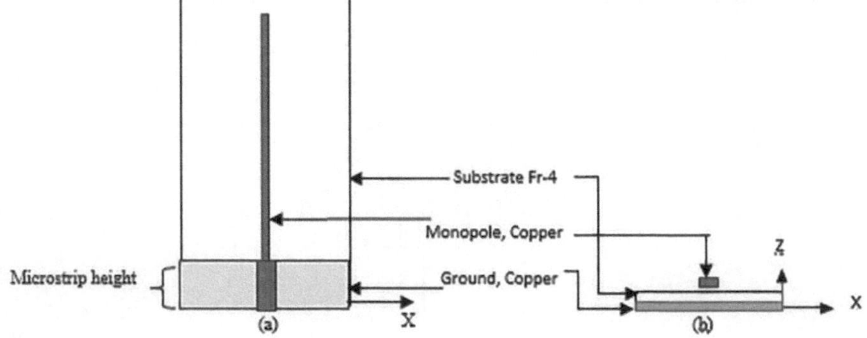

Figure 7.1 A monopole. (a) Top view and (b) cross-sectional.

The modifications in the antenna structure will reveal the design complexity experienced by a designer.

A monopole designed to work for Bluetooth frequency, 2.4 GHz, is of length $\lambda_o/4$ and is 31.25 mm. This portion of the monopole is responsible for the radiation of the EM waves. But this is possible only if it is excited with a time-varying source. A microstrip feed line is added to excite the monopole and thus the overall length of the entire antenna is going to increase by this additional element. Now, a designer should keep in mind that as there may be very limited space for the antenna in a final product The designer cannot add a microstrip line of any height as shown in Figure 7.1. The height of a microstrip line should be good enough to accommodate the connector required to feed the antenna. There are various feeding connectors available in the market and are chosen based on not only their size but the cost factor is also taken into consideration. Figure 7.2 shows an SMA connector and it can easily be understood that this connector requires at least 4.75 mm of space on PCB.

Keeping all these constraints in mind, a monopole is designed on an FR-4 substrate with a thickness of 1.6 mm. The length of the monopole is reduced to 23.5 mm for a width of 1 mm. Since free-space wavelength is 125 mm, the monopole should be 31.25 mm whereas the monopole realized on the Fr-4 substrate reduces to 23.5 mm. This monopole is fed with a 50 Ω microstrip line having a width and height of 3 and 10 mm, respectively. The feed line is supported by a ground plane of a width of 40 mm. The total length of the antenna thus becomes 33.5 mm. The 2D dimension of antenna is $33.5 \times 40\,\text{mm}^2$. To reduce the overall size of the antenna, the width of the ground plane can be reduced from 40 to 24 mm, but this influences the antenna tuning. The antenna length changes from 23.5 to 25 mm for a ground plane of width 24 mm. The reflection coefficients for both are shown in Figure 7.3. The reduction in the width of the antenna reduces the overall dimension of the antenna approximately by 37%.

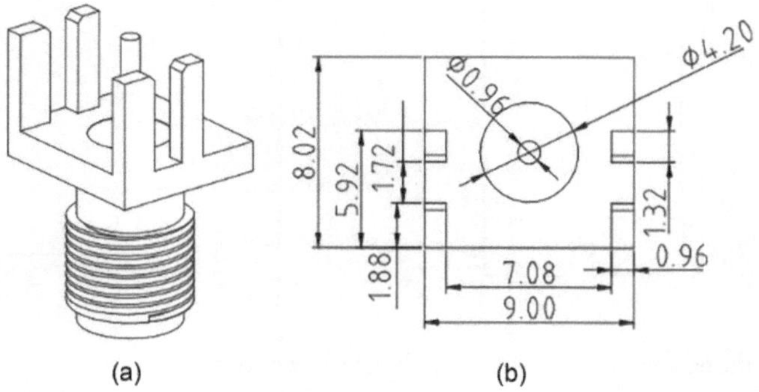

Figure 7.2 (a) 3-D view of a SMA connector and (b) dimensional detail of the connector, all the dimensions are in mm.

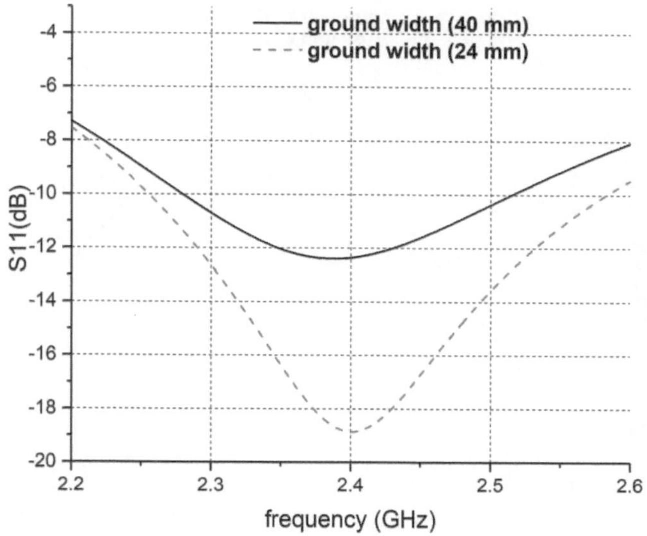

Figure 7.3 Reflection coefficient of monopole antenna for two different dimensions.

The antenna radiation pattern does not change as there is hardly any significant change in the antenna length. The radiation available in its E-plane is shown in Figure 7.4 and agrees with the theoretical radiation pattern.

While designing an antenna, it is equally important to keep track of the antenna efficiency along with gain and directivity. Figure 7.5 shows the antenna efficiency for two different dimensions discussed so far. The improvement in the efficiency is due to improved impedance matching of

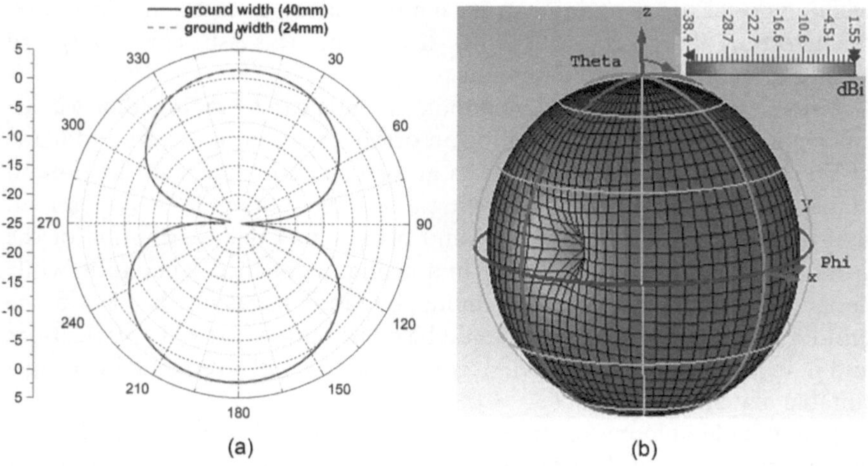

Figure 7.4 Radiation patterns in (a) E-plane, for two different monopoles. (b) 3-D for reduced size monopole.

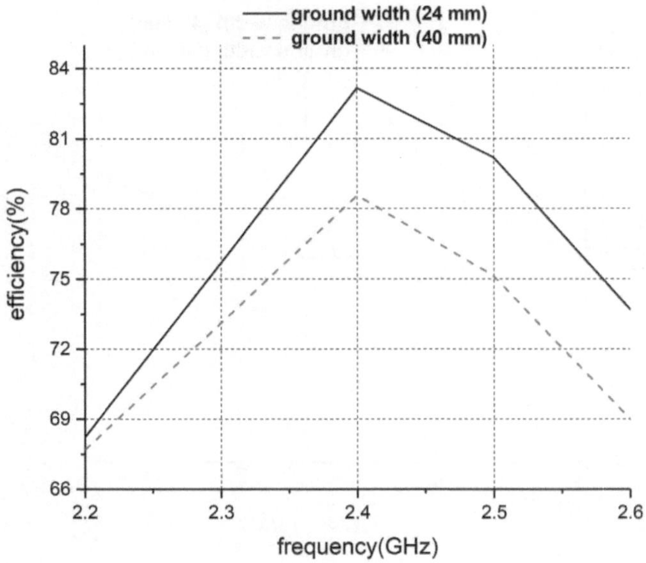

Figure 7.5 Efficiency of monopole for two different dimensions.

the monopole antenna. Thus far, it should be noted that the monopole antenna radiates in all directions in the H-plane and it is directional in the E-plane as shown in Figure 7.4a and b. The efficiency of the antenna is 83.15%, and the antenna gain is about 1.55 dBi, as shown in Figure 7.4b.

A monopole antenna is a linear polarized antenna and this kind of polarization may not be very much useful for industrial purposes, especially for handheld devices. For example, a packed good with a tag enveloped on it when exposed for scanning can have any random orientation. Therefore, the detection of such goods becomes difficult but this issue can be resolved using a circularly polarized antenna.

A simple linearly polarized monopole antenna can be modified into a circular polarized by simply including an orthogonal slot in the ground plane of the antenna. This slot introduces an orthogonal current in the ground plane with respect to current flowing along the monopole. It is observed that the slot introduced in the ground plane alters the distribution of the current and this leads to a change in antenna impedance. Therefore, while designing a circularly polarized monopole antenna using the slot in the ground plane, the entire effort should be on to achieve desired polarization, and if any impedance mismatch occurs, then it can be overcome by using a suitable matching technique.

A monopole antenna of dimensions $33.5 \times 40\,mm^2$ is used to add a slot in the ground plane. Figure 7.6 shows the structure of a circular polarized monopole antenna. It is observed that the height of slot C_h, slot width C_w, and length C_L plays important role in achieving circular polarization.

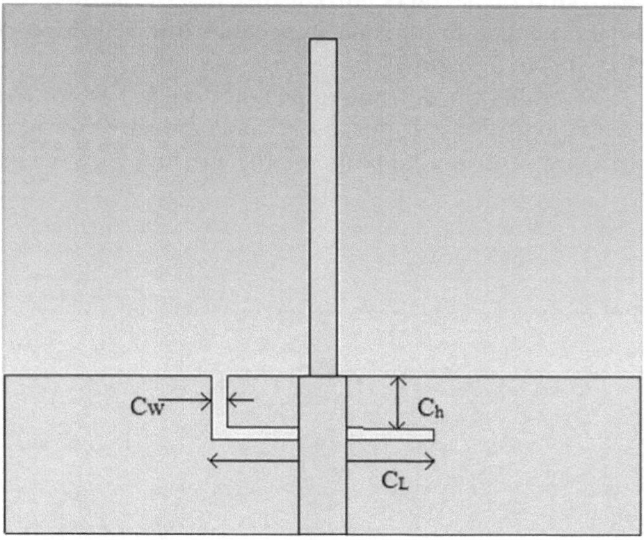

Figure 7.6 Circularly polarized monopole.

The circular polarization is achieved for C_h=2 mm, C_w=1.25 mm, monopole width of 1.7 mm, and C_L=14 mm symmetrically distributed across a microstrip feed line of width 3 mm. Figure 7.7 shows the axial ratio of the antenna to be 0.635 dB at θ=0° and 180°. The axial ratio remains within 3 dB for a beamwidth of 40° around θ=0° and 180°. The introduction of

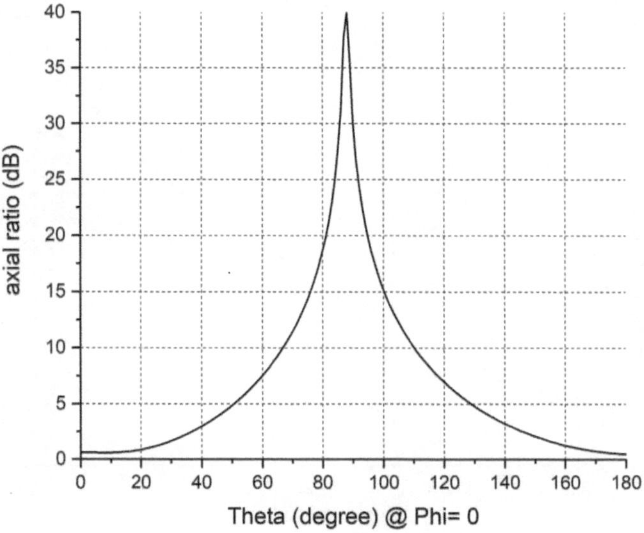

Figure 7.7 Axial ratio of circularly polarized monopole antenna.

a slot in the ground plane alters current distribution in the ground plane which leads to a change in antenna impedance and reflection coefficient degradation as shown in Figure 7.8.

Figure 7.9 shows that the antenna impedance is $45+j\ 58$ for the inserted slot. The antenna impedance is purposely made inductive so that an open-circuited stub can be introduced to match the antenna with the feed.

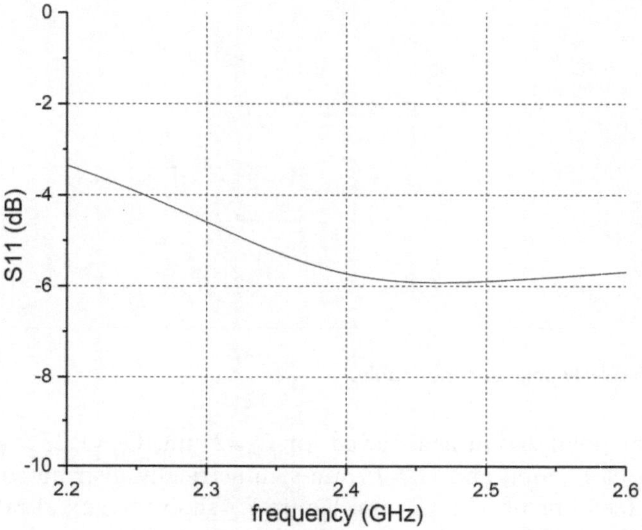

Figure 7.8 Frequency v/s reflection coefficient of circular polarized monopole.

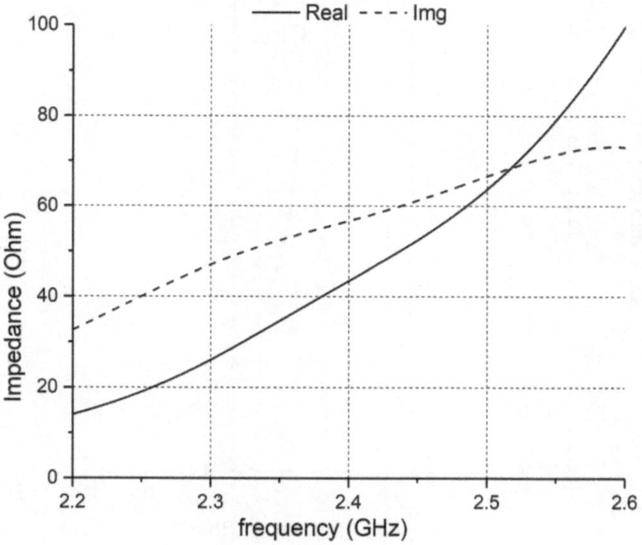

Figure 7.9 Antenna impedance of circularly polarized monopole.

Now, to achieve a match, a stub is introduced along with the feed. An open-circuited stub is a better option than a short circuit as it will cause an increase in the manufacturing cost of the antenna due to the involvement of a short stub with a ground plane.

Figure 7.10 shows that the insertion of an open-circuited stub of length 10 mm and width of 1.5 mm improves antenna impedance to $54+j8.9\ \Omega$ which leads to the excellent matching of the antenna with 50 Ω feed as shown in Figure 7.11.

Figure 7.12 shows the antenna reflection coefficient after loading it with the stub. It can be observed that the antenna is very well-matched and the reflection coefficient improves from –5.7 to –21.33 dB at 2.4 GHz although the antenna is tuned at 2.442 GHz with a reflection coefficient of –38.33 dB.

Figure 7.13 shows the axial ratio of the antenna and it can be observed that the axial ratio of the antenna is within 3 dB for a beam width of 50° about $\theta=0°$ and 180°. Figure 7.14 shows the antenna efficiency which has dropped to 74% as compared to the earlier efficiency of linearly polarized monopole antenna. But it is good enough for any handheld device used for short range.

So far, we have covered the design of a simple monopole antenna with linear and circular polarizations and have noted the maximum length and width of a monopole antenna at 2.4 GHz. It is observed that the antenna length cannot be reduced significantly to have better efficiency and tuning at the desired frequency. But the antenna width (ground plane) can be reduced significantly which leads to an overall size reduction of the antenna. This reduction in the antenna size is not sufficient sometimes and different

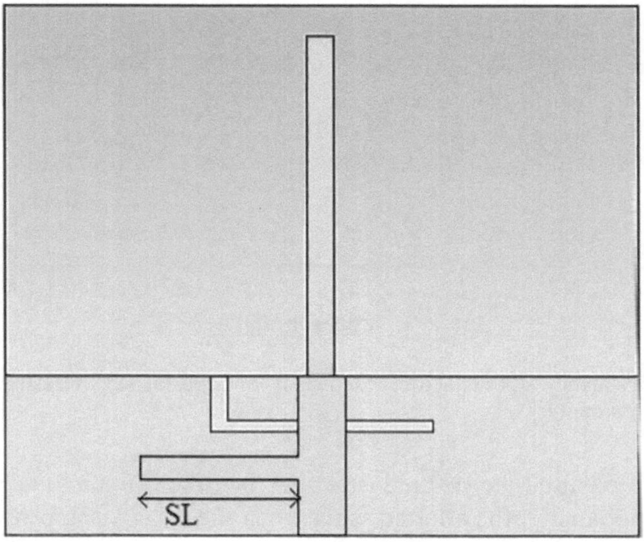

Figure 7.10 Circularly polarized monopole with stub.

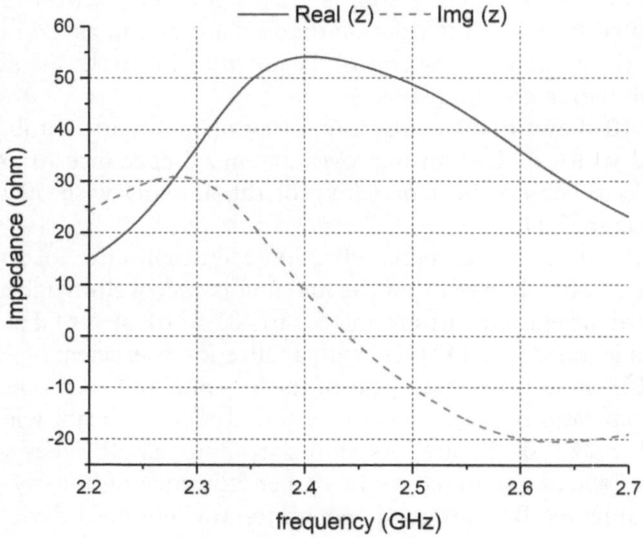

Figure 7.11 Impedance of circularly polarized monopole loaded with stub.

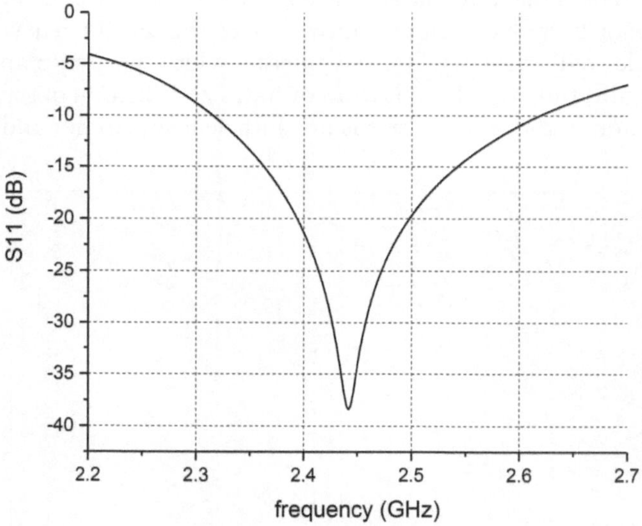

Figure 7.12 Frequency v/s reflection coefficient of stub-loaded circularly polarized monopole.

approaches are used for size reduction. If the designed antenna can easily be accommodated in the allotted space, then the design is approved or else it gets rejected for obvious reasons. A designer faces space constraints while designing an antenna for compact handheld devices. Therefore, a designer

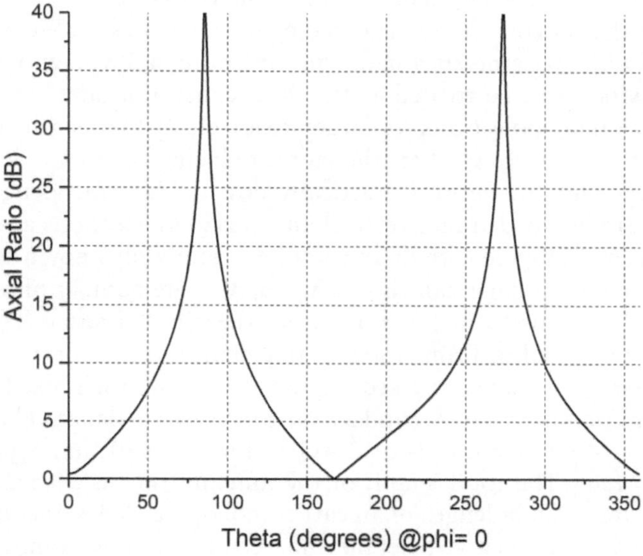

Figure 7.13 Axial ratio of circularly polarized stub-loaded monopole antenna.

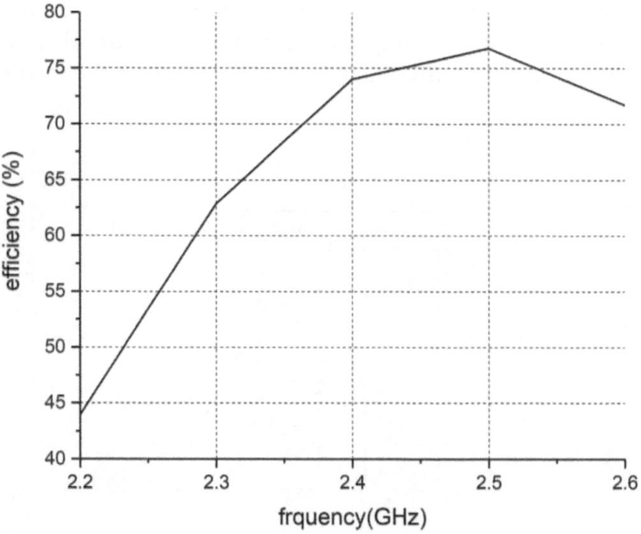

Figure 7.14 Efficiency of circularly polarized stub-loaded monopole antenna.

needs to explore size reduction techniques available in practice, or sometimes a completely new approach is used. Here, we are going to show a conventional approach to designing a compact monopole antenna which can be the first choice for any designer if antenna gain is not a major concern.

There are three dimensions, length, width, and thickness of antenna, which the designer has to consider while reducing the antenna size. We begin with length and width parameters as a design constraint and will ignore antenna thickness which will be treated as the third constraint later. The straight shape of the monopole occupies larger space along the length, while the width is comparatively smaller; therefore, reducing monopole length can bring the antenna dimensions drastically down. The entire length of the monopole can be divided into vertical and horizontal sections as shown in Figure 7.15. A monopole with this structure is commonly known as a meander monopole. A designer can choose an appropriate number of horizontal and vertical sections to fit the antenna in the allotted space. Horizontal length can be varied to bring antenna impedance in a favorable situation once the vertical sections are fixed. Figure 7.16 shows the impact of variations in horizontal sections S_{h1} and S_{h2} on antenna impedance. The antenna impedance is $44.35+j58.35\ \Omega$ at 2.4 GHz for $2{*}S_v=10$ mm, $S_{h2}=7.5$ mm, and $S_{h1}=12$ mm. The total length of the antenna thus is 29.5 mm and is larger than the straight length of an earlier monopole. A designer can introduce inductive or capacitive reactance in the antenna, if it cannot be eliminated at the desired frequency, by varying only horizontal section/s, or both horizontal and vertical sections. But the choice of the vertical section is dependent on the space available for the antenna.

It should also be noted that the antenna impedance contains 44.35 Ω resistance and is very close to the reference impedance of 50 Ω. The inductive reactance can be eliminated using a suitable stub at the feed end.

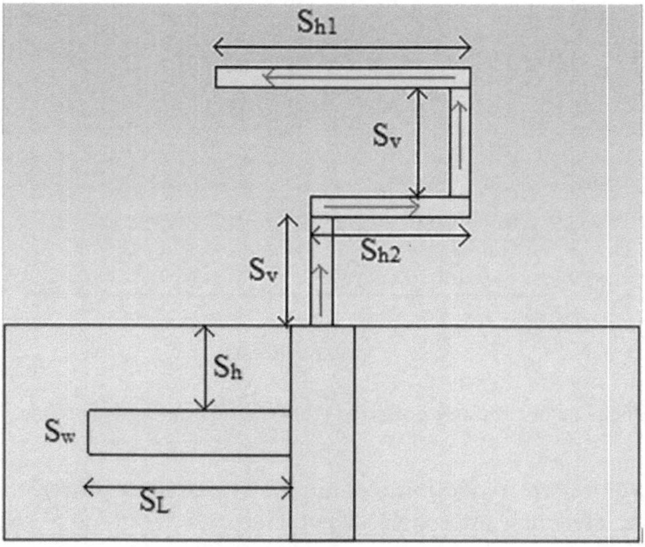

Figure 7.15 Meander monopole antenna.

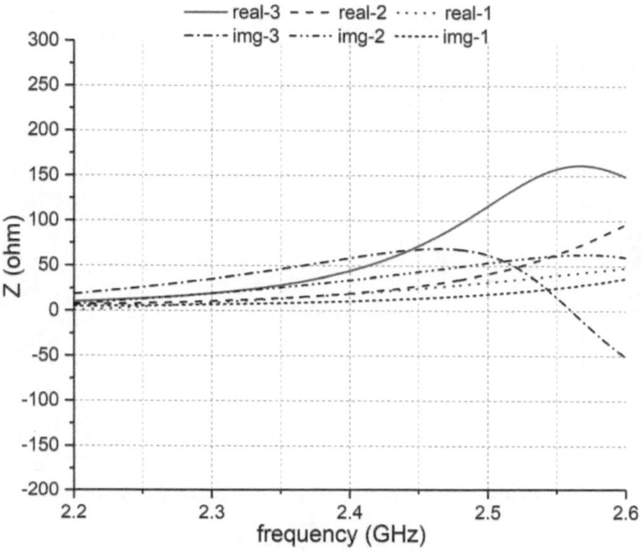

Figure 7.16 Variation of antenna impedance as a function of S_{h1} and S_{h2}.

Table 7.1 Antenna impedance for different lengths of horizontal sections

Case	*1*	*2*	*3*
S_{h1} (mm)	10	11	12
S_{h2} (mm)	5.5	6.5	7.5
Z (ohm)	10.8+*j*19	19.4+*j*34	44.34+*j*58.34

Antenna impedance for three cases of S_{h1} and S_{h2} is summarized in Table 7.1.

Now to match the antenna with a reference impedance of 50 Ω, a stub can be introduced as shown in Figure 7.15. The antenna tuning is completely dependent on length S_L, width S_W, and height S_h. S_L is chosen as 9.5 mm after multiple simulations to attain acceptable matching and then the stub height and width are optimized for improved matching. The variation in antenna impedance can be observed in Figure 7.17 for different widths S_W of stub for S_L=9.5 mm and S_h=3 mm. The stub thickness is increased from 1 to 2 mm in the step of 0.5 mm. The antenna impedance is larger for a stub thickness of 1 mm and decreases for every 0.5 mm increase in its width. All the variations in antenna impedance are summarized in Table 7.2.

The change in impedance results in improved matching and can be observed in Figure 7.18.

Since horizontal segments, S_{h1} and S_{h2}, carry current in the opposite direction and their radiation cancels out while vertical segments, S_v, carry current in the same direction, therefore, contributing to radiation.

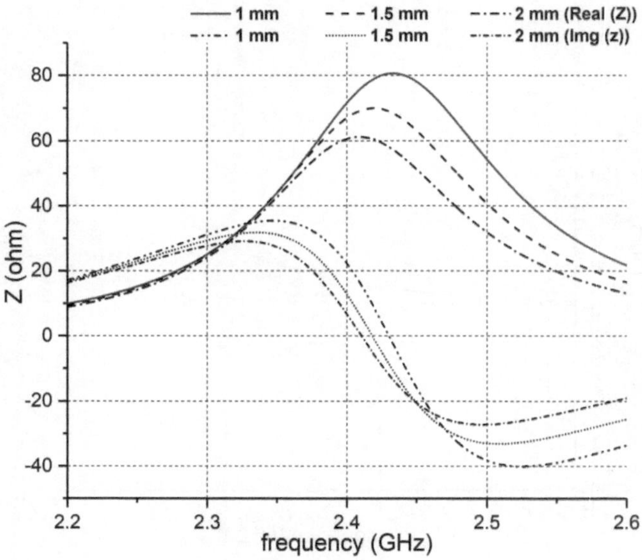

Figure 7.17 Variation of antenna impedance as a function of stub width S_W.

Table 7.2 Antenna impedance for different widths of the stub

S_W (mm)	1	1.5	2
Z (ohm)	71.55+j22.22	67+j12.76	60.6+j6.65

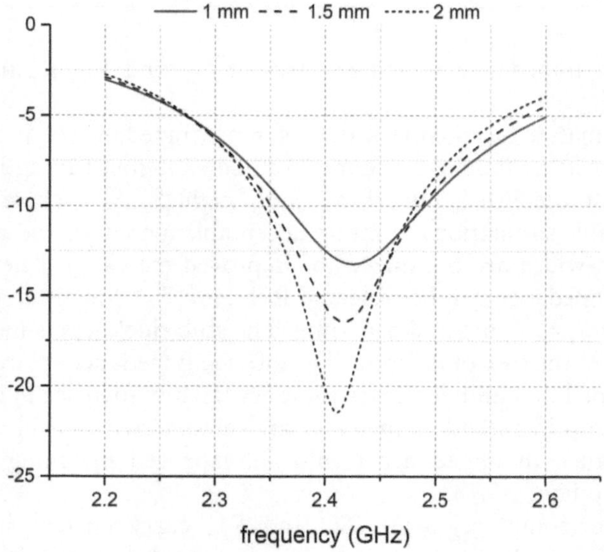

Figure 7.18 S11 v/s frequency as a function of stub width S_W.

This causes degradation in antenna efficiency. A well-matched antenna need not have good efficiency. To have good efficiency, not only the antenna matching but also antenna dimensions with respect to design wavelength λ_o matter a lot. Figure 7.19 shows antenna efficiency for different widths of stub, $S_{W,}$ The maximum efficiency is 73.33% for a stub width of 2 mm. The same monopole for a straight structure has 83.15% efficiency, therefore, it is evident that the size reduction technique leads to a drop in the efficiency and here it is almost a drop of 10%. A designer has to compromise between antenna size and efficiency wherever applicable. Figure 7.20 shows the radiation pattern of the antenna and is similar to the straight monopole antenna with a gain of 0.86 dBi.

Now the two antennas, the straight monopole and the meander monopole, should be compared to check the compactness of the meander antenna. The straight monopole has dimensions of 25×24 $(L\times W)=600\,mm^2$, whereas the meander antenna has $12\times40=480\,mm^2$. The meander antenna occupies 20% lesser space. A further reduction of the meander antenna can be achieved by reducing antenna width. The reduction in width causes a reduction in the ground plane of the antenna which leads to a change in the current distribution in the ground plane and consequently changes antenna impedance. To establish matched condition, antenna length, stub location, width, and length are reconsidered. The modified meander monopole is shown in Figure 7.21; $S_{h1}=14$ mm, $S_{h2}=7.5$ mm, $S_w=1.5$ mm, and $S_h=4.5$ mm. The final matching is achieved by employing two shunt stubs, each of length 7.5 mm, and the antenna shows a reflection coefficient of –17.17 at 2.4 GHz.

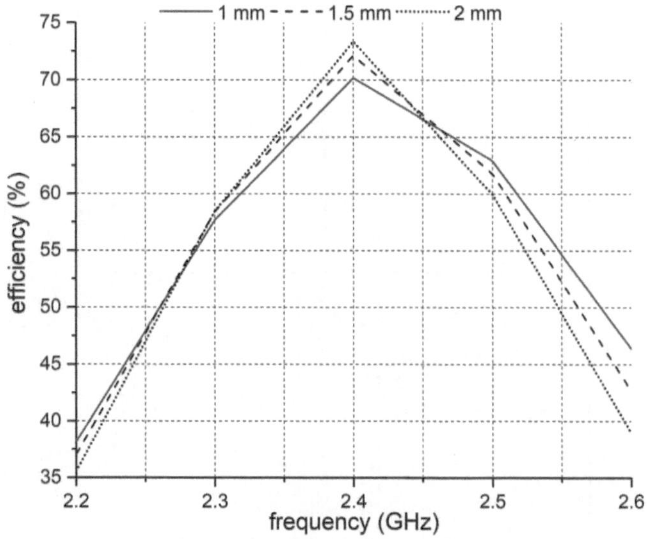

Figure 7.19 Efficiency v/s frequency as a function of stub width S_W.

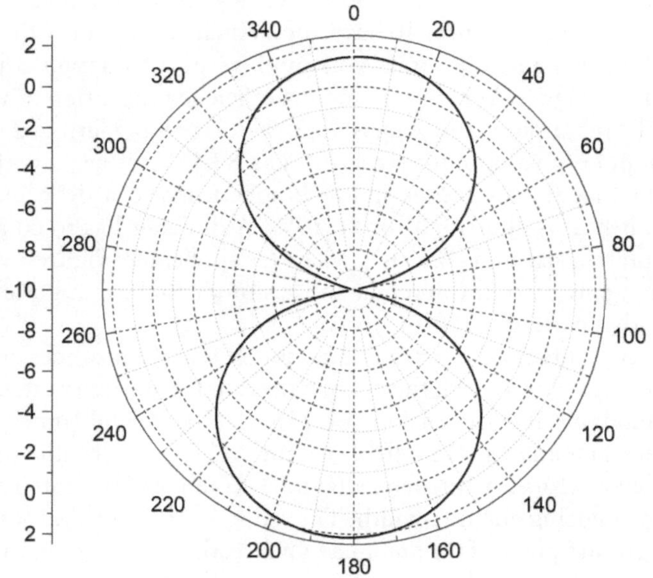

Figure 7.20 Radiation pattern of meander monopole antenna.

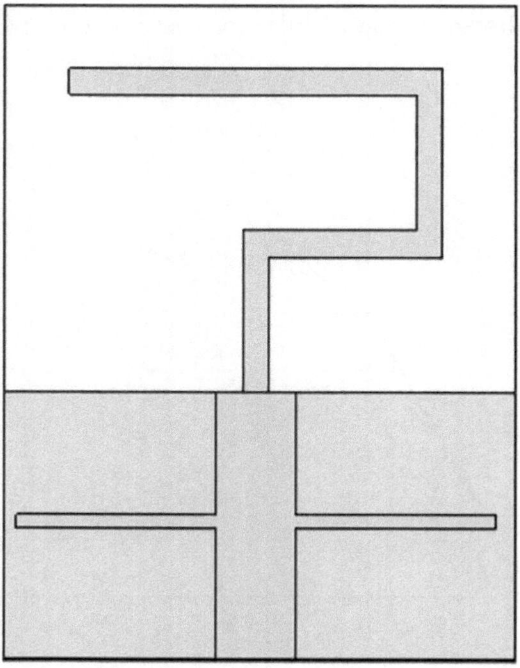

Figure 7.21 Modified meander monopole.

The overall dimensions of the modified monopole are now $12 \times 20 = 240\,mm^2$ and are 60% smaller in size.

It must be noted that the feed length for both antennas is ignored in this calculation.

The two antennas covered so far are omnidirectional and if used for directional applications, such as goods scanning in a godown and toll tax collection on the highway, will lose most of the energy radiated in an unwanted direction.

A simple solution to this problem or, in other words, we can say that to make omnidirectional monopole antenna into a directional antenna, a reflector can be employed. A simple conducting plate can be placed to block the radiation but the separation gap needed to do this will be $\lambda_0/4$, i.e., 31.25 mm. This will make the antenna very bulky despite of reduction in the antenna area. Now, we will investigate a common technique already in practice to make the antenna directional as explained in Ref. [7]. The introduction of a metallic reflector plane will lead to an increase in the overall volume of an antenna and therefore employing a simple metallic reflector to block radiation in an unwanted direction is not a clever choice. An electronic bandgap (EBG) structure will be a better choice. This structure can be used as a lens to focus electromagnetic radiation in one direction. Other than this, the EBG-based reflector need not be placed at a distance of $\lambda_0/4$ which makes the overall antenna size very compact. Investigation of EBG is beyond the scope of this chapter and a reader needs to refer to relevant literature to explore this area. Figure 7.22 shows the geometrical arrangement of the monopole antenna and EBG structure. There is an additional layer of Fr-4 to accommodate the ground plane for the EBG structure, and thus, the entire antenna is now a two-layered structure which increases the thickness of the antenna from 1.6 mm to a larger figure.

Complete dimensional details of the EBG structure are given in Table 7.3. The monopole is of length 22 mm and width 1 mm. The dimensions of the monopole feed and ground plane are kept the same. A rectangular EBG structure is placed below the monopole, along the monopole ground. This EBG structure is open and is extended with an inner bent as shown in Figure 7.22. The length of the inner bent M_L is 12 mm. This length can be varied to tune the antenna at a desired frequency. The gap between the inner bent is 1 mm. Our prime goal is to get unidirectional radiation by employing this EBG structure and to bring antenna impedance close to the reference impedance of 50 Ω.

There are multiple parameters to tune the antenna at the desired frequency, i.e., 2.4 GHz. These parameters are monopole length, M_w, M_h, and M_L. These parameters can be varied to get the desired impedance at 2.4 GHz as shown in Figure 7.23. The antenna impedance at 2.4 GHz is $75.8 + j94$ Ω for the EBG dimensions listed in Table 7.4. The antenna can be matched with the reference impedance using shunt stub/s as was done earlier.

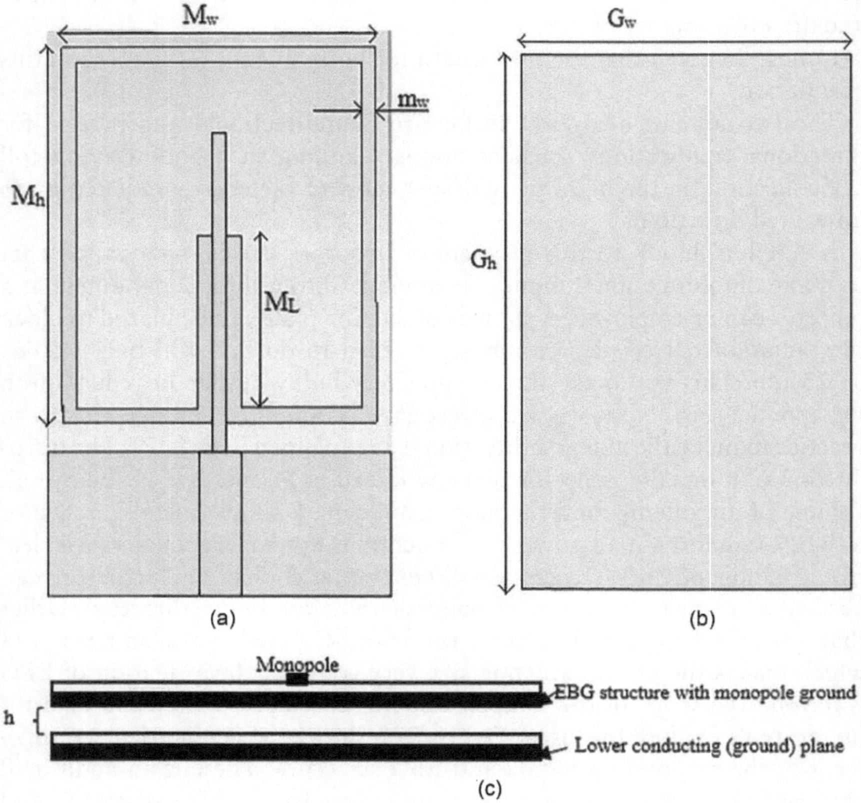

Figure 7.22 (a) Monopole with EBG structure and (b) conducting plane for EBG structure and (c) cross-sectional view of the complete antenna.

Table 7.3 Comparison of monopole antennas

Antenna parameters	*Simple monopole*	*Meander monopole*	*Modified meander monopole*
Area (mm^2)	600	480	240
Efficiency (%)	83.15	73.33	71.2
Gain (dBi)	1.55	0.86	0.87
Thickness (mm)	1.6	1.6	1.6

Figure 7.24 shows the radiation pattern of the antenna. It can easily be concluded that the radiation is through the reflecting plane. Although the radiation is toward the reflecting plane, it solves our purpose of unidirectional radiation and can be utilized in any handheld system where unidirectional radiation is desired.

Now we can move a step ahead to match the antenna while maintaining the directional characteristic of the antenna intact. An open-circuited stub

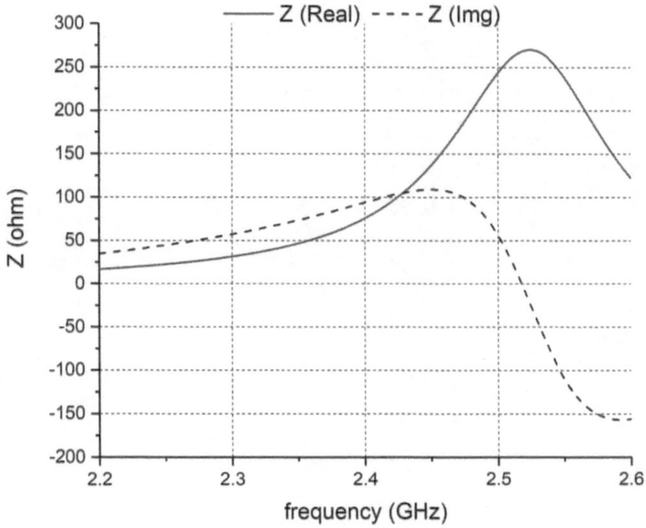

Figure 7.23 Impedance v/s frequency.

Table 7.4 Dimensional details of EBG structure

Parameter	*Dimension (mm)*
M_w	22
M_h	26
m_w	1
M_L	12
G_h	40
G_w	24

is included at the feed of the antenna as shown in Figure 7.25. There are two shunt stubs added to the antenna feed. Each stub is of length S_L=8.5 mm, S_w=2 mm, and its height, S_h, from the lower edge of antenna ground is 3 mm. Other dimensions of the antenna and EBG structure are kept the same.

Figure 7.26 shows antenna impedance after the inclusion of the stub. The antenna impedance has now changed to $68.5+j13.8\ \Omega$ and shows a favorable matched condition as shown in Figure 7.27.

Now, it is important to analyze the gain and beamwidth of the antenna which plays an important role in the acceptance of the designed product. Figure 7.28 confirms the unidirectional nature of the EBG-backed monopole antenna after it is loaded with the stub. The gain of the antenna is 2.85 dBi and is sufficient for handheld devices. The beamwidth of the antenna is 79.4°. Figure 7.28 shows the directivity of the antenna which is around

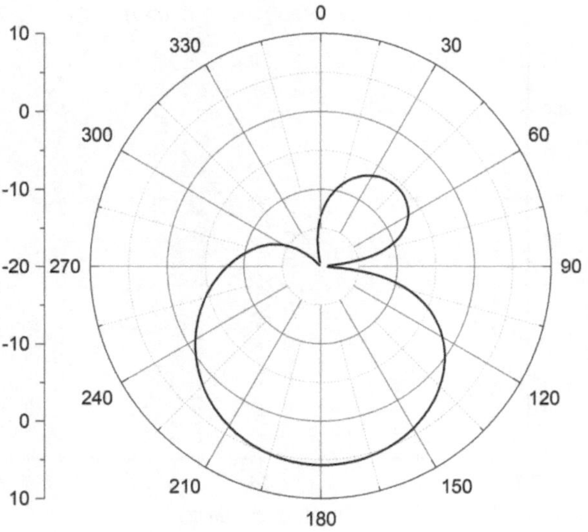

Figure 7.24 Directional radiation pattern of monopole with EBG structure.

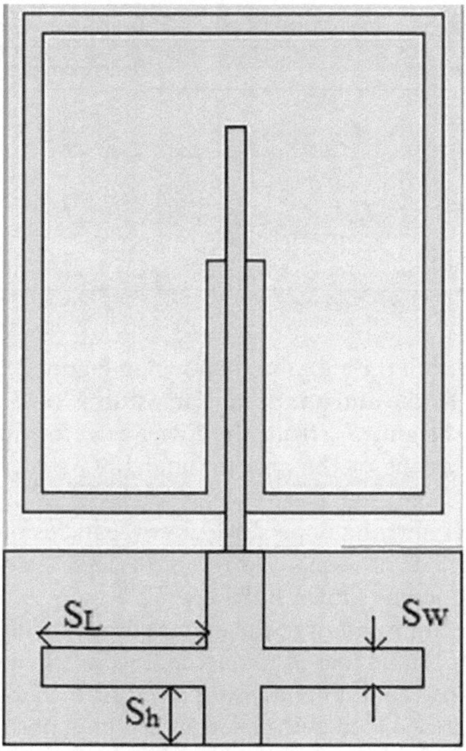

Figure 7.25 Stub-loaded EBG supported monopole.

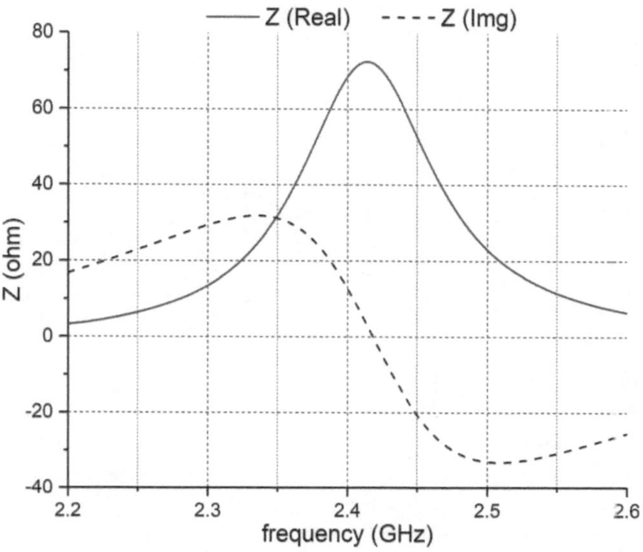

Figure 7.26 Stub-loaded EBG-backed monopole impedance.

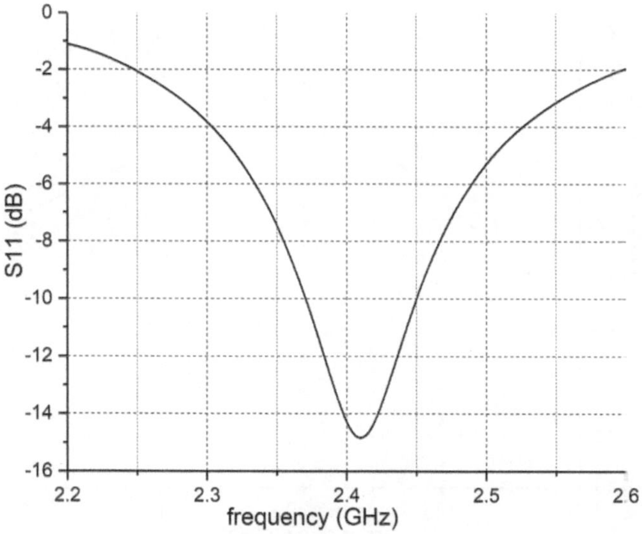

Figure 7.27 Reflection coefficient v/s frequency of EBG-backed stub-loaded monopole.

5.68 dBi but the antenna suffers from poor efficiency as shown in Figure 7.29. The efficiency is dropped due to the introduction of a backplane. The gap h between the monopole and backplane is 5 mm, and antenna efficiency can be improved by increasing this gap but at the cost of compactness of

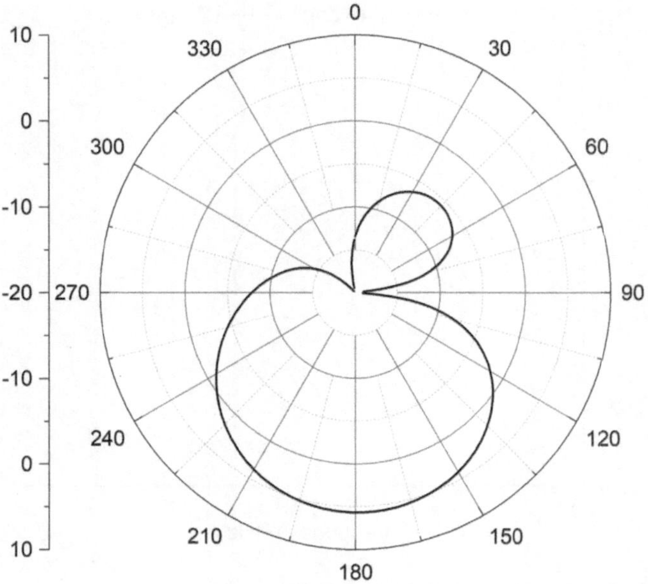

Figure 7.28 Radiation pattern of EBG loaded monopole antenna.

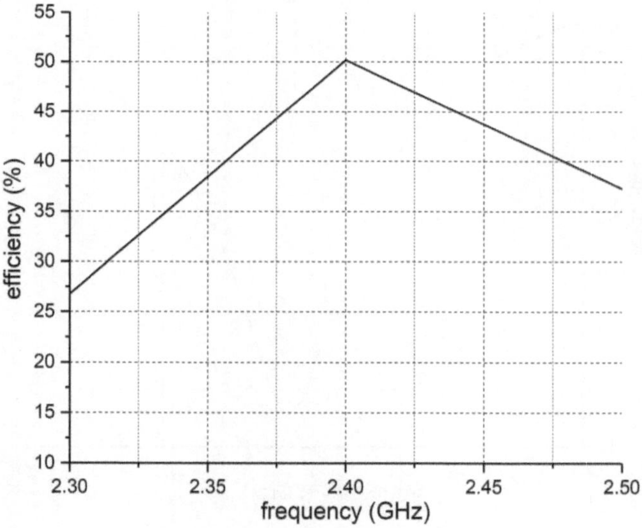

Figure 7.29 Efficiency v/s frequency for EBG-backed monople antenna.

the antenna. It can be observed that the antenna efficiency is 50% at the design frequency, and therefore, antenna gain is limited to 2.85 dBi only.

The volume $L \times W \times h$ of the designed antenna is now $30 \times 24 \times 8.2 = 5904\,mm^3$. All three designs are summarized in Table 7.5. The dimensions

Table 7.5 Performance comparison of all the monopoles

Antenna category	*L (mm)*	*W (mm)*	*h (mm)*	*Efficiency (%)*	*Gain (dBi)*	*Radiation*
Straight monopole	25	24	1.6	83.15	1.55	Omnidirectional
Meander monopole	12	20	1.6	71.12	0.87	Omnidirectional
EBG monopole	30	24	8.6	50	2.85	Directional

calculated in the tableexclude the ground plane of the monopole which is 10 mm in height and increases the length *L* of all the monopoles by 10 mm.

7.4 CONCLUSION

An antenna designer is offered a fixed volume for the antenna while working on compact handheld devices. It is up to the designer to explore and apply various techniques to design the best antenna under given constraints. A designer needs to compromise in the best possible way between antenna volume and gain as obtaining higher gain at a compact size is always a challenge.

REFERENCES

1. G Feng, L Chen, X Shi, A broadband circularly polarized monopole antenna employing parasitic loops and defective ground plane, *Microwave and Optical Technology Letters*, 62, 251–256, 2020, DOI: 10.1002/mop.31998.
2. ZH Jia et al., A compact, low-profile metasurface-enabled antenna for wearable medical body-area, *IEEE Transactions on Antennas and Propagation*, 62, No. 8, August 2014.
3. D Chaturvedi, S Raghavan, SRR-loaded metamaterial-inspired electrically-small monopole antenna, *Progress in Electromagnetics Research C*, 81, 11–19, 2018.
4. B Chen, Y-C Jiao, F-C Ren, L Zhang, Broadband monopole antenna with wide-band circular polarization, *Progress in Electromagnetics Research Letters*, 32, 19–28, 2012.
5. MS Ellis et al., Compact broadband circularly polarized printed antenna with a shifted monopole and modified ground plane, Published by IOP Publishing for Sissa Medialab, November 26, 2019.
6. L Zhang et al., CPW-fed broadband circularly polarized planar monopole antenna with improved ground-plane structure, *IEEE Transactions on Antennas and Propagation*, 61, No. 9, September 2013.
7. MAB Abbasi et al., Compact EBG-backed planar monopole for BAN wearable applications, *IEEE Transactions on Antennas and Propagation*, 65, No. 2, February 2017.
8. *Planar Antenna: Design, Fabrication, Testing, and Application*, Nova Science Publishers Inc., New York, 15 October 2021, ISBN: 9781536198980.

9. P Malik, J Lu, BTP Madhav, G Kalkhambkar, S Amit (Eds.), *Smart Antennas: Latest Trends in Design and Application*, Springer, ISBN: 9783030766368, doi: 10.1007/978-3-030-76636-8.
10. PK Malik, S Padmanaban, JB Holm-Nielsen, *Microstrip Antenna Design for Wireless Applications*, Taylor and Francis, Aug 04, 2021, ISBN: 9780367554385.

Chapter 8

A novel tri-band bandpass filter for 5G applications

[1]Rayaluru Akshay, [2]M. Anas, [3]Kourike Sai Kiran, [4]Patri Upender, [5]B. Yakub, [6]Amarjit Kumar, and [7]B. K. Sharma

[1,2,3,4,5,6]Department of ECE, NIT Warangal, India

[7]Director, Planar Microwave Technologies Ltd, United Kingdom.

CONTENTS

8.1 INTRODUCTION

BPF is the frequency selective network which passes only one band of frequencies while rejecting both higher and lower frequencies. It's a passive component that can sample signals within a particular bandwidth at a certain center frequency and suppress signals in different frequency bands [1,2]. It is necessary for high-performance applications, and is the first choice for accuracy reasons. In addition to a good selectivity and low insertion loss, this classic waveguide filter has a high Q factor.

On the other hand, integrating it between plain and non-plain circuits is difficult. Furthermore, it is large and costly to produce. Rectangular waveguides are mostly employed for frequencies over 8–100 GHz at modest signal levels [3–5]. The SIW method is a unique way of putting this into practice. It can combine the benefits of planar and nonplanar structures, such as microstrip lines and waveguides [6,7]. Waveguide structures can be created in a planar form by employing metalized vias or holes in the dielectric substrate. As a result, it combines the advantages of traditional microstrip circuits, such as ease of manufacture, low cost, and small size, with waveguide capabilities previously only achievable by bulky [8,9]. Substrate-integrated waveguide is a laminated rectangular electromagnetic waveguide created synthetically. It's made from a dielectric substrate with a dense pattern of metalized posts or via-holes connecting the lower and upper conducting plates. It was created in the early 2000s by Ke Wu. It has a simple manufacturing procedure that allows for low-cost mass production employing

DOI: 10.1201/9781003347057-8

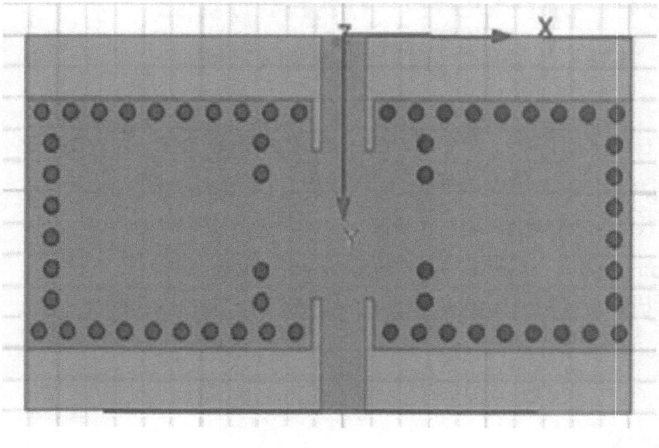

Figure 8.1 Proposed tri-band bandpass filter.

a variety of through-hole techniques, and the post walls are made up of fences. It possesses mode characteristics and a guided wave that are identical to those of a standard rectangular waveguide with an analogous guide wavelength [10–12].

SIW is a rectangular electromagnetic waveguide formed by covering both faces of the dielectric substrate with metal conductors. In addition to that, there are metallic via-holes which connect both the metallic plates. In traditional rectangular waveguides, the propagation of electromagnetic waves includes both the TE and TM modes. But in SIW, there is only the TE mode [13]. This is because in the TM mode, the current is in the direction of the propagation of the wave, and in the case of SIW, it is not possible for the current to move from one via-hole to another because of the presence of a substrate. Therefore, only the TE mode is possible in the case of SIW. SIW has demonstrated that utilizing a normal printed circuit board (PCB) technique can produce a low-cost waveguide filter. The via fence inside the substrate is the structural component of a SIW that consists of the top and bottom metal surfaces of a substrate, and also two parallel rows of via-holes, as shown in Figure 8.1.

The remaining chapter has been structured in the following way: Section 8.2: Design process, Section 8.3: Proposed structure, Section 8.4: Simulation results, and Section 8.5: Conclusion.

8.2 DESIGN PROCESS

A. Design Equations

The equations that are followed in designing a tri-band filter are given in Ref. [6]: Geometry. On a specific substrate, this can integrate passive elements including an antenna with active components such as an amplifier, reducing transitions, and parasitic effects. In recent years, SIW was vastly explored and there is a lot of advancement in this technology [14–16].

$$f_{101} = \frac{c}{2\pi}\sqrt{\left(\frac{\pi}{W_{\text{eff}}}\right)^2 + \left(\frac{\pi}{L_{\text{eff}}}\right)^2} \tag{8.1}$$

$$W_{\text{eff}} = w - \frac{d^2}{0.95p} \tag{8.2}$$

$$d < \frac{\lambda_g}{5} \tag{8.3}$$

$$P \leq 2d \tag{8.4}$$

where P=pitch, d=diameter of the via-hole, w=width of the SIW, l_{eff}=effective length of the SIW, and W_{eff}=effective width of SIW.

These equations 8.1–8.4 govern the dimensions of the substrate-integrated waveguide. In equation 8.1, by keeping the center frequency to 21 GHz and taking the operating mode as TE (101), where m=1 and n=0, we get the dimensions of w from which we can calculate f using equation 8.2, and following equations 8.3 and 8.4, we can calculate the dimension of the pitch and the diameter of the via-holes. The calculated dimensions of the proposed filter are depicted in Table 8.1.

B. Proposed Structure

A tri-bandpass filter has been designed following the design equations mentioned in Section 8.2. The operating mode has been set to the TE101 mode. Initially following these equations, a wide band bandpass filter has been designed with a pass band from 19 to 22 GHz. The wide band bandpass filter so designed has been converted into a triple band filter by the introduction of two stop bands.

Table 8.1 Dimensions of proposed filter

Width	21 mm
Length	10 mm
Height of substrate	0.508 mm
Pitch	1 mm
Diameter	1 mm
Patch length	3.64 mm
Patch width	1.55 mm
Patch slot width	0.3 mm

This has been done by the proper introduction of the via-holes. The introduction of the via-holes on both sides of the center helps the structure in resonating at three different frequencies. The gap between the via-holes on either side of the center acts as a feed for both the resonating parts, which acts as an RLC resonant circuit. Roger Duroid 5880 substrate, which has an electric permittivity of 2.2, has been used and the bottom and top are covered using copper material. Center coupling feed has been used which was realized using two 50 Ω microstrip lines. The structure of the proposed tri-band BPF is shown in Figure 8.1. The dimensions of the proposed BPF are illustrated in Table 8.1.

8.3 SIMULATION RESULTS

The S_{11} and S_{21} plots have been plotted using the HFSS software. The S_{11} plot or return loss has been shown in Figure 8.2. The S_{21} plot or insertion loss is shown in Figure 8.3. It has been observed that the center frequencies for the three bands are 20, 21, and 21.8 GHz. From the S_{11} and S_{21} plots, various parameters like return loss, insertion loss, bandwidth, and fractional bandwidth are among the metrics that have been determined and tabulated below. The total area of the bandpass filter is 1.575 cm^2. The field distribution taken at 21 GHz is shown in Figure 8.4, where we can see that the structure is resonating resulting in minimal losses.

Various parameters of the proposed structure have been calculated and have been tabulated below. From Table 8.2, it can be seen that the return loss is 22, 17.5, and 18.5 dB at the three center frequencies, respectively. The insertion loss is 1.58, 0.9 and 0.85 dB, respectively, and the bandwidths

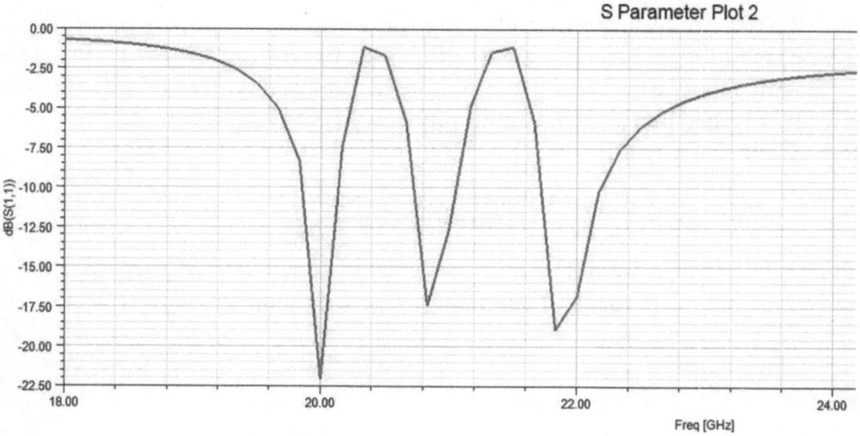

Figure 8.2 S_{11} plot for proposed structure.

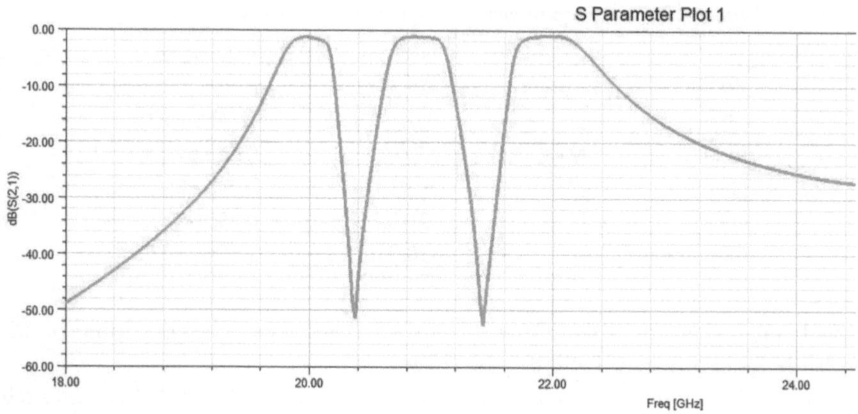

Figure 8.3 S_{21} plot for proposed structure.

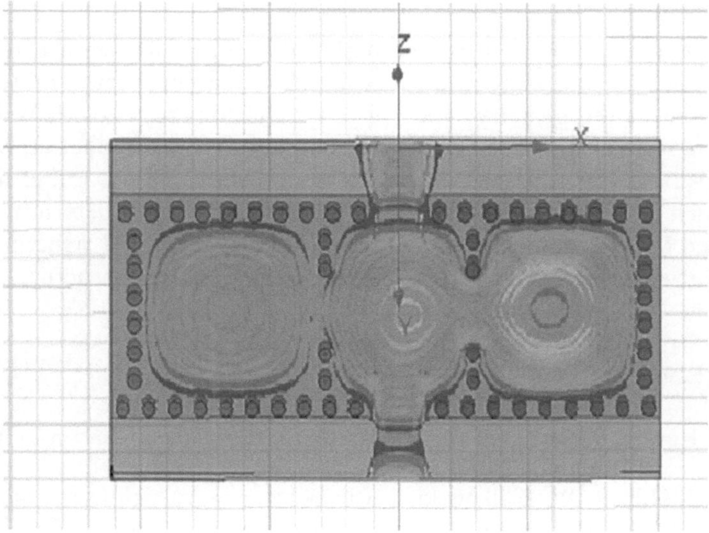

Figure 8.4 Field distribution at 21 GHz.

Table 8.2 Parameters of proposed filter

	Center frequencies		
Parameter (GHz)	20	21	21.8
Return loss (dB)	22	17.5	18.5
Insertion loss (dB)	1.58	0.9	0.85
Bandwidth (GHz)	1	0.8	1.4

Table 8.3 Comparison with other research articles

Works	*Resonate structure*	*Center frequency (GHz)*	*Insertion loss (dB)*
[7]	ML	1.2/1.93/3.5	1.7/0.54/1.16
[8]	SIW+CSRR	3.27/4.75/6.3	3.23/3.69/1.67
[10]	SIW	29.9/34.8/36.8	1.65/1.68/1.79
[11]	SIW	9.72/10.76/11.76	0.33/0.45/0.3
Our work	SIW	20/21/21.8	1.58/0.9/0.85

have been 1, 0.8, and 1.4 GHz at the three frequencies. The fractional bandwidth has been nearly 10% for the three bands.

Table 8.3 shows the comparison between the previously published research articles with the proposed articles.

8.4 CONCLUSION

In this chapter, we have used the design of a BPF filter with three pass bands at 20, 21, and 21.9 GHz using the SIW structure. The TE101 mode was chosen as the primary mode of operation for the whole SIW cavity. The plots for *S*11 and *S*21 have been completed. The distribution of electric fields in the 21 GHz pass band has also been displayed. Other metrics like return loss, insertion loss, and bandwidth have been determined and compared to other filters that have previously used the SIW structure. The HFSS software was used to create all the simulations and structures.

REFERENCES

1. G. Hunter, *Theory and Design of Microwave Filters*, The Institution of Electrical Engineers, 2001.
2. M. J. Hill, R. W. Ziolkowski, J. Papapolymerou, A high-Q reconfigurable planar EBG cavity resonator, *IEEE Microw Wireless Compon Let*, Vol. 11, no. 6, Dec. 2001.
3. U. Patri, K. Amarjit, HEM11δ and HEM12δ-based quad band quad sense circularly polarized tunable graphene-based MIMO dielectric resonator antenna, *Frequenz*, 2022, https://doi.org/10.1515/freq–2021–0145.
4. K. Wu, D. Deslandes, Y. Cassivi, The substrate integrated circuits A new concept for high-frequency electronics and optoelectronics, *Int Conf TELSIKS*, Vol. 1, pp. III–X, 2003.
5. D. Deslandes, K. Wu, Single-substrate integration techniques for planar circuits and waveguide filters, *IEEE Trans Microwave Theory Tech*, Vol. 51, pp. 593–596, Feb. 2003.
6. P. Upender, A. Kumar, Design of a multiband graphene-based absorber for terahertz applications using different geometric shapes, *J Opt Soc Am B*, Vol. 39, pp. 188–199, 2022.

7. D. Deslandes, K. Wu, Integrated microstrip and rectangular waveguide in planar form, *IEEE Microw Wirel Compon Lett*, Vol. 14, pp. 446–448, Sep. 2004.
8. S. Achraou, H. Elftouh, A. Farkhsi, A. Zakriti, S. B. Haddi, Substrate integrated waveguide bandpass filter for mm-wave applications, *Procedia Manuf*, Vol. 46, pp. 766–770, 2020, ISSN: 2351-9789.
9. P. Upender, A. Kumar, Quad-band circularly polarized tunable graphene based dielectric resonator antenna for terahertz applications, *Silicon*, 2021, https://doi.org/10.1007/s12633-021-01336-5.
10. G. Shen, W. Che, Q. Xue, Novel tri-band bandpass filter with independently controllable frequencies, bandwidths, and return losses, *IEEE Microw Wirel Compon Lett*, Vol. 27, no. 6, pp. 560–562, 2017.
11. Y. Dong, C. T. M. Wu, T. Itoh, Miniaturised multi-band substrate integrated waveguide filters using complementary split-ring resonators, *IET Microw Antennas Propag*, Vol. 6, no. 6, pp. 611–620, 2012.
12. W-L. Tsai, T-M. Shen, B-J. Chen, T-Y. Huang, R.-B. Wu, Triband filter design using laminated waveguide cavity in LTCC, *IEEE Trans Compon Packag Manuf Technol*, Vol. 4, no. 6, pp. 957–966, 2014.
13. M. Esmaeili, J. Bornemann, Substrate integrated waveguide triplepass – band dual-stopband filter using six cascaded singlets, *IEEE Microw Wirel Compon Lett*, Vol. 24, no. 7, pp. 439–441, 2014.
14. P. K. Malik, Chapter 4. Mathematical modeling and principle of wireless communication, In: *Energy Harvesting Technologies for Powering WPAN and IoT Devices for Industry 4.0 Up-Gradation*, Nova Science Publishers, Inc., Hauppauge, NY, April 2020, ISBN: 9781536169430.
15. R. Roges, P. K. Malik, Planar and printed antennas for Internet of Things-enabled environment: Opportunities and challenges, *Int J Commun Syst*, Vol. 34, no. 15, p. e4940, 2021, https://doi.org/10.1002/dac.4940, ISSN: 1099-1131.
16. A. Rahim, P. K. Malik, Analysis and design of fractal antenna for efficient communication network in vehicular model, In: *Sustainable Computing: Informatics and Systems*, Elsevier, Vol. 31, 2021, p. 100586, https://doi.org/10.1016/j.suscom.2021.100586, ISSN 2210-5379.

Chapter 9

Microstrip feed half-mode substrate-integrated waveguide loaded with SRR for dual-band applications

Nanda Kumar M.
Sreenidhi Institute of Science and Technology

G. Srihari
Sree Vidyanikethan Engineering College

D. Prasad
Sasi Institute of Technology and Engineering

CONTENTS

9.1 INTRODUCTION

Communication plays a significant role in day-to-day life. It is necessary to provide high data rates to all users to access communication. The trend in communication is growing fast due to applications like industry, medical emergency and academia, while researchers are concentrating on high-frequency applications like microwave and millimetre to fulfil requirements like high gain and broadband applications [1–3].

The microstrip transmission line is not appropriate for high-frequency applications due to its insignificant wavelength and high tolerance. The coplanar waveguide is the next selected device for operating high-frequency applications because of its advantages like more power control and smaller losses, whereas its drawbacks are increased size and cost.

DOI: 10.1201/9781003347057-9

To conquer the drawbacks of CPW and microstrip, a technology named substrate-integrated waveguide (SIW) is derived from the substrate-integrated circuits (SICs) and appears like a waveguide having two horizontal rows of vias connected through a dielectric substrate from the top to bottom ground planes. Planar technology is used for fabrication leading to its simple design and lower cost [4–7].

The generalised formula for the cut-off frequency is represented in equation 9.1. The dominant mode is TE_{10}, and the simplified formula for the width of the rectangular waveguide is mentioned in equation 9.2 [8–10].

$$f_c = \frac{C}{2\pi}\sqrt{\left(\frac{m\pi}{a}\right)^2 + \left(\frac{n\pi}{b}\right)^2} \quad (9.1)$$

$$a = \frac{c}{2^* f_c} \quad (9.2)$$

The dielectric material used to fill is air in the rectangular waveguide called dielectric-filled rectangular waveguide (DFW) and the width is mentioned in equation 9.3.

$$a_R = \frac{c}{2^* f_c^* \sqrt{\varepsilon_r}} \quad (9.3)$$

The width of SIW is

$$a_{\text{siw}} = a_R + \frac{d^2}{0.95^* s} \quad (9.4)$$

where d is the diameter and S is the spacing between vias. The diameter (d) to a width (a_R) is not considered, thereby leading to an error value in equation 9.4. The modified equation without error is represented in equation 9.5 [11,12].

$$a_{\text{siw}} = a_R + 1.08^* \frac{d^2}{s} - 0.1^* \frac{d^2}{a_{\text{SIW}}} \quad (9.5)$$

The condition used to keep up the loss-free radiation is

$$d \le \frac{\lambda_g}{5} \quad \text{and} \quad s \le 2d \quad (9.6)$$

In this research paper, a rectangular-based Sierpinski carpet slot is introduced on the top of the half-mode SIW and the bottom is loaded with metamaterial for dual paper applications [13,14]. Their design equations, antenna geometry and design flow are discussed in Section 9.2. The results are analysed in Section 9.3, and finally, the conclusion is stated in Section 9.4 followed by the Reference section.

9.2 ANTENNA GEOMETRY

The design procedure of the HMSIW is shown in Figure 9.1. Choose to consider a rectangular SIW cavity and a triangular SIW cavity individually cut into half modes. Again, repeat the same procedure to convert into QMSIW individually. Finally, to make an HMSIW, combine the rectangular QMSIW and triangular QMSIW. The parametric representation of the proposed design's top and bottom views is represented in Figure 9.2. The proposed antenna is designed with a Rogers substrate, i.e., RT-Duriod-5880 with a dielectric constant of 2.2. The height of the substrate is 1.57 mm, and the copper thickness chosen is 35 μm. The HMSIW fed by a microstrip, rectangular-based Sierpinski slot, and SRR is introduced in the front- and

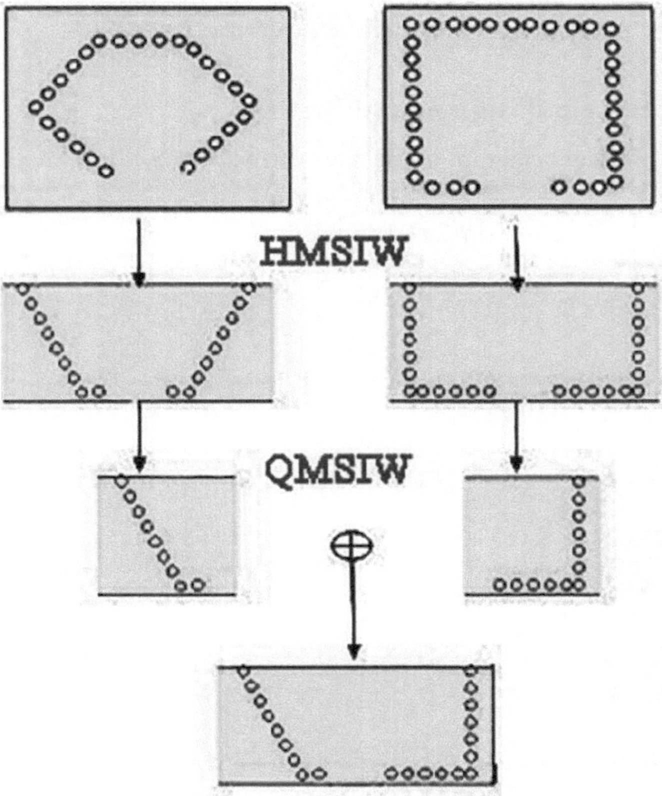

Figure 9.1 Design procedure of HMSIW.

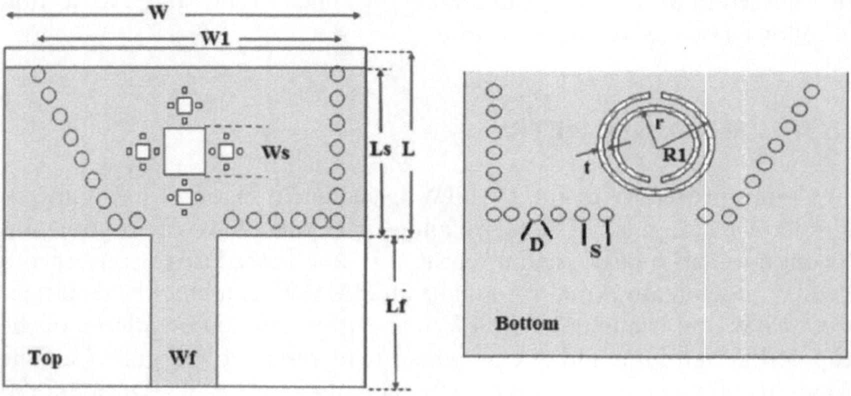

Figure 9.2 Proposed design.

backsides of the antenna, respectively, for k and k_u bands, i.e., dual-band applications [5–7]. The EM tool is used to design and simulate the proposed design and analyse the generalised parameters of the antenna like VSWR, S_{11}, surface current, radiation pattern, efficiencies, etc.

9.2.1 Microstrip design equations

The width (W) and height (h) of the microstrip are [12–14]

$$\frac{W}{h}=\begin{cases}\frac{2}{\pi}\left\{\begin{array}{c}a-1-\ln(2a-1)+\cdots\\ \frac{\varepsilon_r-1}{2\varepsilon_r}\left[\ln(a-1+0.39)-\frac{0.1}{\varepsilon_r}\right]\end{array}\right\} & \frac{W}{h}>2\\ \frac{8e^a}{e^a-2}, & W/h<2\end{cases} \tag{9.7}$$

where

$$a=\frac{377\pi}{2Z_o\sqrt[2]{\varepsilon_r}} \tag{9.8}$$

$$b=\frac{Z_o}{60\sqrt[2]{\varepsilon_r}}\sqrt[2]{\frac{\sqrt[2]{\varepsilon_r}+1}{2}}+\frac{\varepsilon_r-1}{\varepsilon_r+1}\left(0.23+\frac{0.11}{\varepsilon_r}\right) \tag{9.9}$$

The length of the microstrip is [7,12–14]

$$L_m=n^*\lambda_g;\qquad n=1,3,5,7\ldots \tag{9.10}$$

9.2.2 Sierpenski carpet iteration equations

The Sierpinski carpet fractal is constructed with the Sierpinski sieve, where the triangles are replaced by octagonal shapes. It can be constructed with matrix format and described as cell [1], cell [0]. The first iteration is described in equation 9.11 [15].

$$\left\{ 0 \rightarrow \begin{bmatrix} 0 & 0 & 0 \\ 0 & 0 & 0 \\ 0 & 0 & 0 \end{bmatrix}, 1 \rightarrow \begin{bmatrix} 1 & 1 & 1 \\ 1 & 0 & 1 \\ 1 & 1 & 1 \end{bmatrix} \right\} \tag{9.11}$$

Equation 9.3 represents the second iteration, and the same procedure is continued for the nth iteration. The black cells after $n=0, 1, 2, \ldots$ iterations are, therefore, 1, 8, 64, etc.

$$\begin{bmatrix} \begin{bmatrix} 0 & 1 & 0 \\ 1 & 0 & 1 \\ 0 & 1 & 0 \end{bmatrix} & \begin{bmatrix} 0 & 1 & 0 \\ 1 & 0 & 1 \\ 0 & 1 & 0 \end{bmatrix} & \begin{bmatrix} 0 & 1 & 0 \\ 1 & 0 & 1 \\ 0 & 1 & 0 \end{bmatrix} \\ \begin{bmatrix} 0 & 1 & 0 \\ 1 & 0 & 1 \\ 0 & 1 & 0 \end{bmatrix} & 0 & \begin{bmatrix} 0 & 1 & 0 \\ 1 & 0 & 1 \\ 0 & 1 & 0 \end{bmatrix} \\ \begin{bmatrix} 0 & 1 & 0 \\ 1 & 0 & 1 \\ 0 & 1 & 0 \end{bmatrix} & \begin{bmatrix} 0 & 1 & 0 \\ 1 & 0 & 1 \\ 0 & 1 & 0 \end{bmatrix} & \begin{bmatrix} 0 & 1 & 0 \\ 1 & 0 & 1 \\ 0 & 1 & 0 \end{bmatrix} \end{bmatrix} \tag{9.12}$$

The L_n and N_n are the white box side length and number of block boxes, respectively. They are described in the below equations [15–17]:

$$N_n = 8^n \tag{9.13}$$

$$L_n = 3^{-n} \tag{9.14}$$

9.3 DISCUSSION OF RESULTS

Figure 9.3 represents the proposed design S_{11} over the frequency range of 12–14 GHz and resonates five frequencies with respect to –10 dB reference line which are operating in the k band and k_u band applications. The S_{11} values are as follows: –35 dB at 15.07 GHz, –13.36 dB at 15.81 GHz, –13.19 dB at 17.83 GHz, –30.53 dB at 20.27 GHz and –13.06 dB at 21.17 GHz. The

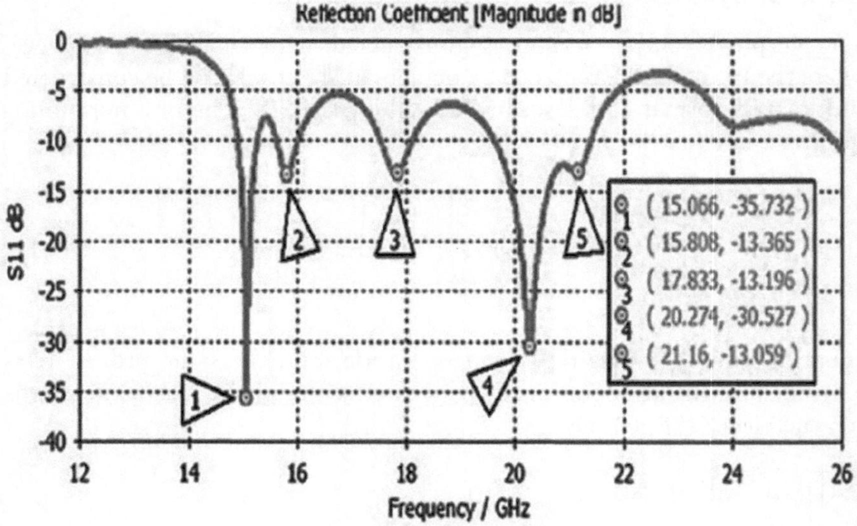

Figure 9.3 Reflection coefficient.

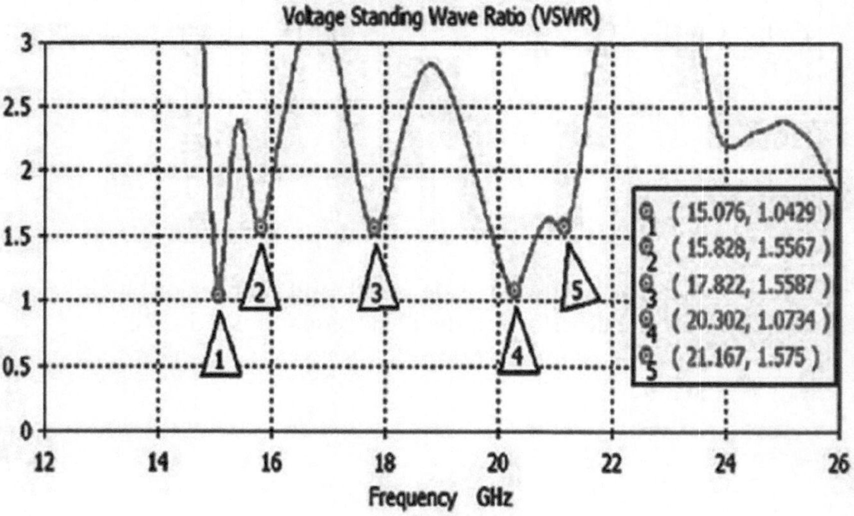

Figure 9.4 Voltage standing wave ratio.

operating frequency of the proposed antenna is 14.908–15.272 GHz (resonant frequency is 15.07 GHz), 15.61–16.02 GHz (resonant frequency is 15.81 GHz), 17.5–18.13 GHz (resonant frequency is 17.83 GHz) and 19.61–21.38 GHz (resonant frequencies are 20.27 and 21.17 GHz). The voltage standing wave ratio of the proposed design explored in Figure 9.4 is (1:2

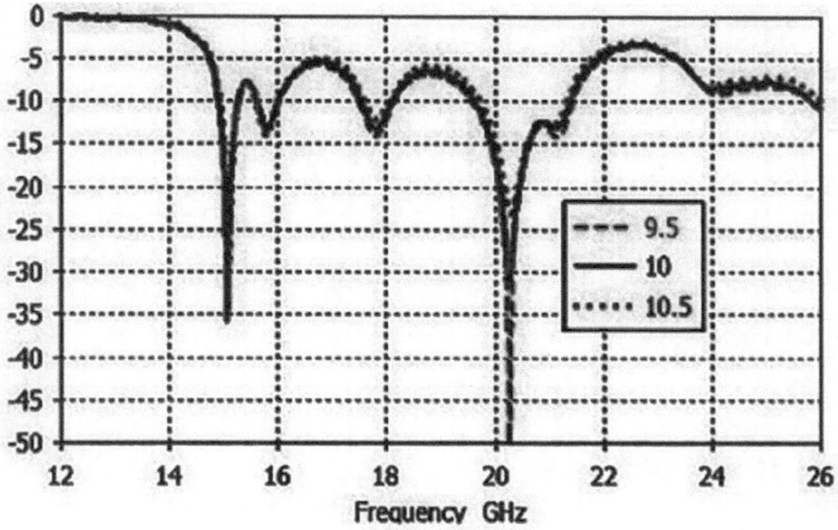

Figure 9.5 Reflection coefficient for different values of feed length (L_f).

ratio) matched with the S_{11} values. The VSWR values are 1.0429, 1.5567, 1.5587, 1.0734 and 1.575 at 15.08, 15.828, 17.822, 20.302 and 21.167 GHz frequencies, respectively.

The analysis of the reflection coefficient for three different values of feed length (L_f) is represented in Figure 9.5, and the values considered for analysis are 9.5, 10 and 10.5 mm. The increasing feed length decreases the accuracy, leading to an increase in the bandwidth and moves towards the –10 dB line. It is observed that 10 mm results are best when compared to other values with respect to accuracy and bandwidth.

The variation of reflection coefficient for different values of feed widths (4.75, 4.8, 4.85 mm) is represented in Figure 9.6 where the blue colour represents 4.85 mm, red colour represents 4.8 mm, and light green is used to represent 4.75 mm. The increasing feed width will decrease the accuracy, and a little bit of variation in the bandwidth and among the three values—4.8 mm—will produce the best results. The reflection coefficient for different values of spacing between vias is represented in Figure 9.7, and these values are 155, 1.57, 1.6, and 1.62 mm. The S = 1.6 mm will give the best results in terms of accuracy and bandwidth.

The S_{11} results for three different values of diameter (D) are observed in Figure 9.8, and those values are 0.9, 1 and 1.1 mm. The decrease in the diameter will move the S_{11} results above the –10dB reference line, increasing accuracy and reducing the bandwidth. The D = 1 mm is the best value chosen as the diameter of SIW vias compared to the other two values. The S_{11} for three different values of the width of the proposed design is represented in

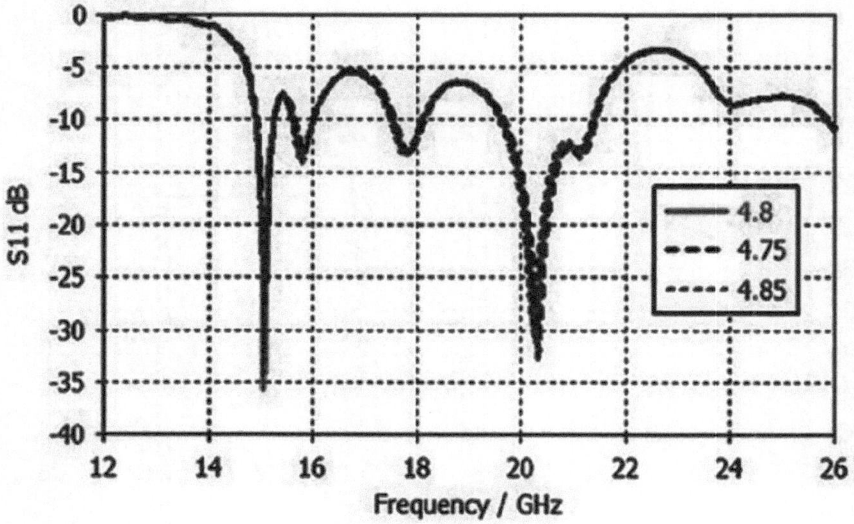

Figure 9.6 S_{11} for different values of feed width (W_f).

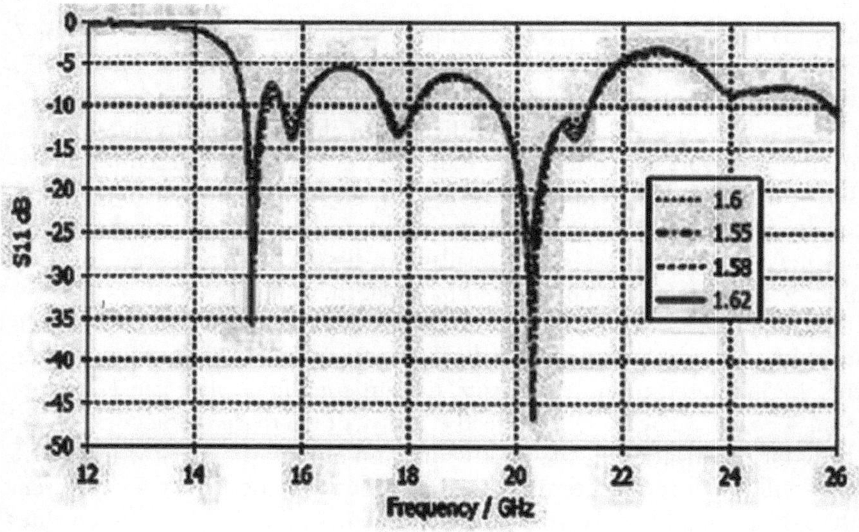

Figure 9.7 S_{11} for different values of spacing between vias (S).

Figure 9.9, and three values of 25, 26 and 27 mm are considered in analysing the results. Increasing the width of the design will decrease the accuracy and provide a slight improvement in the bandwidth. Figure 9.10 represents S_{11} for three different values of SRR widths, and the values are 0.4, 0.5 and 0.6 mm. An increase in the width of SRR will increase the accuracy and

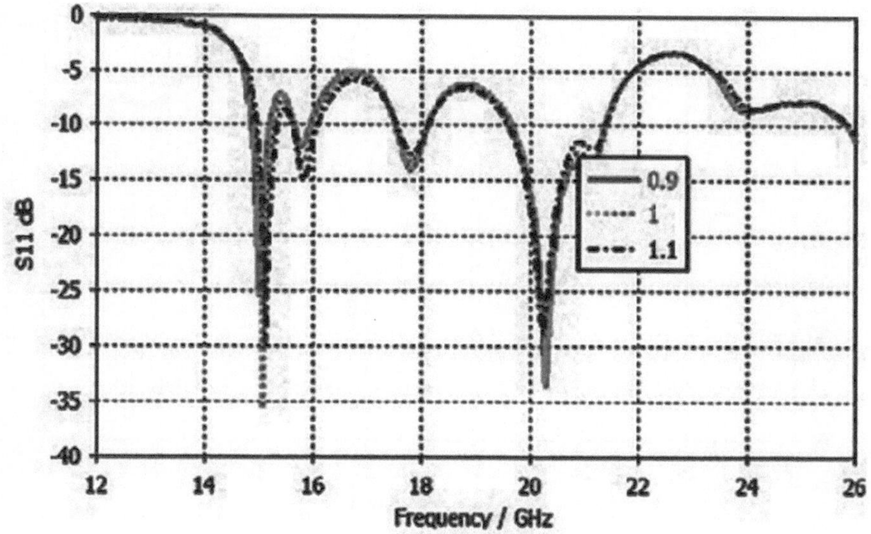

Figure 9.8 S_{11} for different values of diameter (*D*).

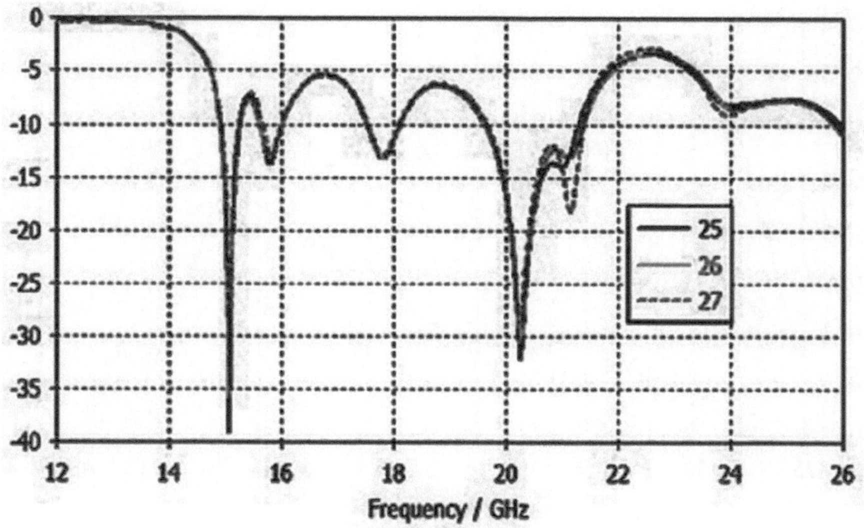

Figure 9.9 S_{11} for different values of width of proposed antenna (*W*).

impedance bandwidth, and $t=0.6$ mm is the best value chosen in the design of the SRR width.

The efficiencies of the proposed design are represented in Figure 9.11 over the frequency range of 15–22 GHz. The radiation and total efficiencies are 87.7%, 87.6% at 15.07 GHz, 93.6%, 89.1% at 15.83 GHz, 91.5%, 87.3%

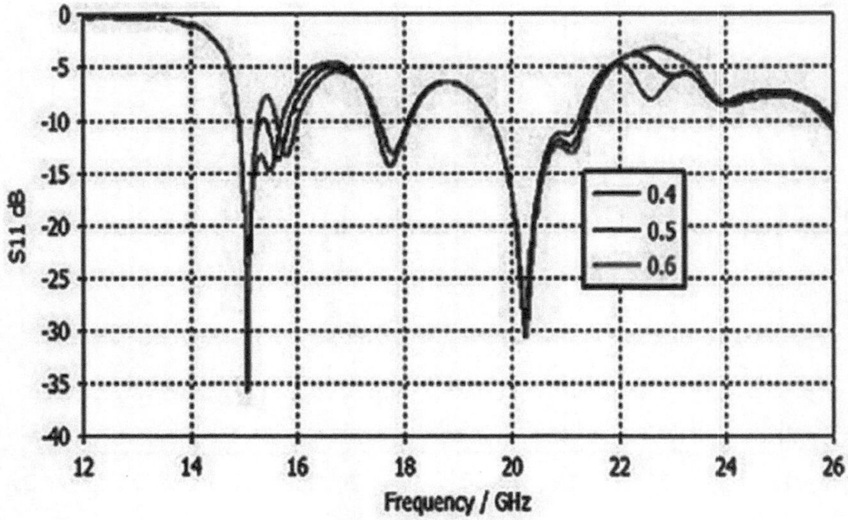

Figure 9.10 S_{11} for different values of SRR width (t).

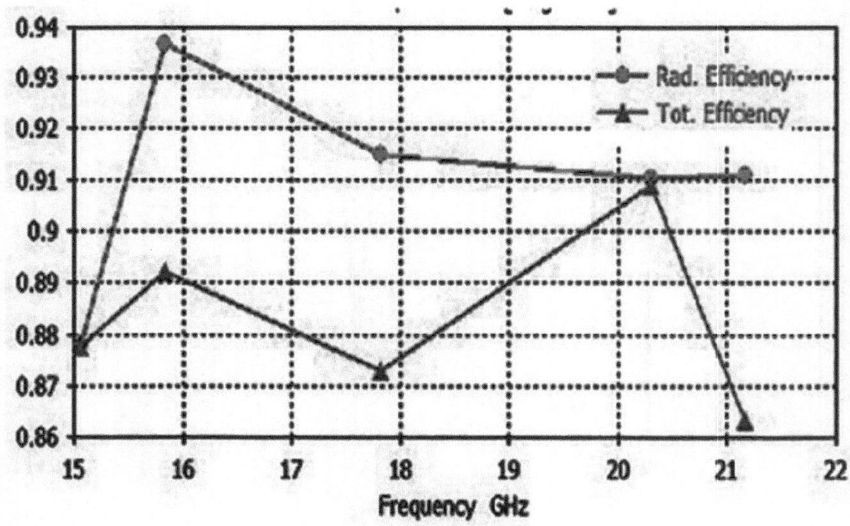

Figure 9.11 Efficiencies of the proposed design.

at 17.8 GHz, 91%, 90.9% at 20.3 GHz, and 91.2%, 86.3% at 21.17 GHz. Figure 9.12 reveals the results of the maximum gain over frequency and approximately more than 6 dBi except at a frequency of 15.1 GHz (4.2 dBi) which is much more useful in wireless communication applications.

The surface current of the proposed antenna is revealed in Figure 9.13. The red colour indicates the maximum flow of the current and it is observed

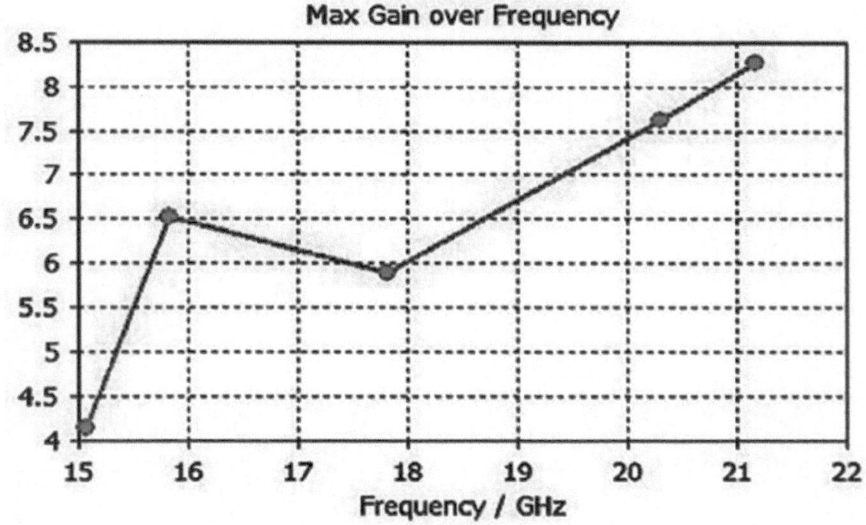

Figure 9.12 Frequency (GHz) vs gain (dBi).

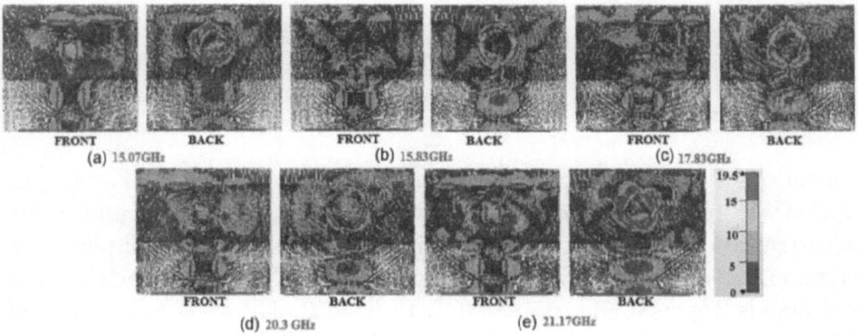

Figure 9.13 Surface current.

that the maximum current flow is at the feed as well as slot positions and SRR. Figure 9.14 represents the radiation pattern of the proposed antenna for five resonant frequencies and is represented by different colours. In that, the left side graph indicates co-polarisation and the right side indicates cross-polarisation and it also radiates bidirectionally.

9.4 CONCLUSION

In this chapter, a half-mode SIW-based Sierpinski rectangular-shaped fractal loaded with SRR has been introduced for dual-band, k and ku bands,

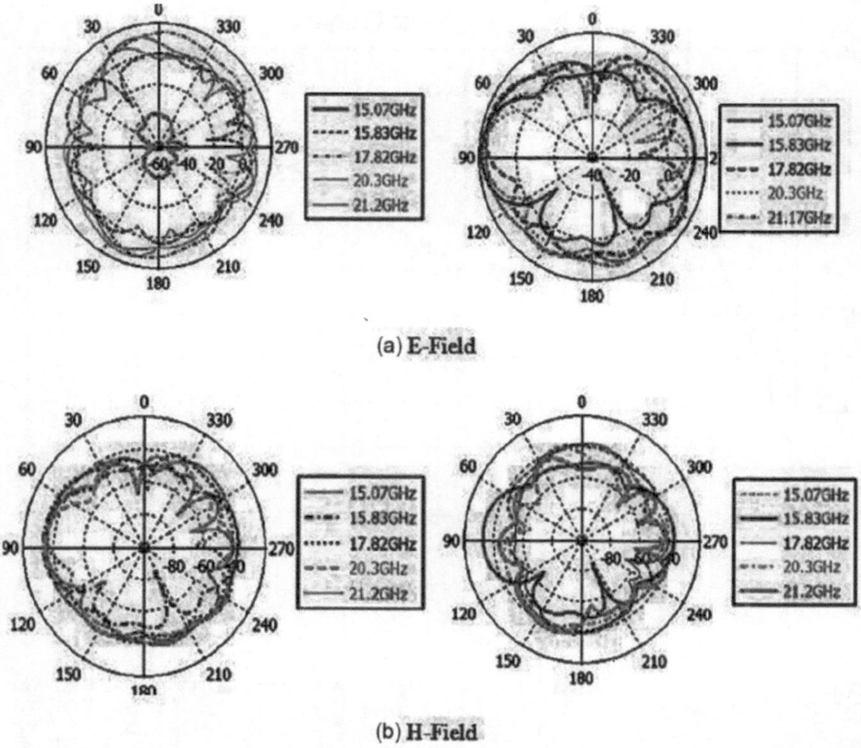

Figure 9.14 Radiation pattern.

applications. The half-mode SIW is merged with two QMSIW. The shapes are QMSIW rectangular cavity and QMSIW hexagonal cavity and feed by microstrip and dimensions are 24 mm×26 mm×1.57 mm. The proposed antenna is designed and simulated using an EM tool, and the material used for design is Rogers RT-Duriod 5880 with E_r=2.2. The antenna is operated in five frequencies: 20.3, 17.8, 15.83, 15.07 and 21.17 GHz, and their *S*11 values are –35, –13.36, –13.19, –30.53 and –13.06 dB. Also, the *S*11 values are matched with VSWR results with a ratio of 1:2. The average gain is approximately 6 dB which is very useful for wireless communication applications and radiated in bidirectional radiation patterns.

REFERENCES

1. Cai Y, Zhang Y, Ding C, Qian Z. A wideband multilayer substrate integrated waveguide cavity backed slot antenna array. *IEEE Transactions on Antennas and Propagation*, Vol. 65, No. 7, 3465–73, 2017.
2. Xu J, Chen ZN, Qing X. CPW center-fed single-layer SIW slot antenna array for automotive radars. *IEEE Transactions on Antennas and Propagation*, Vol. 62, 4528–36, 2014.

3. Mukherjee S, Biswas A. Design of self diplexing substrate integrated waveguide cavity backed slot antenna. *IEEE Antennas Wireless Propagation Letters*, Vol. 15, 1775–8, 2014.
4. Nanda Kumar M, Shanmuganantham T. Broad-band H-spaced head shaped slot with SIW based antenna for 60 GHz wireless communication applications. *Microwave and Optical Technology letters (MOTL)*, Vol. 61, No. 8, 1911–6, 2019.
5. Nanda Kumar M, Shanmuganantham T. Division shaped SIW slot antenna for millimeter wirelesss/automotive radar applications. *Journal of Computer and Electrical Engineering*, Vol. 71, 667–75, 2018.
6. Bozzi M, Georgiadis A, Wu K. Review of substrate integrated waveguide circuits and antennas. *IET Microwave Antennas and Propagation*, Vol. 5, 909–20, 2011.
7. Bozzi M, Perregrini L, Wu K, Arcioni P. Current and future research trends in substrate integrated waveguide technology. *Radio Engineering*, Vol. 18, No. 2, 201–7, 2009.
8. Shaik N, Malik PK. A comprehensive survey 5G wireless communication systems: open issues, research challenges, channel estimation, multi carrier modulation and 5G applications. *Multimedia Tools and Applications*, 2021. https://doi.org/10.1007/s11042-021-11128–z.
9. Tiwari P, Malik PK. Wide band micro-strip antenna design for higher "X" band. *International Journal of e-Collaboration (IJeC)*, Vol. 17, No. 4, 60–74, 2021. http://doi.org/10.4018/IJeC.2021100105 (ISSN: 1548-3673).
10. Wadhwa DS, Malik PK, Khinda JS. High gain antenna for n260- & n261-bands and augmentation in bandwidth for mm-wave range by patch current diversions. *World Journal of Engineering*, 2021. https://doi.org/10.1108/WJE-03-2021-0133 (ISSN: 1708-5284).
11. Nanda Kumar M, Shanmugnantham T. Substrate integrated waveguide cavity backed bowtie slot antenna for 60 GHz applications. *IEEE International Conference on Emerging Technology Trends*, Kollam, Kerala, 20–21 April, 2016.
12. Nanda Kumar M, Shanmuganantham T. Back to back Pi shaped slot with SIW cavity for millimeter wireless applications. *International Journal of Microwave and Optical Technology (IJMOT)*, 2019.
13. Dinesh M, Kumar MN, Balachandra K. Micro-strip feed reconfigurable antenna for wideband applications. *Journal: Lecturer Notes in Electrical Engineering*, 665–71, 2018.
14. Nanda Kumar M, Shanmuganantham T. Broad band I shaped SIW slot antenna for V-band Applications. *Applied Computational Electromagnetic Society (ACES)*, Vol. 34, No. 11, November 2019.
15. Kumar MN, Yogaprasad K, Anitha VR. A quad band sierpenski based fractal antenna fed by CPW. *Microwave and Optical Technology Letters*, Vol. 2, 893–8, 2020.
16. Mukherjee S, Biswas A, Srivastava KV. Broadband substrate integrated waveguide cavity backed bowtie slot antenna. *IEEE Antennas Wireless Propagation Letters*, Vol. 13, 1152–5, 2011.
17. Mukherjee S, Biswas A, Srivastava KV. Substrate integrated waveguide cavity backed dumbbell-shaped slot antenna for dual frequency applications. *IEEE Antennas Wireless Propagation Letters*, Vol. 14, 1314–7, 2011.

Chapter 10

Design of a novel keyhole-shaped multiband MIMO antenna for 5G applications

[1]Anshika Shrivastav, [2]Pradeep Kamal, [3]Yash Deshmukh, [4]Patri Upender, [5]B. Yakub, [6]Amarjit Kumar, and [7]B. K. Sharma

[1,2,3,4,5,6]Department of ECE, NIT Warangal, India.
[7]Director, Planar Microwave Technologies Ltd, United Kingdom.

CONTENTS

10.1 INTRODUCTION

As the number of users and devices increases, there will be a steep rise in traffic. As a result, more antennas (access points) will be required in the future generation of communication networks. Transmission and receiving of signals are done through antennas; hence, they are very important. A good antenna will ensure a good communication system [1]. Thus, designing an antenna plays a dominant role at both ends. Multiple-input-multiple-output antennas are used at both the sending and receiving ends [2,3]. This provides spatial multiplexing leading to higher data rates, link reliability and spectral efficiency. Hence, the Multiple input multiple output (MIMO) technology is an essential requirement for the development of 5G technology [4,5]. The key features of MIMO are spatial multiplexing, enhanced data throughput and high data integrity due to path diversity and beam steering/beam forming techniques [6,7]. Consequently, the multiple-input-multiple-output antenna plays a major role in making 5G a reality [8]. The characteristics of an ideal 5G antenna are high gain, compact designs, very low return loss, high efficiency and very low ECC to meet the 5G technology requirements, which include massive throughput, blazing speeds, capability and ultra-low latency [9]. Since the 5G systems

DOI: 10.1201/9781003347057-10

will use millimetre wave bands, the antennas will be required to have high gain to compensate for atmospheric losses due to absorptions and diminutions [10]. Moreover, they will support a wider bandwidth. The path loss is significantly high in the millimetre wave spectrum; thus, MIMO is essential to provide improved gain [4]. Since MIMO uses multiple antennas for both transmission and reception, they need to be designed and placed in such a way that the mutual coupling between the antennas is the least. A high mutual coupling will affect the radiation pattern of the antenna. Reducing mutual coupling becomes more challenging when the size is required to be compact. Additionally, just reducing the size of the antenna is also not possible as it will affect antenna performance such as efficiency and operating bandwidth. All these will affect the diversity performance of the MIMO systems. Thus, there is a trade-off between the antenna size and performance metrics. The challenge is to optimise all parameters for the best performance keeping in mind the design constraints [8]. Good isolation between the antennas of MIMO configuration is required to provide independent paths for communication channels. The most difficult challenge thus remains to develop MIMO in a compact size with good isolation [11].

We have used circular microstrip antennas to obtain multiband MIMO antennas for 5G applications. Patch antenna contains a dielectric substrate, ground, microstrip feed and patch which can take many shapes like triangular, circular, rectangular, etc. Addition of slots in original structures can add additional operating frequency bands, improve bandwidth and return loss. Moreover, it helps in reducing the size [12–14].

10.2 ANTENNA DESIGN

A. Antenna Design Procedure

The first step to designing a multiband MIMO antenna is to design its single-element structure. We selected a circular patch antenna. First, the dimensions of this patch antenna are determined. Once the basic circular antenna is designed, the next step is to load it with slots to obtain multiband frequencies of operation and improved bandwidth with improved return loss. The slots are loaded in a circular antenna and simulation is carried out.

At this step, we tried many different slot shapes such as T-shaped, U-shaped, star-shaped, and rectangular slots. Finally, the antenna with the best parameters was selected. The next step involved utilising this single element in MIMO configuration. The main requirements are good isolation between the antennas; otherwise, the radiation pattern gets affected. Various configurations were analysed to find the one having the maximum isolation between the single elements. Once the configuration is done, simulations run to check if they give the required parameters. This is the final step. Thus, if all requirements are satisfied, the design is finalised.

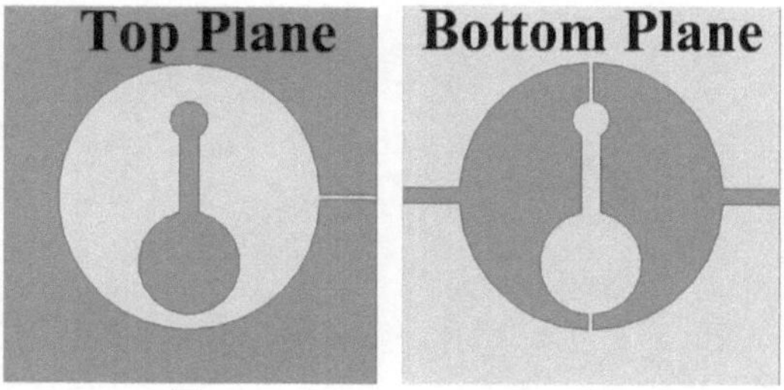

(a) (b)

Figure 10.1 Proposed antenna single-element design. (a) Top view and (b) ground plane (bottom view).

While designing, it is necessary to choose a substrate of lesser relative permittivity for good radiation efficiency. A thicker dielectric may result in a wider bandwidth. Thus, Rogers RT-5800 with a dielectric permittivity of 2.2 and a loss tangent of 0.0013 was chosen as our substrate. The radius was calculated with the below formula for our desired frequency. Since we are looking to design an antenna for a 5G application, we have to choose a frequency in the mm-Wave spectrum.

B. Single-Element Design

The top view and bottom defected ground structure of a single element are depicted in Figure 10.1.

The radius of the circular microstrip antenna can be found using the following formula [15, 16]:

$$r = F \Big/ \left(\sqrt{\left(1+(2/\pi rF)\right)}\right)\left(\left(LN\left(\pi f/(2*h)\right)+1.7726\right)\right) \tag{10.1}$$

Here, $F=\left(8.791\times 10^{9}\right)\Big/\left(f\times\sqrt{\varepsilon}\right)$, f: resonant frequency, and ε: relative permittivity of the dielectric. The thickness of the substrate is 0.72 mm. The dimensions of the ground plane are 10 mm×10 mm, and the radius of the circle is 3.2 mm.

C. MIMO Configuration

Once the single-element design is finalised, it is placed in the MIMO configuration. We have used four single-element antennas in the MIMO antenna. Different configurations were designed and analysed to find the best one. Various configurations tested included linearly arranging all elements side by side, placing two ports down and other two up opposite to them and one where all ports were in different directions [17–19]. The configuration where all ports were in different directions gave the best results in terms of the isolation of antennas. The circular patch antenna is loaded with keyhole slots as shown in Figure 10.2. The proposed antenna design has been made on a substrate of 18.5 mm×18.5 mm and a width of 0.74 mm. The circular patch has a radius of 6.475 mm. In this structure, one side of

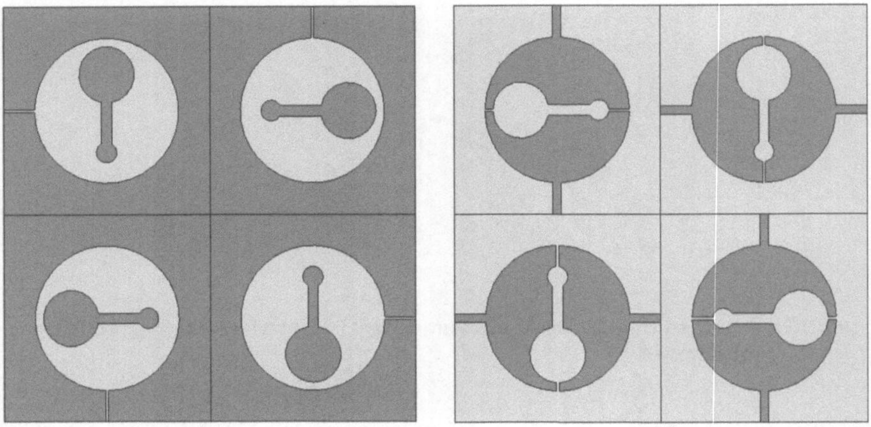
(a) (b)

Figure 10.2 Proposed MIMO antenna configuration. (a) Top view and (b) bottom ground plane.

the keyhole has a bigger circle with a radius of 1.85 mm and a smaller circle with a radius of 0.925 mm. The defected ground structure is represented as dumble shaped but complementary to the patch design of the antenna. The top view and bottom ground plane are depicted in Figure 10.2a and b.

A slit has been cut from both ends to give a surface current. Figure 10.2 shows the defective ground at the bottom of the design.

10.3 MIMO CONFIGURATION

Thus, the radiation pattern of individual elements was not disturbed. Figure 10.2 shows the configuration of four antennas which provided the best results in terms of isolation. As mentioned earlier, all ports here are in the opposite direction from each other.

10.4 SIMULATION RESULTS

A. Return loss

Figure 10.3 shows the plot of return loss or reflection coefficient of the MIMO antenna. The reflection coefficient (S_{11} or S_{22} or S_{33} or S_{44}) is expressed in decibel (dB) and is the ratio of incident to the reflected power. For good performance, its value must be a minimum of –10 dB, or for the best case, greater than –15 dB.

We obtained four frequencies in the mm-Wave spectrum where the reflection coefficient values were below –15 dB. These were 40.52 GHz

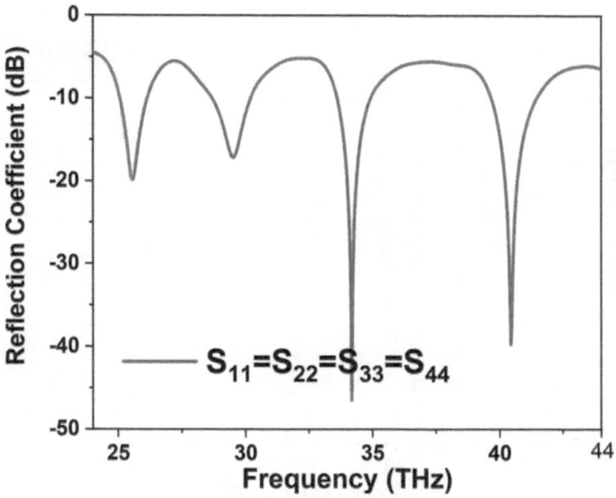

Figure 10.3 Plot of return loss versus frequency.

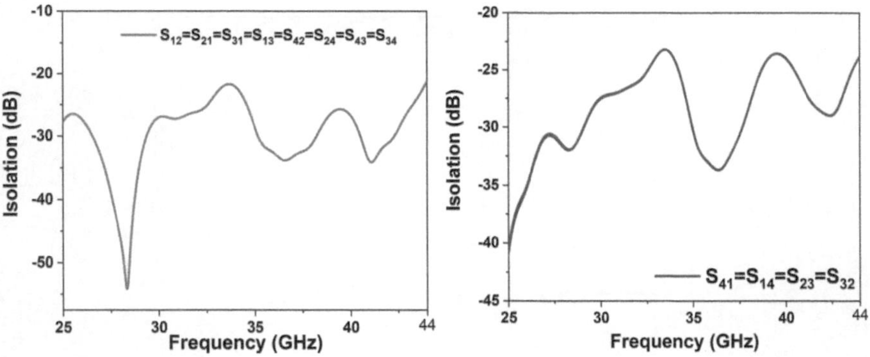

Figure 10.4 Plot of isolation versus frequency.

with a return loss of −41.721 dB, 34.212 GHz with a return loss of −43.464 dB, 29.579 GHz with a return loss of −16.899 dB and 25.623 GHz with a return loss of −20.726 dB. The bandwidths obtained were 40.066–41.003 GHz, 25.351–25.891 GHz, 33.91–34.584 GHz and 29.32–29.797 GHz.

B. Isolation

For good MIMO performance, the isolation must be < −15 dB. The isolation achieved between the individual antennas is < −20 dB as shown in Figure 10.4.

C. VSWR (Voltage standing wave ratio)

Reflection of power from the antenna is indicated by the VSWR value. The VSWR must range between 1 and 2. From Figure 10.5, it can be

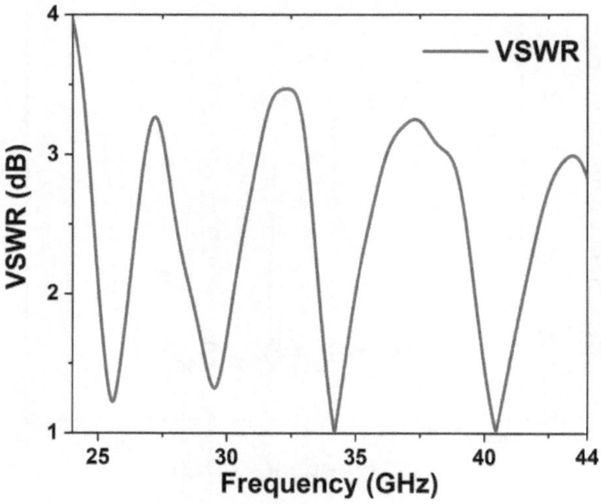

Figure 10.5 Plot of VSWR.

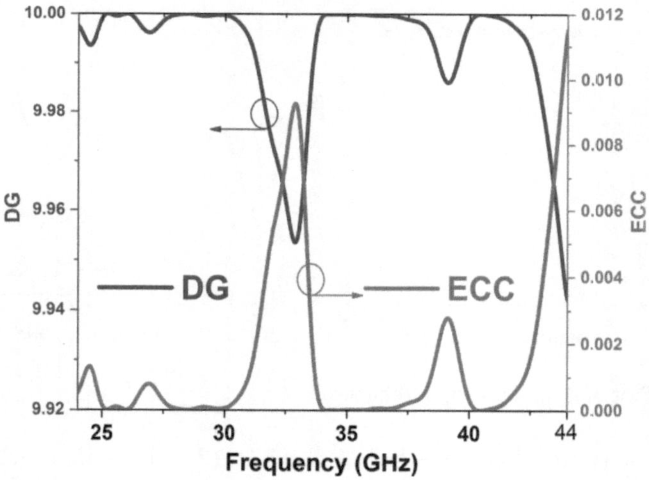

Figure 10.6 Plot of ECC and DG versus frequency.

noted that the VSWR value ranges between 1 and 2, thus giving good results. The VSWR was 1.023 at 40.512 GHz, 1.0221 at 34.206 GHz, 1.3318 at 29.569 GHz and 1.2026 at 25.623 GHz.

D. ECC (Envelope correlation coefficient)

In Figure 10.6, the plot of the envelope correlation coefficient (ECC) and diversity gain (DG) is shown in Figure 10.6 and is evaluated using the method reported in Ref. [3]. The ECC values are 0.000001 at 40.529 GHz, 0.0000006 at 34.215 GHz, 0.00006 at 29.578 GHz and 0.00008 at 25.631 GHz. It is desirable for 5G antennas to have very

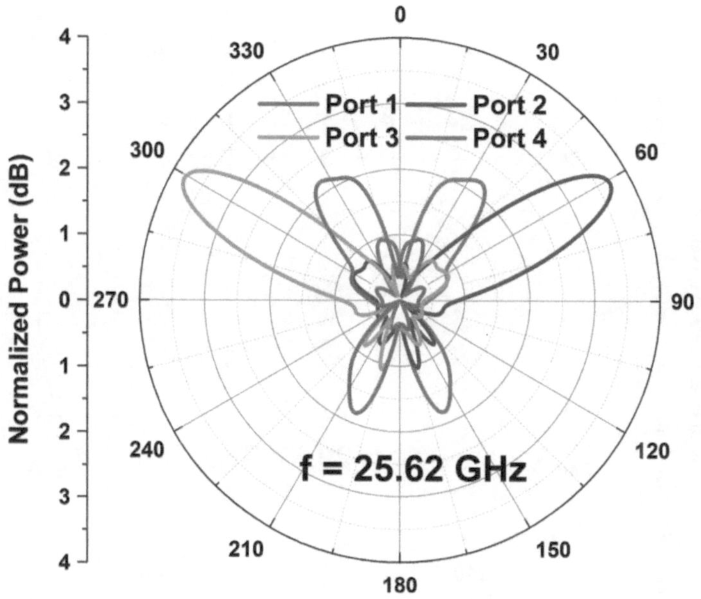

Figure 10.7 Radiation pattern at $f = 25.62$ GHz.

low ECC. It is a good way to analyse the diversity performance of the MIMO antenna.

E. Radiation pattern

Radiation patterns of all four antennas are shown in Figure 10.7 at different frequencies. The amount of power radiated by the antenna is given by the radiation pattern. We can observe that each antenna at a particular frequency has a radiation pattern different from each other. We thus obtain a good ECC for the designed MIMO antenna. At 25.538 GHz, the directivity of all four antennas was 6.93, 6.86, 6.9, 6.83 dBi, respectively, and the realised gains were 6.22, 6.13, 6.2, 6.11 dB, respectively.

At 25.538 GHz, the directivity of all four antennas was 6.93, 6.86, 6.9, 6.83 dBi, respectively, and the realised gains were 6.22, 6.13, 6.2, 6.11 dB, respectively. At 29.72 GHz, the directivity of all four antennas was 7.29, 7.19, 7.27, 7.17 dBi, respectively, and the realised gains were 6.57, 6.45, 6.56, 6.44 dB, respectively (Figures 10.8–10.10). At 37.475 GHz, the directivity of all four antennas was 6.56, 6.54, 6.59, 6.57 dBi, respectively, and the realised gains were 5.83, 5.81, 5.86, 5.84 dB, respectively. A comparison of the proposed antenna with the previously published research is shown in Table 10.1.

F. Gain

The plot of gain is depicted in Figure 10.11. A peak gain of 8 dB is shown for the proposed MIMO antenna at a frequency of 40.52 GHz.

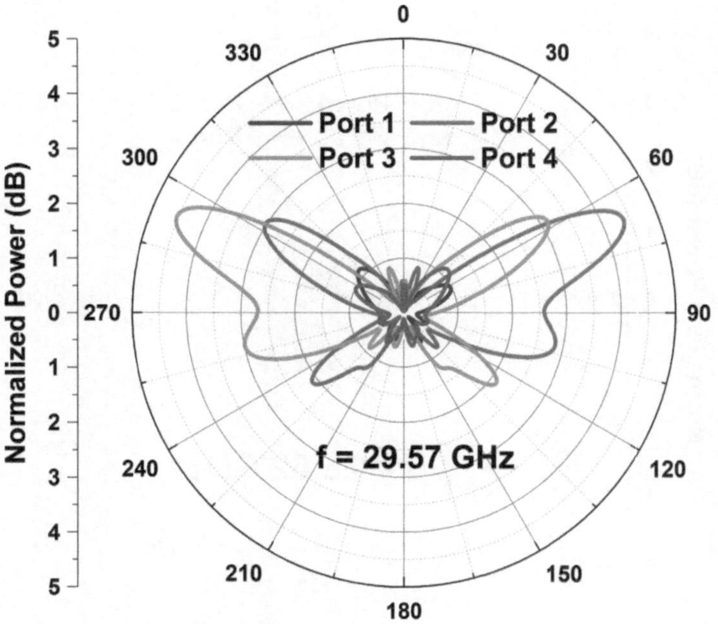

Figure 10.8 Radiation pattern at $f = 29.57$ GHz.

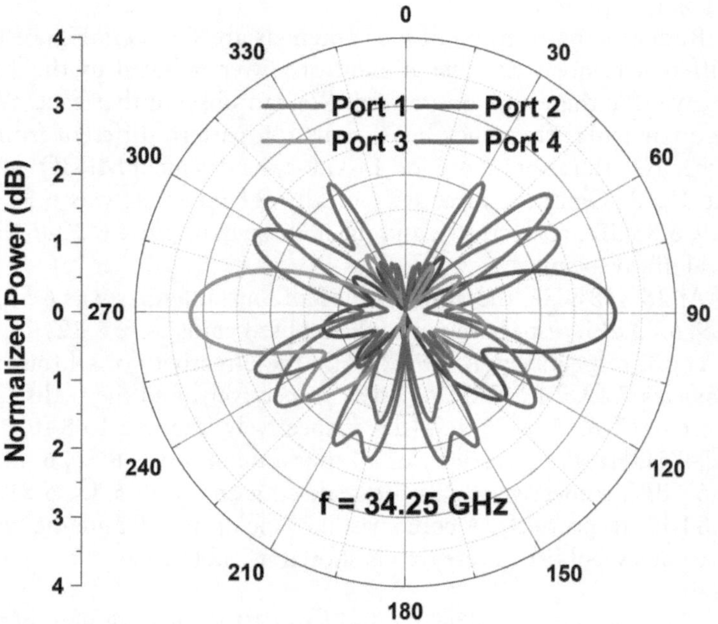

Figure 10.9 Radiation pattern at $f = 34.25$ GHz.

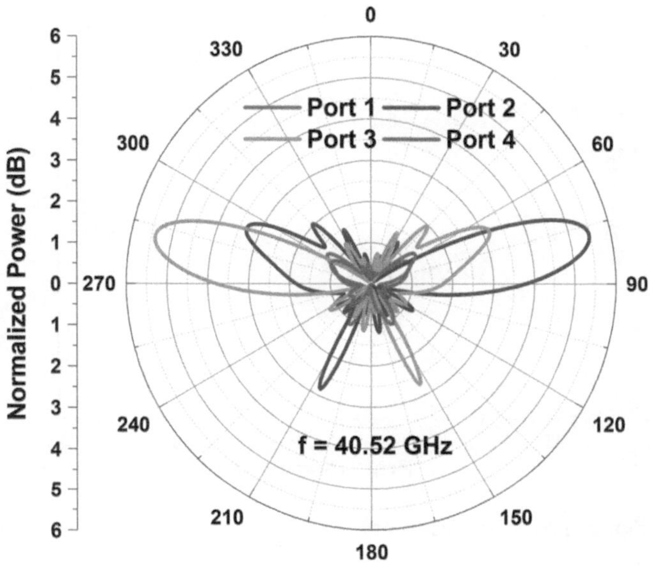

Figure 10.10 Radiation pattern at $f = 40.52$ GHz.

Table 10.1 Comparison with other published articles

Ref.	*MIMO*	*Mm-wave support?*	*Frequency*	*Gain (dB)*	*Isolation (dB)*	*Both 4G and 5G?*	*ECC*
[2]	Yes	Yes	5G -28, 37, 39 GHz 4G -2 GHz	7.2 at mm wave	25 db at mm wave 16 db at 2 GHz	Yes	<0.001
[4]	Yes	Yes	1870–2530 MHz mm-Wave – 28 GHz	3.8 dbi	>10 db	Yes	<0.5
[12]	Yes	Yes	25.5–29.6 GHz	8.3 dbi	Below −10 db	No	<0.01
[13]	Yes	Yes	28 GHz	9.5 dbi	< −30 db	No	0.015
[19]	Yes	Yes	WAN -2.45, 5.2 GHz LTE -2.6 GHz 5G -24, 28 GHz	11 dbi	>16 db	Yes	0.16
Our Work	Yes	Yes	mm-Wave – 25.62, 29.57 GHz, 34.25 GHz 40 GHz 52 GHz	6.59 dBi	< −25 dB	Yes	<0.008

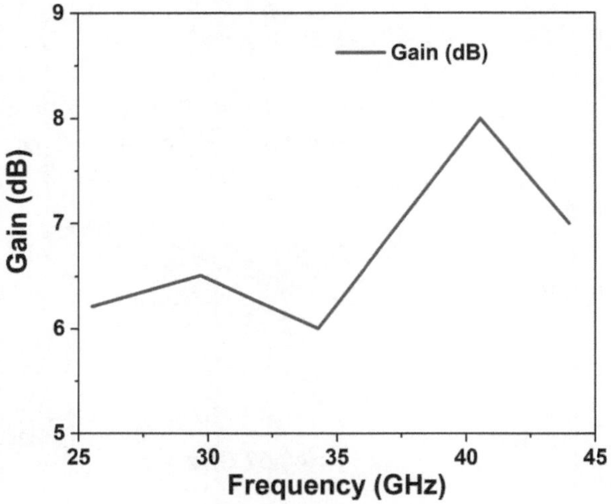

Figure 10.11 Plot of gain versus frequency.

Table 10.1 shows the comparison of the proposed MIMO antenna with other research articles. Our antenna shows excellent performance in terms of return loss, isolation, gain and ECC.

10.5 CONCLUSION

The MIMO antenna proposed in this chapter has four elements. These four are placed opposite to each other for good isolation. On carrying out simulations, we obtained four bands of operations. The bandwidths obtained were 40.066–41.003 GHz, 25.351–25.891 GHz, 33.91–34.584 GHz and 29.32–29.797 GHz. In these regions, the return loss obtained was less than –15 dB, which is required for 5G operations. Moreover, a low value of ECC was obtained, which indicates good isolation between the antennas. The far-field directivity is of all frequencies and different ports in different directions. Eventually, this will help in diversifying data in different directions in different power and data rates.

REFERENCES

1. J. G. Andrews et al., What Will 5G Be? *IEEE Journal on Selected Areas in Communications*, vol. 32, no. 6, pp. 1065–1082, June 2014, doi: 10.1109/JSAC.2014.2328098.

2. E. Al Abbas, M. Ikram, A. T. Mobashsher and A. Abbosh, MIMO Antenna System for Multi-Band Millimeter-Wave 5G and Wideband 4G Mobile Communications, *IEEE Access*, vol. 7, 181916–181923, 2019, doi: 10.1109/ACCESS.2019.2958897.
3. P. Upender and A. Kumar, HEM11δ and HEM12δ-based Quad band Quad Sense Circularly Polarized tunable Graphene-based MIMO Dielectric Resonator Antenna, *Frequenz*, 2022, https://doi.org/10.1515/freq–2021–0145
4. T. S. Rappaport, *Wireless Communications: Principles and Practice*, 2nd ed. Englewood Cliffs, NJ: Prentice-Hall, 2002.
5. R. Hussain, A. T. Alreshaid, S. K. Podilchak and M. S. Sharawi, Compact 4G MIMO Antenna Integrated with a 5G Array for Current and Future Mobile Handsets, *IET Microwaves, Antennas Propagation*, vol. 11, no. 2, 271–279, 2017.
6. P. Upender and A. Kumar, Design of a Multiband Graphene-Based Absorber for Terahertz Applications Using Different Geometric Shapes, *Journal of the Optical Society of America*, vol. 39, 188–199, 2022.
7. M. S. Sharawi, Printed Multi-Band MIMO Antenna Systems and Their Performance Metrics [Wireless Corner], *IEEE Antennas and Propagation Magazine*, vol. 55, 218–232, 2013.
8. Estimated Spectrum Bandwidth Requirements for the Future Development of IMT-2000 and IMT-Advanced, ITU-R, Geneva, Switzerland, Rep. M.2078, 2006.
9. P. Upender and A. Kumar, Quad-Band Circularly Polarized Tunable Graphene Based Dielectric Resonator Antenna for Terahertz Applications, *Silicon*, 2021, https://doi.org/10.1007/s12633-021-01336–5.
10. A. Tikhomirov, E. Omelyanchuk and A. Semenova, Recommended 5G Frequency Bands Evaluation, 2018 Systems of Signals Generating and Processing in the Field of on Board Communications, 1–5, 2018, doi: 10.1109/SOSG.2018.8350639.
11. I. Dioum, K. Diallo, M. M. Khouma, I. Diop, L. Sane and A. Ngom, Miniature MIMO Antennas for 5G Mobile Terminals, *2018 6th International Conference on Multimedia Computing and Systems (ICMCS)*, 1–6, 2018, doi: 10.1109/ICMCS.2018.8525870.
12. A. Kaur and P. K. Malik, Multiband Elliptical Patch Fractal and Defected Ground Structures Microstrip Patch Antenna for Wireless Applications, *Progress in Electromagnetics Research B*, vol. 91, 157–173, 2021, doi: 10.2528/PIERB20102704, ISSN: 1937–6472.
13. N. Shaik and P. K. Malik, A Retrospection of Channel Estimation Techniques for 5G Wireless Communications: Opportunities and Challenges, *International Journal of Advanced Science and Technology*, vol. 29, no. 05, 8469–8479, 2020, ISSN: 2005–4238.
14. P. K. Malik and M. Singh, Multiple Bandwidth Design of Micro strip Antenna for Future Wireless Communication, *International Journal of Recent Technology and Engineering*, vol. 8, no. 2, 5135–5138, July 2019, doi: 10.35940/ijrte.B2871.078219, ISSN: 2277–3878.

15. P. Upender, P. A. H. Vardhini, and V. Prakasam, "Performance Analysis and Development Of Printed Circuit Microstrip Patch Antenna with Proximity Coupled Feed at 4.3 GHz (C-band) with Linear Polarization for Altimeter Application," Int. J. Comput. Digit. Syst., vol. 10, no. 1, pp. 1293–1304, 2021, doi: 10.12785/IJCDS/1001116.
16. P. Upender and P. A. H. Vardhini, "Design and Development of a Printed Circuit Microstrip Patch Antenna at C-Band for Wireless Applications with Coaxial Coupled Feed Method," Planar Antennas Des. Appl., pp. 31–48, Jan. 2021, doi: 0.1201/9781003187325-2.
17. P. Upender and A. Kumar, "Circularly Polarized 2×2 MIMO Dielectric Resonator Antenna for Terahertz Applications," 2021 IEEE Indian Conference on Antennas and Propagation (InCAP), 2021, pp. 283-286, doi: 10.1109/InCAP52216.2021.9726292.
18. P. Upender and P. A. Harsha Vardhini, "Design Analysis of Rectangular and Circular Microstrip Patch Antenna with coaxial feed at S-Band for wireless applications," 2020 Fourth International Conference on I-SMAC (IoT in Social, Mobile, Analytics and Cloud) (I-SMAC), 2020, pp. 274–279, doi: 10.1109/I-SMAC49090.2020.9243445.
19. U. Patri and A. Kumar, "A novel ultra wideband circularly polarized stacked cylindrical dielectric resonator antenna with modified ground plane," Int. J. RF Microw. Comput. Eng., p. e23221, May 2022, doi: 10.1002/MMCE.23221

Chapter 11

Design of a compact multiband fractal antenna using ANN and firefly algorithm for wireless body area network

Satheeshkumar Palanisamy and T. Balakumaran
Coimbatore Institute of Technology

E. Suganya
Sri Eshwar College of Engineering

T. Prabhu
Presidency University

Osamah Ibrahim Khalaf
Al-Nahrain Nanorenewable Energy Research Center

CONTENTS

11.1 INTRODUCTION

The swift growth in wireless communications technology and commercial production has been significant in the enhanced use of both wireless and user computers. In the end, the customer needs higher capacity quickly to support and secure, anywhere and anytime, wireless communications. The RF spectrum is thus overcrowded and radio resources are emitted. One of the most promising technologies is the multiband technology which offers

DOI: 10.1201/9781003347057-11

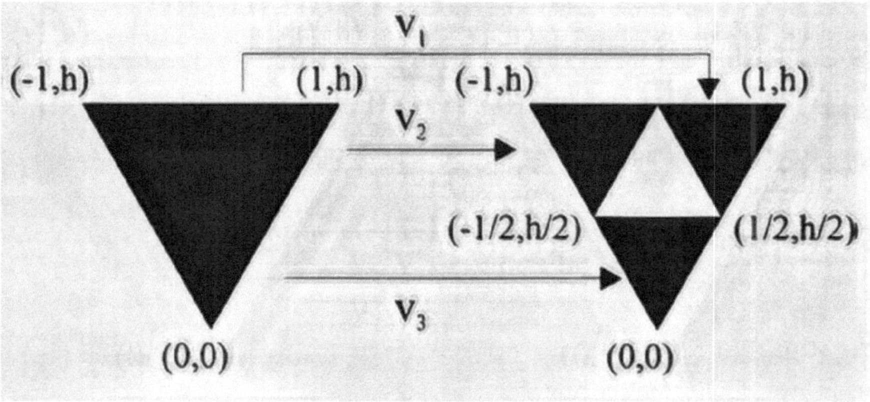

Figure 11.1 Modeling of fractal geometry.

a hopeful solution to the RF spectrum scarcity and enables the coexistence of new services with the existing, minimally, or no interference [1,2] wireless systems. Multiband varies from other communication techniques mainly because the pulses of the width of nanoseconds are extremely narrow. The power emitting limit has been set by the FCC at –41.3 dBm/MHz [3,4], allowing multiband systems under the noise floor of a traditional narrowband receiver. One of the major issues in designing and developing multiband antennas is to possess omnidirectional indoor radiation patterns which requires a return loss of less than 10 dB (S_{11}<–10 dB) over the 7.5-GHz impedance bandwidth. It suits indoor short-range wireless communication [5–7]. Multiband signals can successfully penetrate materials. Multiband signals are beneficial in the low-frequency radio spectrum [8,9]. These signals' material penetration capacity is vital. In the indoor case, the building's design and construction materials influence radio wave propagation [10–13]. The building's geometry determines the coverage degree (Figure 11.1).

Iterated Function Systems (IFS)

$$w\begin{pmatrix} p \\ q \end{pmatrix} = \begin{pmatrix} w & x \\ y & z \end{pmatrix}\begin{pmatrix} p \\ q \end{pmatrix} + \begin{pmatrix} a \\ b \end{pmatrix} \tag{11.1}$$

where w, x, y, z, a, b are the fractal object's space motion controllers in real numbers:: x, y –revolution by φ_1, φ_2 , w, z - scaling, with respect to the coordinating axis, and a, b – the vector (a, b) lends to linear scaling and translation and is expressed [14–23] as follows:

$$w = \delta_1 \cos\varphi_1 \; ; \; z = \delta_2 \cos\varphi_2 \; ; \; x = \delta_2 \sin\varphi_2 \; ; \; y = \delta_1 \sin\varphi_1 \tag{11.2}$$

11.2 ANTENNA DESIGN

The equations used to design antennas are listed here. Transmitter gain and directionality are controlled by the middle-layer substrate thickness. An antenna's resonant frequency determines the substrate's thickness [24,25]. The substrate's thickness is chosen to fit within the parameters of the antenna [26–29]. The thickness of the substrate affects the antenna's gain. The resonating frequency of an antenna determines the substrate thickness. The substrate height is calculated based on $0.003\lambda < h < 0.05\lambda$.

$$\lambda = c/f_r \tag{11.3}$$

where c is the velocity of light and f_r=resonant frequency.

A 1.6 mm substrate thickness is required for the 2.4 GHz operation [30–32]. The substrate material affects the relative permittivity, which is a variable. The ε_r=4.4 for FR4 material, length (L_{sub}) and width (W_{sub}) of the substrate [33–36] should be

$$W_{\text{sub}} = W + 6h \tag{11.4}$$

$$L_{\text{eff}} = 2\Delta L + L \tag{11.5}$$

The dielectric constant (effective) [37–39] is

$$\varepsilon_{\text{eff}} = \frac{\varepsilon_r + 1}{2} + \frac{\varepsilon_r - 1}{2}\left(1 + 12\frac{h}{w}\right)^{-0.5} \tag{11.6}$$

The additional length due to fringing consequence is

$$\Delta L = h(0.412)\frac{(\varepsilon_{\text{eff}} + 0.3)\left(\frac{w}{h} + 0.264\right)}{(\varepsilon_{\text{eff}} - 0.258)\left(\frac{w}{h} + 0.8\right)} \tag{11.7}$$

ΔL=Length extension due to fringing effect.

Fractal geometries are one of the potential candidates for antenna miniaturization and bandwidth of the fractal geometry with the aid of the properties for spatial filling and self-similarity [40–44]. Therefore, we built a multiband and compact size antenna with fractal geometry. We have taken a circular monopoly in our design and have produced a smiley face that is iterated twice to make the bandwidth wider. Table 11.1 shows the design considerations for designing a conventional patch antenna. Feeding to the antenna structure is through the microstrip line method Figure 11.2 shows the proposed antenna's geometry,

Table 11.1 Design considerations

Antenna	*Fractal patch antenna*
Substrate	FR4
Feeding method	Microstrip line
Tangent δ	0.019
Substrate height	1.6 mm
Relative permittivity	4.7
Polarization	Linear
Frequency of operation	2.45, 4.2 and 5.8 GHz

Figure 11.2 Geometry of the antenna.

11.3 ARTIFICIAL NEURAL NETWORK

Artificial neural networks (ANNs) were designated as influential tools in several disciplines in prediction, design, and simulation. Designing and analyzing antenna using ANN is highly advantageous. Overall, three key layers are related to the ANN architecture. The receiving layer is considered as (consideration) input layer. The output (realized) layer generates a complete network output. The concealed (hidden) layer is created between the consideration and realized layers. For achieving direct communication links, each neuron is interlinked with successive neurons in each layer and all the communication links are related with a weight value. The weights

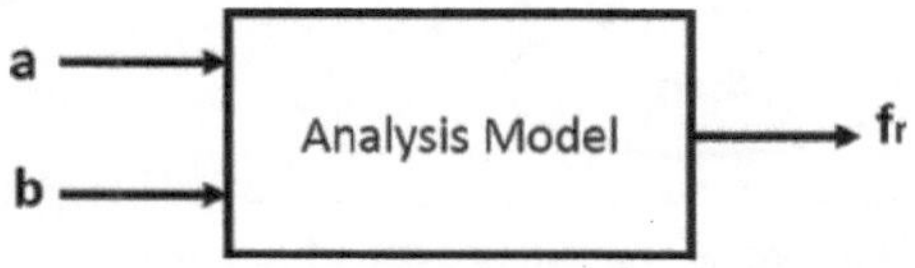

Figure 11.3 Artificial neural network-based analysis model for proposed fractal antenna.

deliver the evidence about an input signal. The error message is created when the network output is subtracted from the anticipated output. An ANN thus has the same potential as that of the human brain.

The ANN-based system is shown in Figure 11.3

11.4 CURVE FITTING EMPLOYMENT

Curve modification is a geometrical approach to deriving scientific interrelationships for dependent and independent curve parameters or the mathematical equation for the given data collection. The assessment of a suitable model for the mathematical relationship between the variables is used. This technology defines a model which best fits in accordance with the data package provided. A total of 37 models are created on erratic feed positions along the *x*-axis. If these values are used, the suggestion is amid the microstrip line feed location (y) and frequency of resonance (f_r).

$$f_r = 2.461\cos(0.1259x + 6.949) + 0.0306\cos(1.124x + 9.128) + 0.002287\cos(7.366x - 13.29) + 0.002519\cos(8.732x - 7.318) + 0.001725\cos(11.08x - 0.5975) - 0.0009745\cos(14.5x - 9.16) \quad (11.8)$$

Thus, light strength $L(d)$ is given by

$$L(d) = e^{-\gamma d^2} L_o \quad (11.9)$$

A microstrip line is used to forage the antenna structure [45–47].

11.5 ANN CONFIGURATION

Antenna Placement (Transmitter and Receiver)

The built fractal multiband antenna is the sender and antenna receiver. It is an all-way fractal of the entire ultra-wideband antenna (3.1 – antenna

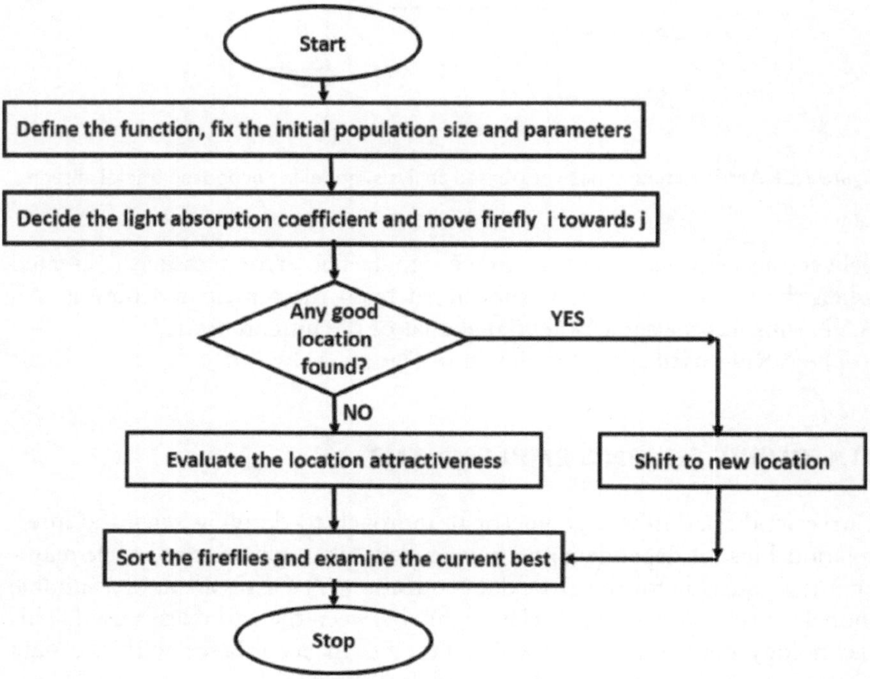

Figure 11.4 Flowchart of firefly algorithm.

running GHz 10.6). With a chirp signal 731, the antenna is excited. Frequency width is from 3.1 to 5.15 GHz (bandwidth) [33–36]. 1.368 GHz). The radio wave spread is usually very widespread and environment-influenced, i.e., construction and mobile terminal use (pedestrian/structural) structures. Figure 11.4 shows the flow diagram for the firefly algorithm.

11.6 RESULTS AND DISCUSSION

For both linear arrays and linear bands, multiband impedance bandwidth and gain values were analyzed in detail. Using parametric analysis, the impact of the component size and inter-component distance can be determined. Simulated measurements were taken of impedance variance, response, and bandwidth. There is an additional 47 MHz of bandwidth generated for the dual 4.2 and 5.8 GHZ channels in the linear array. Array geometry and advanced wireless technology applications are two of the many uses for the proposed approach. Figure 11.5 demonstrates the effect of antenna designs on the return loss parameter as a function of frequency. Table 11.2 summarizes the comparative analysis of various

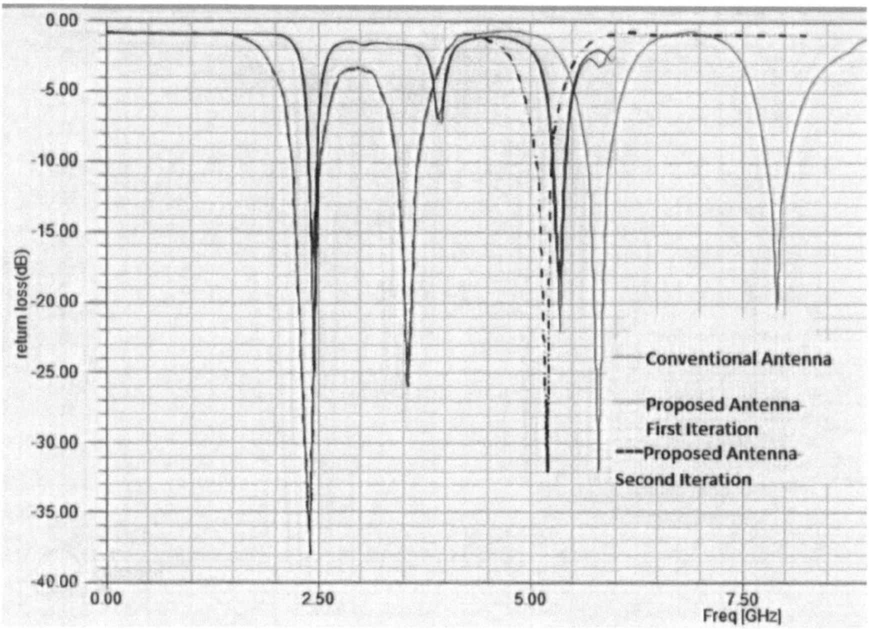

Figure 11.5 Comparative analysis of return loss of different antennas for conventional antenna, proposed fractal for first iteration and second iteration.

Table 11.2 Performance analysis of proposed fractal geometry for the different iterations

Iterations	*Resonating frequency (GHz)*	*Reflection coefficient, S_{11} (dB)*	*Gain (dBi)*
First	2.44	−36	8.39
	3.6	−25	6.5
	5.8	−32	7.17
Second	2.43	−38	14.84
	3.6	−26	8.4
	5.2	−32	9.4

radiation characteristic parameters for two iterations of the proposed fractal antenna.

The proposed on-body antenna is a printed monopole, it is expected to offering the radiation pattern as omnidirectional pattern as shown in Figures 11.6 and 11.7, where the regulated H- and E-plane radiation patterns are shown for the resonating case at f=2.4 GHz.

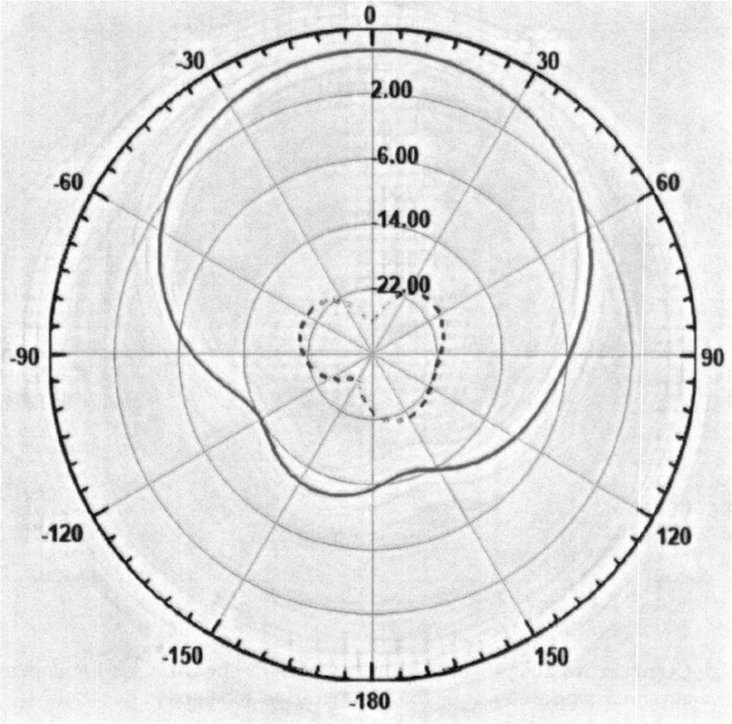

Figure 11.6 Radiation patterns at 2.4 GHz in the horizontal magnetic field plane (solid line) and the electric plane (dashed line).

11.7 FABRICATION PROCESS

The software used to fabricate is Design Pro which can generate a .dxf file. This .dxf file is taken from the HFSS software from the modular option. The fabricating machine named PCB prototype machine from MITS Company understands the .dxf file information and does fabrication. The fabrication process involves three steps namely milling, etching, and routing. In the milling process, the outline shape of an antenna using a needle which is attached to a spindle head is drawn. After completion of the milling process, the etching of an unwanted portion of an antenna can be done. Finally, the antenna is routed out from the module. Figure 11.8 shows the fabricated fractal antenna of the third iteration,

11.8 CONCLUSION

An innovative ultra-thin multiband geometry on-body fractal antenna is developed, simulated, and fabricated for multiple healthcare applications. For on-body contact, microwave characteristics of evolved geometry are presented. The simulated radiation characteristics comprising the return

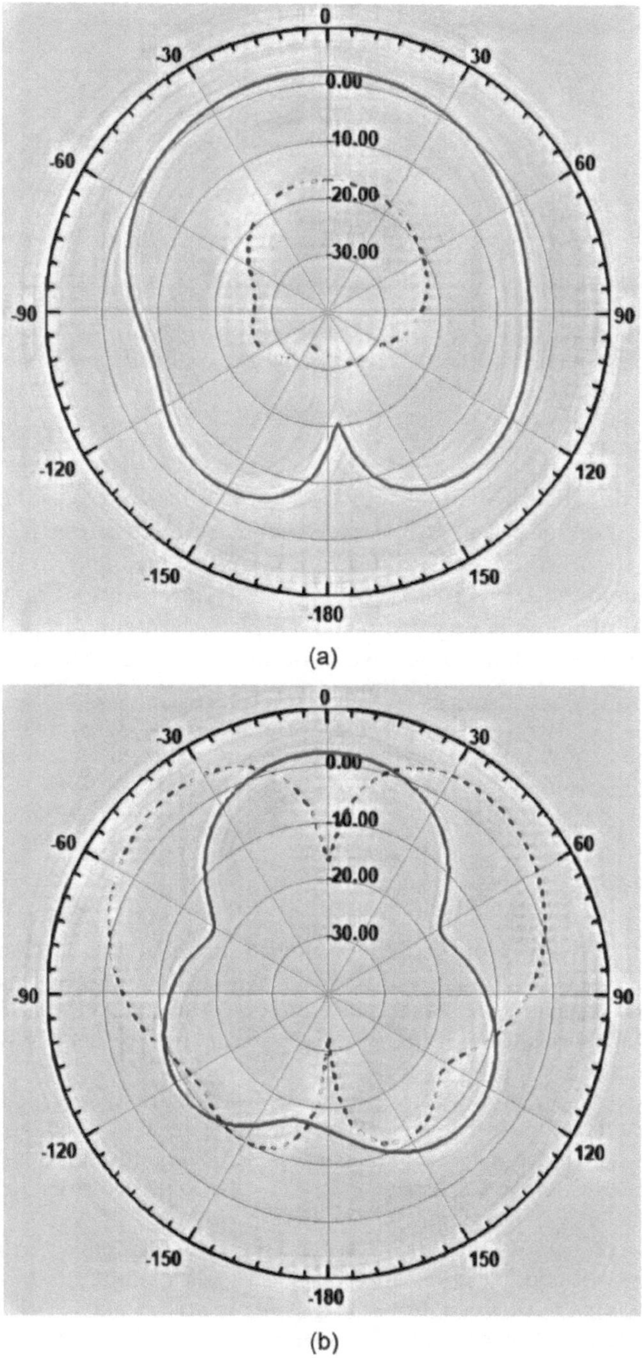

Figure 11.7 (a) Radiation pattern at 3.6 GHz in the horizontal magnetic field plane (solid line) and the electric plane (dashed line) in far field region. (b) Radiation pattern at 5.8 GHz in the horizontal magnetic field plane (solid line) and the electric plane (dashed line) in far field region.

Figure 11.8 Fabricated antenna.

loss of the designed radiator using HFSS was –38 dB at 2.4 GHz, with a 380 MHz bandwidth of –10 dB (15.83%). The simulated return loss transceiver was found to be –26 dB at 3.6 GHz and –32 dB at 5.2 GHz with a 620 MHz bandwidth of –10 dB (25.83%). With a return loss of 12.67 and 10 dB, the return loss is very promising (13.75%). Antenna impedance and thickness, as well as bandwidth, were found to be negatively correlated. In comparison to bone tissue, the human body's fatty tissues demonstrated a higher degree of absorption coefficient for a given energetic stage.

The results show a ramp, as predicted, relationship of function between levels of excitation power and SAR average values.

REFERENCES

1. Sridevi, S., Karpagam, G. R., & Vinoth Kumar, B. (2022). Genetic algorithm – optimized gated recurrent unit (GRU) network for semantic web services classification, *Malaysian Journal of Computer Science*, 35(1), 70–88. Retrieved from https://ejournal.um.edu.my/index.php/MJCS/article/view/25988
2. Adamu Kakudi, H., Chu Kiong, L., Moy, F. M., Chee Kau, L., & Pasupa, K. (2021). Diagnosis of metabolic syndrome using machine learning, statistical and risk quantification techniques: a systematic literature review, *Malaysian Journal of Computer Science*, 34(3), 221–241. Retrieved from https://ejournal.um.edu.my/index.php/MJCS/article/view/31484
3. Lin, W. & H. Wang. (2016). Polarization reconfigurable circular patch antenna with multiple probes for biomedical applications, *IEEE International Symposium on Antennas and Propagation (APSURSI)*, ISSN: 1947-1491.
4. Prasad Jones Christydass, S., Kusuma Kumari, E., Sowjanya, A., Satheesh Kumar, P., Selvam, N., & Murali, K. (2020). Microstrip metamaterial bandpass filter For 5G application, *Solid State Technology*, 63(3).

5. Hall, P. S., & Hao, Y., *Antennas and Propagation for Body-centric Communications*, ArtechHouse, London and Boston, 2006.
6. Kaur, G., Kaur, A., Toor, G. K., Dhaliwal, B. S., & Pattnaik, S. S. (2015). Antennas for biomedical applications, *Biomedical Engineering Letters*, 5(3), 203–212.
7. Sabban, A. (2013). New wideband printed antennas for medical applications, *IEEE Transactions on Antennas and Propagation*, 61(1), 84–91.
8. Mandelbrot, B. B., *The Fractal Geometry of Nature*, W. H. Freeman, New York, 1983.
9. Ali, J. K., Yassen, M. T., Hussan, M. R., & Salim, A. J. (2012). A printed fractal based slot antenna for multiband wireless communication applications, *Proceedings of PIERS*, 618–622, Moscow, Russia, Aug. 19–23.
10. Palanisamy, S., Balakumaran, T., Khalaf, O. I., Alotaibi, Y., Alghamdi, S., & Alassery, F. (2021). A novel approach of design and analysis of a hexagonal fractal antenna array (HFAA) for next-generation wireless communication, *Energies*, 14(19), 6204. https://doi.org/10.3390/en14196204
11. Sundaram, A., Maddela, M., & Ramadoss, R. (2007). Koch-Fractal folded-slot antenna characteristics, *IEEE Antennas and Wireless Propagation Letters*, 6, 219–222.
12. Oraizi, H., & Hedayati, S. (2011). Combined fractal geometries for the design of wide band microstrip antennas with circular polarization, *PIERS Proceedings*, 1262–1267, Suzhou, China, Sept. 12–16.
13. Sharma, N., Sharma, V., & Bhatia, S. S. (2018). A novel hybrid fractal antenna for wireless applications, *Progress in Electromagnetics Research M*, 73, 25–35.
14. Dhaliwal, B. S., & Pattnaik, S. S. (2017) BFO-ANN ensemble hybrid algorithm to design compact fractal antenna for rectenna system, *International Journal on Neural Computing and Applications*, 28(1), 917–928.
15. Nivethitha, T., Palanisamy, S. K., Mohana Prakash, K., & Jeevitha, K. (2021). Comparative study of ANN and fuzzy classifier for forecasting electrical activity of heart to diagnose Covid-19. *Materials Today Proceedings*, 45, 2293–2305. https://doi.org/10.1016/j.matpr.2020.10.400
16. Kaur, K., & J. S. Sivia. (2017). A compact hybrid multiband antenna for wireless applications, *International Journal on Wireless Personal Communications*, 97(4), 5917–5927.
17. Sharma, N., & Bhatia, S. S. (2018). Split ring resonator based multiband hybrid fractal antennas for wireless applications, *International Journal of Electronics and Communications*, 93, 39–52.
18. Choukiker, Y. K., & Behera, S. K. Design of wideband fractal antenna with combination of fractal geometries, *International Conference on Information, Communications and Signal Processing*, Singapore, Dec. 13–16, 2011.
19. Oraizi, H., & Hedayati, S. (2012). Circularly polarized multiband microstrip antenna using square and Giuseppe Peano fractals, *IEEE Transactions on Antennas and Propagation*, 60(7), 3466–3470.
20. Satheesh Kumar, P., Jeevitha, & Manikandan, Diagnosing COVID-19 virus in the cardiovascular system using ANN. In: Oliva D., Hassan S.A., Mohamed A. (eds) *Artificial Intelligence for COVID-19. Studies in Systems, Decision and Control*, vol 358, Springer, Cham, 2021. https://doi.org/10.1007/978-3-030-69744-0_5.

21. Prabhu, T., et al. *An intensive study of dual patch antennas with improved isolation for 5G Mobile communication systems. Future Trends in 5G and 6G*. CRC Press, 205–217.
22. Satheesh, K. P., Chitra, P., & Sneha, S. (2021). *Design of Improved Quadruple-Mode Bandpass Filter Using Cavity Resonator for 5G Mid-Band Applications. Future Trends in 5G and 6G: Challenges, Architecture, and Applications*, p. 219.
23. Rawat, S. S., Alghamdi, S., Kumar, G., Alotaibi, Y., Khalaf, O. I., & Verma, L. P. (2022). Infrared small target detection based on partial sum minimization and total variation, *Mathematics*, 10, 671. https://doi.org/10.3390/math10040671
24. Bangi, I. K., & Sivia, J. S. (2018). Minkowski and Hilbert curves based hybrid fractal antenna for wireless applications, *International Journal of Electronics and Communications*, 85, 159–168.
25. Brar, A. S., Sivia, J. S., & Bharti, G. (2016). A compact hybrid Minkowski fractal antenna for C and Xband applications, *International Journal of Computer Science and Information Security (IJCSIS)*, 14(12), 349–352.
26. Mohan, P., Subramani, N., Alotaibi, Y., Alghamdi, S., Khalaf, O. I., & Ulaganathan, S. (2022). Improved metaheuristics-based clustering with multihop routing protocol for underwater wireless sensor networks, *Sensors*, 22(4), 1618. https://doi.org/10.3390/s22041618
27. Mohammed, H. J., Abdullah, A. S., Ali, R. S., Abd-Alhameed, R. A., Abdulraheem, Y. I., & Noras, J. M. (2014). Design of a unipolar printed triple band-rejected ultra-wideband antenna using particle swarm optimization and the firefly algorithm, *IET Microwaves, Antennas & Propagation*, 10(1), 31–37.
28. Dhaliwal, B. S., & Pattnaik, S. S. (2016). Performance comparison of bio-inspired optimization algorithms for Sierpinski gasket fractal antenna design, *Neural Computing and Applications*, 27, (3), 585–592.
29. Kaur, R., & Rattan, M. (2015). Optimization of the return loss of differentially fed microstrip patch antenna using ANN and firefly algorithm, *Wireless Personal Communications*, 80(4), 1547–1556.
30. Prabhu, T., & Pandian, S. C. (2021). Design and development of planar antenna array for mimo application, *Wireless Networks*, 27(2), 939–946.
31. Prabhu, T., & Pandian, S. C. (2021). Design and implementation of T-shaped planar antenna for MIMO applications, *CMC-Computers Materials & Continua*, 69(2), 2549–2562.
32. Prabhu, T., Pandian, S. C., & Suganya, E. Contact feeding techniques of rectangular microstrip patch antenna for 5 GHz Wi-Fi, *2019 5th International Conference on Advanced Computing & Communication Systems (ICACCS)*, pp. 1123–1127. IEEE.
33. Sam, P. J. C., Surendar, U., Ekpe, U. M., Saravanan, M., & Satheesh Kumar, P., A Low-profile compact EBG integrated circular monopole antenna for wearable medical application. In: Malik P.K., Lu J., Madhav B.T.P., Kalkhambkar G., Amit S. (eds) *Smart Antennas. EAI/Springer Innovations in Communication and Computing*, Springer, Cham, 2022. https://doi.org/10.1007/978-3-030-76636-8_23
34. Bhushan, B., & S. S. Pillai. Particle swarm optimization and firefly algorithm: Performance analysis, *IEEE International Advances Computing Conference (IACC)*, Feb. 22–23, 2013, pp. 746–751.

35. Dhaliwal, B. S., & S. S. Pattnaik. (2013). Artificial neural network analysis of Sierpinski Gasket fractal antenna: A low-cost alternative to experimentation, *Advances in Artificial Neural Systems*, 2013, 7 pages, Article ID 560969.
36. Gil, I., & Fernandez-Garcia, R. (2017). Wearable PIFA antenna implemented on jean substrate for wireless body area network, *Journal of Electromagnetic Waves and Applications*, 31(11–12), 1194–1204.
37. Satheesh Kumar, P., & Manikandan, J., Diagnosing COVID-19 virus in the cardiovascular system using ANN. In: Oliva D., Hassan S.A., Mohamed A. (eds) *Artificial Intelligence for COVID-19, Studies in Systems, Decision and Control*, vol. 358, Springer, Cham, 2021.
38. Satheesh Kumar, P., & Valarmathy, S. Development of a novel algorithm for SVMBDT fingerprint classifier based on clustering approach, *IEEE-International Conference on Advances in Engineering, Science and Management (ICAESM-2012)*, 2012, pp. 256–261.
39. Li, Y., Yang, X., Liu, C., & Jiang, T. (2017). Miniaturization cantor set fractal ultrawideband antenna with a notch band characteristic, *Microwave and Optical Technology Letters*, 54(5), 1227–1230.
40. Palanisamy, S., Thangaraju, B, Khalaf, O. I., Alotaibi, Y., & Alghamdi, S. (2021). Design and synthesis of multi-mode bandpass filter for wireless applications, *Electronics*, 10(22), 2853. https://doi.org/10.3390/electronics10222853
41. Kumar, P. S., Boopathy, S., Dhanasekaran, S., & Anand, K. R. G. Optimization of multi-band antenna for wireless communication systems using genetic algorithm, *2021 International Conference on Advancements in Electrical, Electronics, Communication, Computing and Automation (ICAECA)*, 2021, pp. 1–6, doi: 10.1109/ICAECA52838.2021.9675686.
42. *Planar Antenna: Design, Fabrication, Testing, and Application*, Nova Science Publishers Inc, New York, 2021, ISBN: 9781536198980.
43. Malik, P., Lu, J., Madhav, B.T.P., Kalkhambkar, G., Amit, S. (Eds.). *Smart Antennas: Latest Trends in Design and Application*, Springer, ISBN: 9783030766368, doi: 10.1007/978-3-030-76636-8.
44. Malik, P. K., Padmanaban, S., & Holm-Nielsen, J. B., *Microstrip Antenna Design for Wireless Applications*, Taylor and Francis, 2021, ISBN: 9780367554385.
45. Sivia, J. S., Pharwaha, A. P. S., & Kamal, T. S. (2013). Analysis and design of circular fractal antenna using artificial neural networks, *Progress in Electromagnetics Research B*, 56, 251–267.
46. Salim, M., & Pourziad, A. (2015). A novel reconfigurable spiral-shaped monopole antenna for biomedical applications, *Progress in Electromagnetics Research Letters*, 57, 79–84.
47. Oraizi, H., & Hedayati, S. (2011). Miniaturized UWB monopole microstrip antenna design by the combination of Giuseppe Peano and Sierpinski Carpet fractals, *IEEE Antennas and Wireless Propagation Letters*, 10, 67–70.

Chapter 12

Effects of metamaterial on bioinspired microstrip patch antenna

P. R. Satarkar and Rajesh B. Lohani
Goa College of Engineering (Goa university)

CONTENTS

DOI: 10.1201/9781003347057-12

12.1 INTRODUCTION

With the introduction of the Internet, the demand for and popularity of mobile terminals (laptops, cell phones, etc.) have skyrocketed, propelling the communications industry to greater heights. Society has benefited from the development of digital information systems and wireless communication technology. Various wireless services, such as Wi-Fi, Bluetooth, GPS, NFC, 4G, and 5G systems that operate in multiple frequency bands, necessitate the use of portable devices for broadband applications. The antennas are having a good scope to be used in this portable terminal which should satisfy some important characteristics, such as miniaturization and system integration. Microstrip patch antennas (MPA) are ideal for wireless communication applications due to several important characteristics. For further advances and suitability for the construction of antennas, the researchers concentrated on using bioinspired geometrical shapes. An important one is bioinspired plant geometries, which employ plant parts such as leaves, fruits, and flowers. The study of MPA based on models inspired by leaves (leaf-shaped antennas) has piqued researchers' curiosity due to promising outcomes. We propose hibiscus and *Kaju* leaf-shaped geometries for patch antennas. The second category is bioinspired antennas in animals, which strive to mimic the antenna operation used in communication systems by using the shape of the exterior components or the organs of the animals [1,2]. Antennas of these types are loaded with metamaterials for improvement of their performance. A metamaterial is an artificial material that is constructed using an array of metallic resonant cells. Metamaterials' lattice constants should be much smaller than their wavelengths. Double-negative metamaterials are artificial materials with negative epsilon and mu. Zero-index metamaterials are those with both indices close to zero but less than zero [3,4]. Material cells with only a negative permeability are called mu-negative metamaterials. A square split-ring resonator (SSRR) is simulated at a resonance frequency of 5.8 GHz to find the double negativity of the cell. A cell made of SSRR is also simulated to determine the negativity of a cell. The proposed double-negative and mu-negative metamaterial cells are designed and tested using MATLAB and CST microwave software. A conventional antenna and a proposed 5.8 GHz novel bioinspired MPA (BMPA) and substrate loaded with SSRR are simulated using the FEKO software. The analysis and comparison of parameters such as gain, return loss, bandwidth, and VSWR are presented. This patch antenna is built on an FR-4 substrate with a permittivity of 4.4 and 1.6 mm thickness. The FR-4 material is cheaper and more easily available. Surface wave propagation is decreased by a metamaterial-loaded substrate, which improves antenna performance. Many researchers have discussed about split-ring resonators (SRRs) [5,6]. This antenna with improvement can be used for WiMAX applications. It is a wireless technology based on the IEEE 802.16 standard that provides fast data over a widespread area. The WiMAX is a point to multipoint wireless networking technology [7,8]. It works in three layers: the physical layer, MAC layer, and convergence layer. The bandwidth required for the band with a frequency of 5.8 GHz is 300 MHz. This chapter covers the design of circular microstrip patch antennas (CMPA), double-negative and mu-negative metamaterials, and innovative bioinspired microstrip patch antennas (BMPA) with hibiscus and *Kaju*-shaped leaves that are optimized for wireless applications. The other section presents the metamaterial-loaded BMPA and its characteristics. Simulations of antennas are performed using FEKO.

12.2 MICROSTRIP PATCH ANTENNA

12.2.1 Microstrip planar antenna introduction

The MPA was patented in 1955. But around the 1970s, this started receiving considerable attention. In MPA a substrate of dielectric material sandwiched between a very thin circular disk (patch) and the ground plane. The pattern on the microstrip patch is made to be as normal to the patch as possible (broadside radiator). There are a variety of substrates that are utilized to make MPA with dielectric constants ranging from 2.2 to 12. On the dielectric substrate, the main radiating elements of any shape and microstrip feed lines are photo-etched. Various geometries are used as a radiating patch of MPA. Square, rectangular, and circular shapes of MPA are most common due to their ease of fabrication and analysis and to their appealing radiation properties.

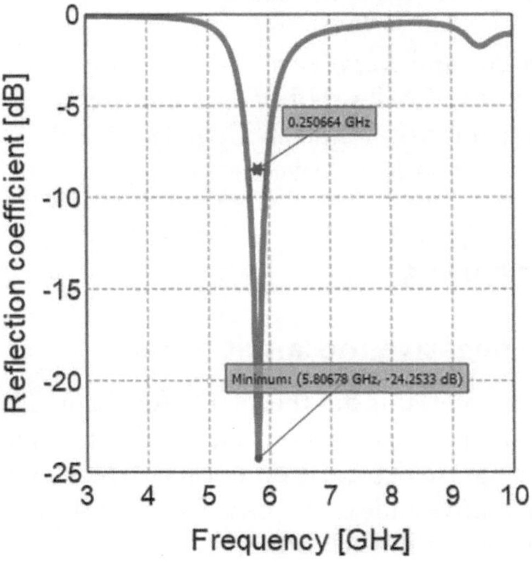

Figure 12.1 Reflection coeff. v/s frequency (CMPA).

12.2.2 Circular shape microstrip patch antenna (CMPA)

The design of various high-performance antennas is essential to cater to the need of society. A CMPA is an initiator for the bioinspired patch antennas (BMPA) like hibiscus and *Kaju* leaf-shaped antenna. Equations 12.1 and 12.2 show the design equations for CMPA. The radius of a circular shape patch (r) is 6.48 mm for a 5.8 GHz resonance frequency [9,10]. The diameter of the substrate is 19.2 mm. The feed length is 2.4 mm. The diameter of CMPA is 12.96 mm. Then the resultant CMPA is perturbed to achieve the proposed BMPA structure [1] where the length of the axis of the proposed BMPA with a resonance frequency of 5.8 GHz is in line with the diameter of the CMPA.

$$r = \frac{F}{\left\{1 + \frac{2h}{\pi \varepsilon_r F}\left[\ln\left(\frac{\pi F}{2h}\right) + 1.7726\right]\right\}^{\frac{1}{2}}} \tag{12.1}$$

$$F = \frac{8.791 \times 10^9}{f_r \sqrt{\varepsilon_r}} \tag{12.2}$$

The antenna is having a circular shape so the radius is denoted by r and h is the height of the substrate. The FR4 substrate is having permittivity equal to 4.4, loss tangent equal to 0.02, and 1.6 mm height. The reflection coefficient v/s frequency graph of CMPA is depicted in Figure 12.1.

12.3 HIBISCUS AND *KAJU* LEAVES

The shape of hibiscus and *Kaju* leaves is very close to a circle; hence, the designed initiator is the CMPA. Hibiscus is a flowering plant, and *Kaju* is an important dry fruit crop found in Goa. Figure 12.2 presents the real images of the hibiscus and *Kaju* leaves.

12.4 METAMATERIALS

12.4.1 Double-negative metamaterials

A left-handed metamaterial is another name for them. They are manufactured materials with negative epsilon, mu, and refractive index. Natural materials found in the earth's crust do not have these characteristics. Due to the negative refractive index, the phase velocities and group velocities are in opposite directions to each other, resulting in an energy flow in the opposite direction. When both μ_r and ε_r are negative, negative refraction

Figure 12.2 Hibiscus leaf and *Kaju* leaf.

can occur [10–13]. Two of Maxwell's equations, 12.3 and 12.4, are used to comprehend the metamaterial.

$$\nabla \times E = -\mu_0 \mu_r \frac{\partial H}{\partial t} \tag{12.3}$$

$$\nabla \times H = -\varepsilon_0 \varepsilon_r \frac{\partial E}{\partial t} \tag{12.4}$$

Equation 12.5 is a wave equation, and ϵ_r represents relative permittivity and μ_r denotes relative permeability.

$$\nabla^2 \cdot E = -\varepsilon_0 \mu_0 \varepsilon_r \mu_r \frac{\partial^2 E}{\partial t^2} \tag{12.5}$$

If ϵ_r and μ_r are real numbers, the wave equation remains unchanged when the signs of ϵ_r and μ_r are altered. Equation 12.6 gives the obtained refractive index.

$$\eta = \pm\sqrt{|\varepsilon_r||\mu_r|} \tag{12.6}$$

These resources are referred to as left-handed materials due to the foregoing facts. In classical electromagnetics, the Drude–Lorentz model is used to characterize material properties. Epsilon and mu effective can be calculated using simple equations 12.7 and 12.8.

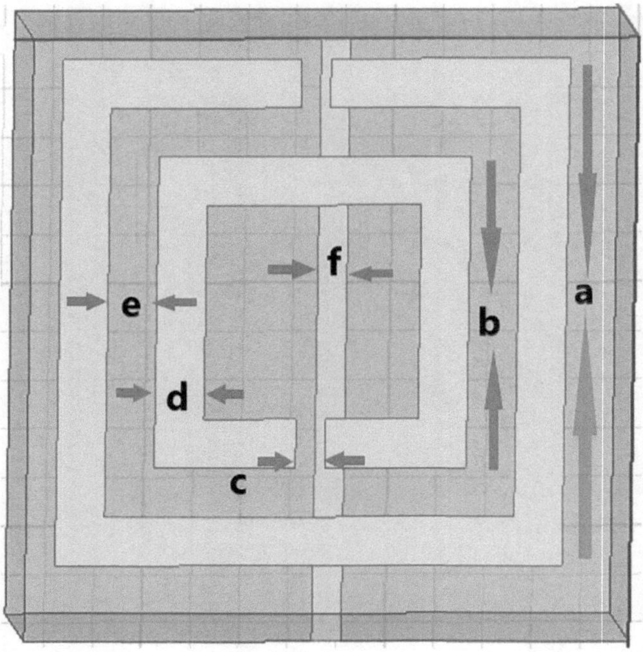

Figure 12.3 Square split-ring resonators (SSRR).

$$\varepsilon_{\text{eff}} = 1 - \frac{f_p^2}{f^2 - j\gamma f} \tag{12.7}$$

$$\mu_{\text{eff}} = 1 - \frac{f_{\text{mp}}^2 - f_0^2}{f^2 - f_0^2 - j\gamma f} \tag{12.8}$$

In equations 12.7 and 12.8, plasma freq. is denoted by f_p and f_{mp}, respectively, f_0 is the resonance frequency and f is a signal frequency. The damping factor γ is a material loss [14–16].

The SSRR is depicted in Figure 12.3 with a copper strip on the substrate ground plane to obtain negative μ and negative ε. The magnetic component of the field vector of the plane wave is normal to the SSRR, which induces currents in the SSRR that produce magnetic moments, and finally, yield negative permeability. The MATLAB script file is used to get the CST platform [17].

12.4.2 Design of the SSRR cell

The SSRR and strip are designed, respectively, on the top and bottom parts of the FR-4 substrate, which is shown in Figure 12.3. Its permittivity is 4.4, and its thickness is 1.6 mm. The dimensions of 4.8×4.8 mm SSRR are

a=3.9 mm, b=3.2 mm, c=0.22 mm, d=0.35 mm, and e=0.45 mm. The strip width f=0.6 mm and strip length=4.8 mm.

12.4.3 Parameters of the SSRR metamaterial cell

S_{11} and S_{21} of the SSRR cell are found by simulating the cell. The SSRR cell is found to be resonating at a 5.8 GHz frequency. The real magnitude of S_{11} and S_{21} at the 5.8 GHz frequency is 0.2 and 0.79, respectively. The epsilon that is permittivity and mu (permeability) of the cell is calculated using both the S parameters. The frequency v/s epsilon and mu are plotted.

The real permittivity and permeability of double-negative metamaterial cells, which are important parameters presented in Figure 12.4, are negative. In the band of frequency between 5.5 and 6.2 GHz, the permittivity and permeability are negative. Therefore, the metamaterial-based antenna must be designed to operate within a range of frequencies. From equation 12.8, in the lossless condition, when $f_{\mathrm{mp}}^2 - f_0^2 = f^2 - f_0^2$, that is, $f=f_{\mathrm{mp}}$ (magnetic plasma frequency) $\mu_{\mathrm{eff}}=0$. The curve diverges at a resonance frequency of the cell. The losses are represented by γ. At 5.8 GHz, the permeability is −4, and the permittivity is −3.6. In Figure 12.4, we have also presented the real refractive index and impedance of SSRR. It is found that the real refractive index is negative and close to the resonance frequency. At the resonance frequency, the imaginary impedance is equal to almost zero and the real impedance is equal to 50 Ω.

The proposed cell is 4.8 mm×4.8 mm. The dimensions of the cell are inversely proportional to the resonance frequency. Some authors have

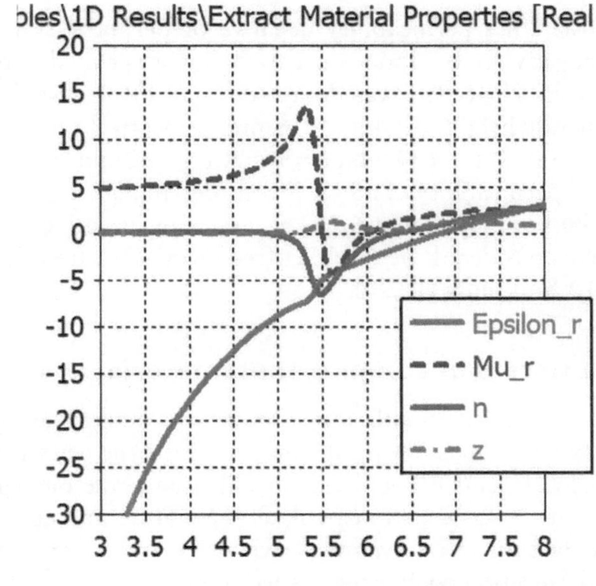

Figure 12.4 Permittivity, permeability refractive index and real impedance of SSRR.

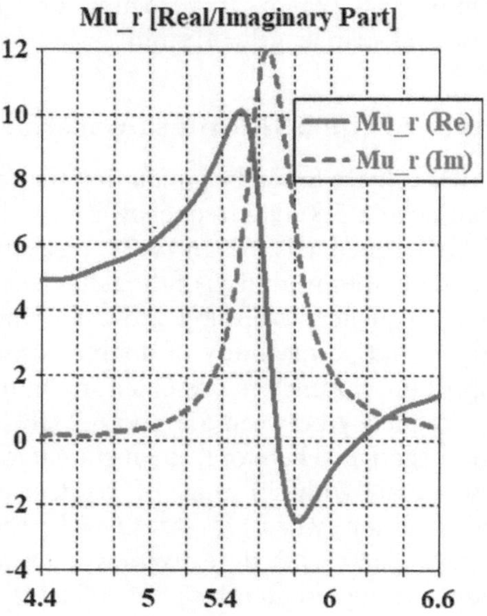

Figure 12.5 Mu of SSRR at 5.8 GHz.

designed such a cell. The 2.2 mm × 2.2 mm cell has a negative parameter at around 10 GHz [18].

12.4.4 Single negative metamaterial cell

Cells that show either permeability negative or permittivity negative are called single negative cells. If mu is negative then it is called a mu-negative metamaterial cell. The SSRR was designed on an FR-4 base. The permittivity and substrate height is 4.4 and 1.6 mm, respectively. The dimensions of 4.8×4.8 mm SSRR are a=4 mm, b=3 mm, c=0.22 mm, d=0.5 mm, and e=0.4 mm. The permeability of SSRR is recorded in Figure 12.5. The band of frequency between 5.65 and 6.2 GHz is having negative permeability. Hence, we have considered the 5.8 GHz frequency. This frequency band is used for WiMAX in India [19,20].

12.4.5 Parameters of complex metamaterial cell

In these cases, numerical simulations are used to determine the parameters of complex structures. We must first determine the cell's transmission and reflection coefficients using algorithms like the Finite Element Method (FEM). Equation 12.9 is used to calculate impedance, and equation 12.10 is used to calculate the refractive index, where S_{21} and S_{11} are transmission coefficient and reflection coefficient, respectively.

$$Z_{\text{eff}} = \pm\sqrt{\frac{(1+S_{11})^2 - (S_{21})^2}{(1-S_{11})^2 - (S_{21})^2}} \tag{12.9}$$

$$\eta_{\text{eff}} = \pm\frac{1}{\text{kod}}\left[\left\{\left[\text{In}\left(e^{\text{inkod}}\right)\right] + 2m\pi\right\} - i\left[\text{In}\left(e^{\text{inkod}}\right)\right]\right] \tag{12.10}$$

$$e^{\text{inkod}} = \frac{S_{21}}{1 - S_{11}\left[\frac{(Z-1)}{(Z+1)}\right]}$$

where k_0 is the wave vector in a vacuum, $k_0 = 2\pi/\omega_o$, the substrate thickness is denoted by d, and m is an integer. From the equations above, we can find Z_{eff} and η_{eff}. Because metamaterials are passive media, they indicate the positive real impedance and negative imaginary impedance. Further, $\epsilon_{r\text{eff}} = \eta_{\text{eff}}/Z_{\text{eff}}$ and $\mu_{\text{reff}} = \eta_{\text{eff}}\, Z_{\text{eff}}$.

12.5 DESIGN OF HIBISCUS LEAF-SHAPED MPA (HLMPA)

12.5.1 Design of HLMPA for WiMAX and other wireless applications

The hibiscus leaf shape is somewhat like a circle; hence, to begin with, a CMPA with a resonating frequency of 5.8 GHz is taken. Its radius is 6.47 mm. The proposed bioinspired HLMPA is designed in a similar shape just like the hibiscus leaf and is presented in Figure 12.6. The metamaterial-loaded hibiscus leaf-shaped antenna (MHLMPA) is presented in Figure 12.6. The circular patch is designed to determine the approximate length of the

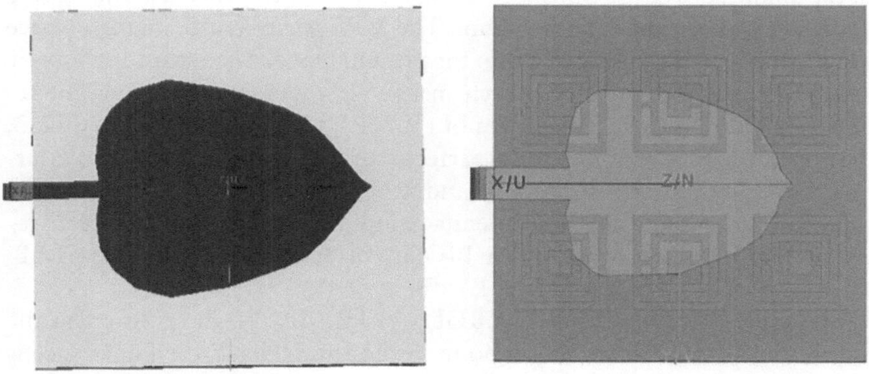

Figure 12.6 Hibiscus leaf microstrip patch antenna without metamaterial and with metamaterial.

Table 12.1 Dimensions of HLMPA

	Dimensions (mm)
The major axis of the patch	14.1
The minor axis of the patch	10
Substrate width	18
Substrate length	20
Feed length	3.3
Feed width	0.8

major axis of the hibiscus leaf MPA (HLMPA). Then, the resultant CMPA is perturbed to acquire the proposed HLMPA structure. [1,2,21,22]. This antenna resonates at frequencies of 5.8, 10, and 13.4 GHz. The HLMPA design is shown in Table 12.1.

12.5.2 Parameters of HLMPA

The gain is an important metric that describes an antenna's performance. Although the antenna gain is closely related to directivity, it is a metric that considers the antenna's efficiency as well as its directional capabilities. The pattern controls the directional qualities, and hence, the directivity of the antenna. The gain of any type of antenna in a particular direction is defined as the ratio of the directional intensity to the isotropically radiated antenna's radiation intensity. The second significant antenna parameter is bandwidth. It can be stated as the frequency range in which the antenna's properties support the set standard values when the antenna input impedance, pattern, beamwidth, polarization, gain, and efficiency are all satisfactory [9]. The reflection coefficient v/s frequency graph of HLMPA is shown in Figure 12.7. It resonates at 5.8, 10 and 13.4 GHz. These parameters determine the ratios of the amplitudes of the transmitted and reflected wave when a plane wave is an incident on the dielectric medium. The MPA performance increases once a split-ring resonator is loaded into the ground plane. The return loss is very low at resonance frequency, which means an antenna radiates maximum power [23–25]. All the parameters of HLMPA are presented in Table 12.2. The return loss is very low for a perfectly matched antenna ffrom the far-field radiating pattern in the UN and VN plane at a 5.8 GHz resonance frequency. It is found that the antenna gain is 3.74 dBi at 0° and directivity is equal to 5.9 dBi at 5.8 GHz. At 10 GHz, the gain of HLMPA is 3.51 dBi at −48°. At 13.4 GHz, the gain is 1.9 dBi at −64°.

The impedance matching at 5.8 GHz of HLMPA is observed using the Smith chart. At the intersection point in the Smith chart, the impedance is almost 50 Ω at 5.8 GHz. The reflection loss is low for a perfectly matched antenna indicating transmission of maximum power. The value of VSWR at 5.8 GHz is 1.01. For a perfectly matched antenna with a source impedance

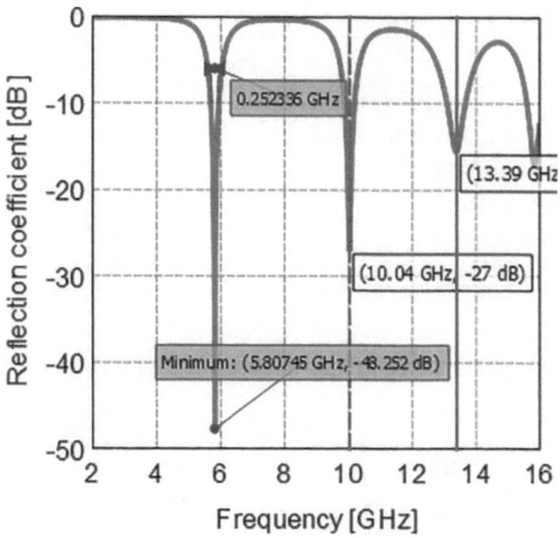

Figure 12.7 Reflection coeff. v/s freq. of HLMPA.

Table 12.2 Parameters of HLMPA

Resonance frequency (GHz)	5.8	10	13.4
Reflection coefficient (dB)	−39.1	−24.1	−14
Bandwidth at (−10 db) (MHz)	267.7	280.5	450.8
Gain (dBi)	3.74	3.51	1.9
VSWR	1.01	1.04	1.12

of 50 Ω, the value of VSWR is 1. The gain of the antenna also can be found from a 3D view of the far-field of HLMPA at 5.8 GHz.

12.5.3 Design of metamaterial-loaded HLMPA (MHLMPA) for WiMAX and other wireless applications

Figure 12.6 shows the MHLMPA and the design of MHLMPA is presented in Table 12.3. The graph of the reflection coefficient v/s frequency of HLMPA with metamaterial (MHLMPA) is presented in Figure 12.8. Return loss is one of the factors responsible for the radiation of power. The lower the return loss, the higher the radiated power.

12.5.4 Parameters of MHLMPA

An improvement in the parameters of patch antenna has been observed after loading metamaterial cells on HLMPA. The reflection coefficient

Table 12.3 Dimensions of MHLMPA

	Optimized dimensions (mm)
The major axis of the patch	11
The minor axis of the patch	9.4
Substrate width	18
Substrate length	19.2
Feed length	3.8
Feed width	1.6

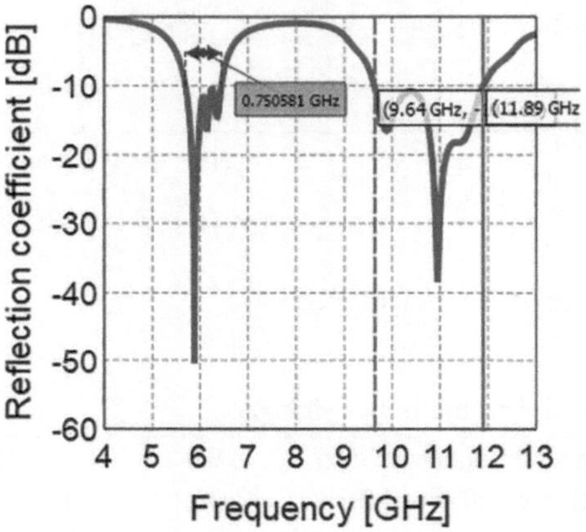

Figure 12.8 Reflection coeff. v/s freq. of MHLMPA.

Table 12.4 Parameters of MHLMPA

Resonance frequency (GHz)	5.8	11
Reflection coefficient (dB)	−50	−38.6
Bandwidth at (−10 db)	750 MHz	2.25 GHz
Gain (dBi)	3.1	1.19
VSWR	1.02	1.04

v/s frequency of MHLMPA is depicted in Figure 12.8. It resonates at 5.8 and 11 GHz. The low return loss of the antenna at resonance frequency is responsible for the maximum power radiated by an antenna. The BW, gain, return loss, and VSWR are presented in Table 12.4. A very low return loss is found for a perfectly matched antenna. The far-field radiating pattern in the UN and VN planes at 5.8 GHz resonance frequency are used to find the gain of antennas. At 11 GHz, the antenna gain is 1.8 dBi at −44°.

Impedance matching at 5.8 GHz of MHLMPA is observed using the Smith chart. At the point in the Smith chart at 5.8 GHz, the ohmic impedance is almost 50 Ω. VSWR at 5.8 GHz is 1.02.

12.6 DESIGN OF *KAJU* LEAF-SHAPED MPA (KLMPA)

12.6.1 Design of geometrical structure of *Kaju* leaf-shaped antenna

The shape of the *Kaju* leaf also matches the geometry of the circle to a great extent. Hence, the design has been initiated from a CMPA with a radius equal to 6.47 mm. The proposed bioinspired *Kaju* leaf-shaped antenna is presented in Figure 12.9. It resonates at frequencies of 5.8 and 9.67 GHz. Then, the resultant CMPA is perturbed to achieve the proposed KLMPA structure [1,2,21]. Also, MKLMPA is shown in Figure 12.9. The proposed antenna is built on an FR4 substrate. Its dielectric constant is 4.4, loss tangent 0.02, and 1.6 mm substrate height, as shown in Figure 12.9. The KLMPA is simulated using FEKO. The design of KLMPA is shown in Table 12.5.

12.6.2 Parameters of KLMPA

The reflection coefficient (dB) v/s frequency (GHz) graph of KLMPA is presented in Figure 12.10. It resonates at 5.8 and 9.67 GHz. Most of the

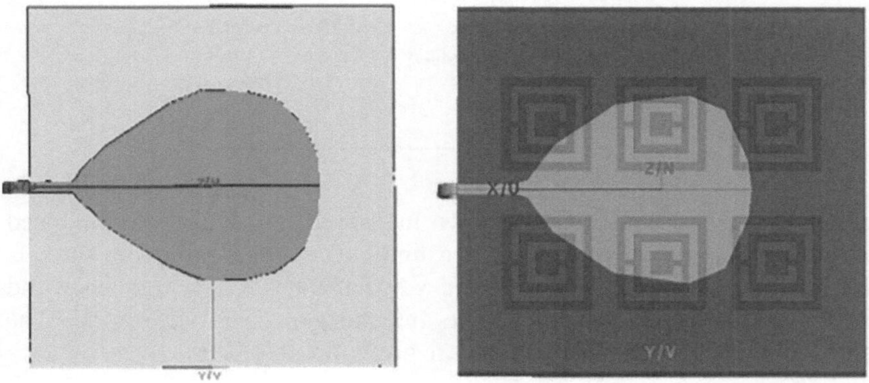

Figure 12.9 *Kaju* leaf microstrip patch antenna without metamaterial and with metamaterial.

Table 12.5 Dimensions of KLMPA

	Dimensions (mm)
The major axis of the patch	14.2
The minor axis of the patch	10.6
Substrate width	20
Substrate length	21
Feed length	2.5
Feed width	0.6

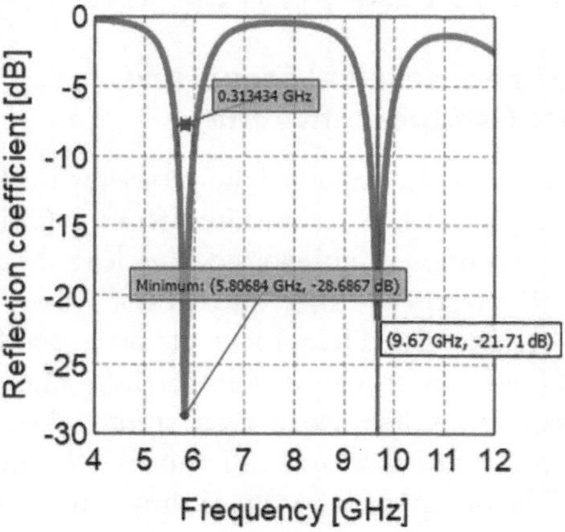

Figure 12.10 Reflection coeff. v/s freq. of KLMPA.

Table 12.6 Different parameters of KLMPA

Resonance frequency (GHz)	5.8	9.67
Reflection coefficient (dB)	−28.6	−21.7
Bandwidth at (−10 db) (MHz)	313.9	390.5
Gain (dBi)	3.1	2.11
VSWR	1.02	1.04

parameters of the antenna are shown in Table 12.6. A perfectly matched antenna has low return loss. We can find out antenna gain from the far-field radiating pattern in the UN and VN plane at 5.8 GHz frequency and 9.67 GHz resonance frequency. The antenna gain is 3.1 dBi at 0°. The antenna gain and other parameters are given in Table 12.6.

12.6.3 Design of metamaterial-loaded KLMPA (MKLMPA) for wireless application

Figure 12.9 depicts the proposed bioinspired metamaterial-loaded *Kaju* leaf antenna (MKLMPA) and the conventional KLMPA. Table 12.7 presents the design of MKLMPA. It resonates at frequencies of 5.8 and 9.18 GHz.

12.6.4 Parameters of MKLMPA

The coefficient of reflection v/s frequency in the GHz graph of MKLMPA is shown in Figure 12.11. The frequencies at which it resonates are 5.8 and 9.18 GHz. The BW, gain, return loss, and VSWR are depicted in Table 12.8.

Table 12.7 Dimensions of MKLMPA

	Dimensions (mm)
The major axis of the patch	12.7
The minor axis of the patch	9.8
Substrate width	19
Substrate length	22
Feed length	3.1
Feed width	0.6

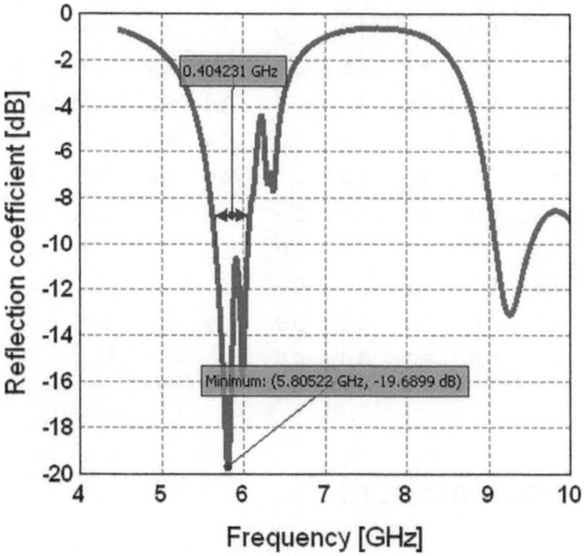

Figure 12.11 Reflection coeff. v/s freq. of MKLMPA.

Table 12.8 Different parameters of MKLMPA

Resonance frequency (GHz)	5.8	9.18
Reflection coefficient (dB)	−19.68	−14
Bandwidth at (−10 db) (MHz)	404.4	480.5
Gain (dBi)	5	4.7
VSWR	1.04	1.14

The gain of the antennas can be obtained from the far-field radiating patterns in the UN and VN planes at 5.8 and 9.18 GHz frequencies. The gain of an antenna is 5.0 dBi at 0°. The gain of an antenna and other parameters are shown in Table 12.8. Table 12.9 presents the *x-y-z* coordinates of each point of the proposed bioinspired MPA.

Table 12.9 The design data (*x*-*y*-*z* coordinates) of each point of the proposed bioinspired microstrip patch antennas

	HLMPA coordinates (mm)			*MHLMPA coordinates (mm)*			*KLMPA coordinates (mm)*			*MKLMPA coordinates (mm)*		
Points	*x*	*Y*	*z*	*x*	*y*	*z*	*x*	*y*	*z*	*x*	*y*	*z*
1	10.0	0.00	0.00	9.60	0.00	0.00	10.5	0.00	0.00	11.0	0.00	0.00
2	10.0	0.4	0.0	9.6	0.8	0.0	10.5	0.3	0.0	11	0.3	0.0
3	6.7	0.4	0.0	5.8	0.8	0.0	8.0	0.3	0.0	7.9	0.3	0.0
4	6.8	1.5	0.0	5.8	1.5	0.0	7.6	0.8	0.0	7.2	0.5	0.0
5	6.6	2.8	0.0	5.6	2.8	0.0	6.0	2.3	0.0	5.8	2.3	0.0
6	6.0	3.8	0.0	5.2	3.8	0.0	2.5	4.5	0.0	2.5	4.4	0.0
7	4.8	4.6	0.0	4.4	4.6	0.0	−0.5	5.3	0.0	−0.5	4.8	0.0
8	3.0	5.0	0.0	3.0	4.7	0.0	−2.0	5.2	0.0	−1.5	4.9	0.0
9	0.0	4.9	0.0	0.0	4.6	0.0	−3.6	4.8	0.0	−3.0	4.3	0.0
10	−3.0	4.0	0.0	−3.0	3.6	0.0	−4.0	4.6	0.0	−3.5	4.1	0.0
11	−4.0	3.3	0.0	−4.2	2.9	0.0	−4.6	4.1	0.0	−3.5	4	0.0
12	−5.0	2.6	0.0	−5.0	1.4	0.0	−5.3	3.4	0.0	−3.9	3.2	0.0
13	−6.0	1.4	0.0	−5.1	0.5	0.0	−6.1	1.5	0.0	−4.7	1.5	0.0
14	−6.8	0.4	0.0	−5.2	0.2	0.0	−6.2	0.0	0.0	−4.8	0.0	0.0
15	−7.4	0.0	0.0	−5.2	0.0	0.0						

12.7 SIMULATION RESULTS

The impact of the metamaterial-based antenna is studied, and its performance is compared with the conventional antenna. The SSRR cell is simulated using MATLAB and the CST software. The CST platform is being invoked through a MATLAB file. Also, only SSRR is simulated using the software. It has been found that the cell with SSRR and strip depicts negative permittivity and permeability, which indicates that it is a double-negative or left-handed metamaterial cell. A metamaterial cell with only SSRR shows negative permeability to state it as a single or mu-negative metamaterial cell. The initiator CMPA structure with a radius equal to 6.48 mm resonates at 5.8 GHz. These bands are useful for the WiMAX and WLAN systems. Hence, the perturbation of the initiator (CMPA) is used to realize the bioinspired MPA. Conventional and metamaterial-loaded hibiscus and *Kaju* leaf-shaped patch antennas have been simulated using the FEKO software. All the proposed antennas resonate at 5.8 GHz, and some other frequencies as presented in Table 12.10. The parameter improvement of metamaterial-loaded antenna is observed. The comparison of the parameters of HLMPA and KLMPA is listed in Table 12.10.

Table 12.10 Parameters of HLMPA, MHLMPA, KLMPA, and MKLMPA

	HLMPA			*MHLMPA*		*KLMPA*		*MKLMPA*	
1. Resonance frequency (GHz)	5.8	10	13.4	5.8	11	5.8	9.67	5.8	9.18
2. Reflection coefficient (dB)	−39.1	−24.1	−14.1	−50	−38	−28.6	−21.7	−19.7	−14
3. Bandwidth (MHz)	267.7	280.5	450.8	750	2.25	313.9	390.9	404.4	480.5
4. Gain (dBi)	3.74	3.51	1.9	3.1	1.19	3.15	2.11	5	4.7
5. VSWR	1.01	1.04	1.12	1.02	1.03	1.02	1.04	1.04	1.14

12.8 CONCLUSION

We have covered various design concepts for unique bioinspired microstrip antennas in this chapter. The hibiscus and *Kaju* leaf-shaped bioinspired conventional and metamaterial-based antennas are designed and simulated using the FEKO software. Both antennas are multi-band. The HLMPA resonates at 5.8, 10 and 13.4 GHz, and the *Kaju* leaf-shaped MPA resonates at 5.8 and 9.67 GHz. Both antennas have considerable bandwidth and gain so that they can be used for WiMAX and other wireless applications. The metamaterial-loaded antenna has shown considerable improvement in parameters. Also, it has shown a reduction in dimensions. Based on the output, it can be stated that the hibiscus leaf-shaped antenna has shown a better performance than the *Kaju* leaf-shaped antenna. In comparison to other conventional designs, bioinspired antennas depict a higher concentration of surface current in a smaller area.

REFERENCES

1. J. O. Abolade, D. B. O. Konditi, V. M. Dharmadhikary, *Bio-inspired Wideband Antenna for Wireless Applications Based on Perturbation Technique*, Elsevier, 2020.
2. A. J. R. Serres, G. K. de Freitas Serres, P. F. da Silva Júnior, R. R. S. Freire, J. do Nascimento Cruz, T. C. de Albuquerque, M. A. Oliveira, P. H. da Fonseca Silva, Bio inspired microstrip antenna, *Trends in Research in Microstrip Antenna*, 2017.
3. K. Gangwar, Paras, R.P.S. Gangwar, Metamaterials: Characteristics, process and application, *Advances in Electronic and Electric Engineering*, Vol. 4, No. 1, pp. 97–106, 2014.
4. Y. Liu, X. Guo, S. Gu, X. Zhao, Zero index metamaterial for designing high gain patch antenna, *International Journal of Antennas and Propagation*, Vol. 25, pp. 1–12, 2013.
5. R. Marques, F. Mesa, J. Martel, F. Median, Comparative analysis of edge and broadside coupled split-ring resonators for metamaterial design theory and experiment, *IEEE Transactions on Antennas and Propagation*, Vol. 51, pp. 2572–2581, 2003.

6. J. D. Baena, J. Bonache, F. Martin, R. M. Sillero, F. Falcone, J. Garcia, I. Gil, M. F. Portillo, M. Sorolla, Equivalent circuit models for split-ring resonators and complementary split-ring resonators coupled to planar transmission lines, *IEEE Transactions on Antennas and Propagation*, Vol. 53, No. 4, pp. 1451–1461, 2005.
7. S. Nelaturi, N. Sarma, CSRR based patch antenna for Wi-Fi and WiMAX applications, *Advance Electromagnetics*, Vol. 7, No. 3, pp. 40–45, 2018.
8. A. Boutejdar, B. I. Halim, Design of multiband microstrip antenna using two parasitic ring resonators for WLAN/WiMAX and C/X/Ku- band application, in *Proceedings of IEEE International Conference on Electromagnetics and Antenna (IEMANTENNA)*, Canada, 2019, pp. 46–50.
9. C. A. Balanis, *Antenna Theory: Analysis and Design*, 3rd ed., John Wiley & Sons Inc., Hoboken, NJ, 2005.
10. P. R. Satarkar, R. B. Lohani, Design and characterization of coaxial feed circular patch antenna on metamaterial substrate, in *Proceedings of 4th IEEE Conference on Smart Trends in System, Security and Sustainability (worlds 4)*, London, UK, Oct. 2020, pp. 424–427.
11. M. A. Hindy, R. Elsagheer, M. S. Aseen, Circular split-ring resonator (CSRR) left handed metamaterial (LHM) having simultaneous negative permeability and permittivity, *International Journal of Hybrid Information Technology*, Vol. 10, pp. 171–178, 2017.
12. N. Gupta, J. B. Saxena, K. S. Bhatia, Design of metamaterial loaded rectangular patch antenna for satellite communication applications, *Iranian Journal of Science and Technology-Transaction of Electrical Engineering*, Vol. 43, pp. 85–90, 2018.
13. P.R. Satarkar and R.B. Lohani, Design and extraction of parameters of metamaterial cell and its applications to enhance the performance of patch antenna, *Presented at the IETE and ISVE International Science Exhibition Congress Symposium (SECS-2020)*, Ranchi, Sept. 12–13, 2020, Paper SECS 20044.
14. S. E. Mendhe, Y. P. Kosta, Metamaterial properties and application, *International Journal of Information Technology and Knowledge Management*, Vol. 4, pp. 85–89, 2011.
15. Y. Liu, X. Zhang. Metamaterials: A new frontier of science and technology, *Chemical Society Reviews*, Vol. 40, pp. 2494–2507, 2011.
16. N. Engheta, R. W. Ziolkowski, *Metamaterials Physics and Engineering Explorations*, John Wiley &Sons, Inc., 2006.
17. M. D. Nazmul, H. B. Monk, MATLAB to CST interfacing method. Metamaterial critics & alternatives, INC publ, 2000.
18. A. B. Numan, M. S. Sharawi, Extraction of material for metamaterials using a full-wave simulator, *IEEE Antenna and Propagation*, Vol. 55, No. 5, pp. 202–211, 2013.
19. S. S. Islam, M. R. Faruque, T. Alam, A new mu negative metamaterials, in *Proceedings of 2nd International Conference on Electrical Information and Communication Technology*, Bangladesh, Jan. 2016.
20. F. Falcone, T. Lopetegi, J. D. Baena, M. Sorolla, Effective negative stopband microstrip lines based on complementary split-ring resonators, *IEEE Microwave and Wireless Components*, Vol. 14, No. 6, pp. 280–282, 2004.

21. V. Latha, D. Rajagopal, R. Kiruthika, L. Reshma, Bioinspired red bay leaf shaped antenna for narrowband application, *International Journal of Applied Engineering Research*, Vol. 14, No. 6, 2019, ISSN: 0973-4562.
22. P. F. da Silva Jr., E. E. C. Santana, C. A. M. Cruz, V. S. Aquino, L. S. O. Castro, A. J. R. Serres, R. C. S. Freire, P. H. F. Silva, *Compact Bioinspired Antenna for WLAN 5 GHz Application*, Spriger Nature, 2021.
23. A. Rahim, P. K. Mallik, V. A. Sankar Ponnapalli, Fractal antenna design for overtaking on highways in 5G vehicular communication ad-hoc networks environment, *International Journal of Engineering and Advanced Technology (IJEAT)*, Vol. 9, No. 1S6, pp. 157–160, 2019, doi: 10.35940/ijeat.A1031.1291S619, ISSN: 2249-8958.
24. P. K. Malik, H. Parthasarthy, M. P. Tripathi, Axisymmetric excited integral equation using moment method for plane circular disk, *International Journal of Scientific and Engineering Research*, Vol. 3, No. 3, pp. 1–3, 2012, ISSN: 2229-5518.
25. P. K. Malik, H. Parthasarthy, M. P. Tripathi, Analysis and design of Pocklingotn's equation for any arbitrary surface for radiation, *International Journal of Scientific and Engineering Research*, Vol. 7, No. 9, pp. 208–213, 2016, ISSN: 2229-5518.

Chapter 13

Design and study of compact bio-inspired-shaped smart MIMO array antenna for 5G-enabled healthcare systems, IoT systems, and environmental care systems

John Colaco and Rajesh B. Lohani
Goa College of Engineering

Dac-Nhuong Le
Haiphong University

CONTENTS

13.1 INTRODUCTION

Microstrip antennas are critical components of healthcare communication systems, remote sensing, satellite communication, mobile communication, and a variety of other 5G-enabled systems. 5G communication is critical in human healthcare systems today due to the various real-time high-quality video-based medical consultations and telemedicine taking place around the world during the ongoing COVID-19 crisis. However, the existing 4G network has limited bandwidth, and real-time medical consultations take an inordinate amount of time to review patients' conditions, thereby delaying patient treatment. This will be accomplished through a fifth-generation

DOI: 10.1201/9781003347057-13

wireless communication operating in the millimetre-wave band. The millimetre-wave band is intended to increase the data transmission bandwidth [1]. While the millimetre-wave band is an advanced wireless technology, it does have some drawbacks, including high signal losses, security concerns, and cost [1,2].

The evolution of botanical structures has enabled bio-inspired microstrip patch antennas to achieve compact and optimal performance in a variety of electronic and communication systems applications.

It has aided researchers in biological and engineering fields in combining a new bioengineering technique comparable to soft-computing techniques such as artificial neural networks and genetic algorithms. The bio-inspiration comes from the properties and shapes of botanical structures that are used as microstrip patch antennas. Using the Gielis formula and the polar transformation defined in vector form as $v(t) = (x), y(t), t_0$, one can design patch antennas by representing the geometry of plants and living organisms. This Gielis formula mathematically describes a variety of abstract and natural forms, such as the shapes of flowers and leaves [3]. By modifying the concept of superellipse, the author transformed natural form equations into modified circles and obtained super formulas using trigonometric polar coordinates [3].

MIMO microstrip patch array antennas, in conjunction with their radio frequency network systems, are critical components of 5G designs. Thus, using the antenna design software to assist in designing, modelling, implementing and analysing will help to minimise the cost, time and risk associated with this complex workflow. Communication engineers are concentrating their efforts on developing advanced microstrip antenna systems capable of beam-steering, phase shifting and multiple data transmission to meet 5G communication requirements such as return loss of at least –10 or –15 dB, the bandwidth of at least 1 GHz and gain of at least 3 dBi. This was made possible by upgrading the antenna system to a MIMO configuration. MIMO antennas are a critical component of the 5G communication network's innovation as the system architecture relies on a greater number of microstrip antennas at base transceiver stations than the frameworks used in 4G communication [4]. Multiple input and multiple output (MIMO) technology has been demonstrated to have enormous potential for meeting the persistently growing demand for transmission-based communication frameworks capable of higher data rates [4]. However, MIMO configurations introduce the possibility of fading and signal multipath when attempting to achieve high data rates in channels with limited bandwidth. On the transmission path, fading and signal multipath occur. The desire for higher data rates over longer distances was a primary motivation for the MIMO orthogonal frequency division multiplexing (OFDM) framework to be upgraded. MIMO communications channels provide an intriguing solution to the multipath challenge by requiring multiple signal paths and acquiring information on the communications channel via a combination of numerous microstrip antennas and multiple signal paths [5].

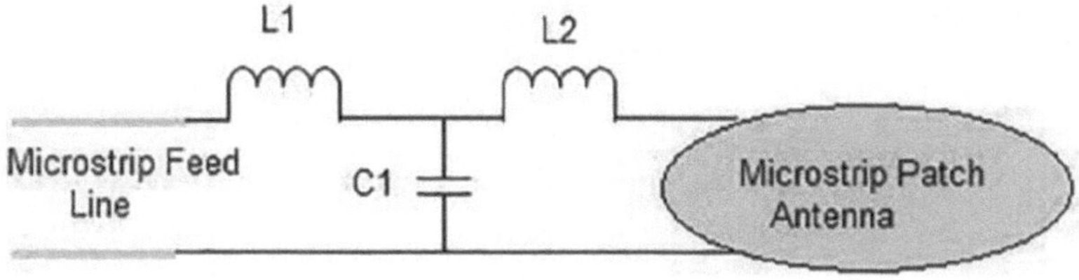

Figure 13.1 Equivalent circuit diagram.

Microstrip patch antennas are composed of a radiating element made of a high-conducting material such as copper in the shape and size of a patch placed on a dielectric material substrate with a ground plane. Numerous feeding or coupling techniques, including microstrip transmission feed line, electromagnetic coupling coaxial probe, and coplanar waveguide feed, can be used to excite these antennas [6]. The equivalent circuits for the proposed elliptical patch antenna are shown in Figure 13.1, indicating that the microstrip patch antenna can be thought of as an extension of the microstrip feed line via edge coupling, with both the microstrip patch antenna and microstrip feedline being easily fabricated on the same substrate [7]. The excitation of a microstrip patch antenna by a microstrip feed line via edge coupling can be expressed in terms of the electric current density, J-z, associated with the magnetic field of the microstrip feed line at the junction plane [7]. Equation 13.1 can be used to determine this coupling constant [7].

$$\text{Coupling} = \iiint_v E_z J_z \, dv \tag{13.1}$$

This research focuses on designing and analysing the bio-inspired-based microstrip patch antennas using the concept of botanical structures such as rose leaves as illustrated in Figure 13.2. This is accomplished by first designing and analysing a single elliptical patch (single leaf) with a 50 Ω transmission line serving as a branch, and then, increasing the number of patches (leaves) to three as illustrated in Figure 13.3 and forming a bio-inspired baffle. This photograph of rose leaves on a branch was taken in one of the authors' gardens. Additionally, to increase the bandwidth capacity of data rates over longer communication distances, the said bio-inspired rose leaves antenna is converted into a uniform linear MIMO microstrip array with equally spaced 0.5 mm elements and its performance is analysed. The array with a uniformly linear spacing of d will have an array factor of [8],

$$AF(\varnothing) = \sum_n w_n e^{jkd(n-1)\sin\varnothing} \tag{13.2}$$

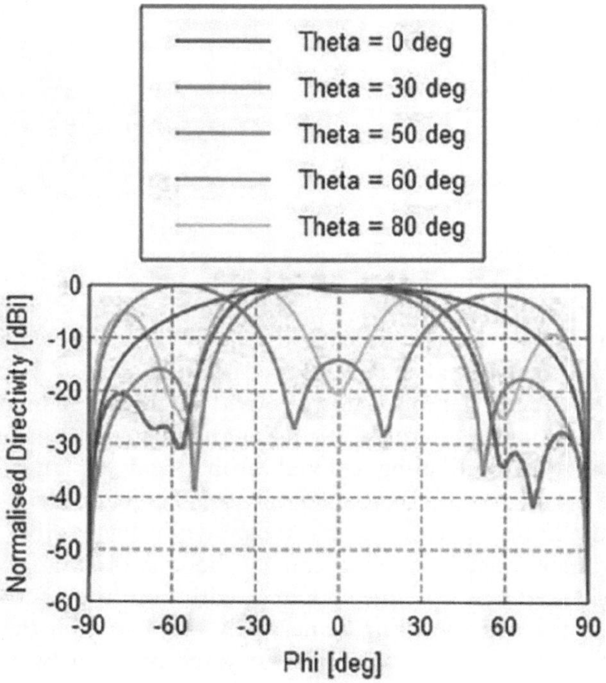

Figure 13.2 Proposed microstrip patch array factor with respect to phi.

Figure 13.3 Rose leaves.

which repeats itself in the sin∅ domain when:

$$kd\sin\varnothing_1 - kd\sin\varnothing_2 = \pm m2\pi \quad (13.3)$$

$$\sin\varnothing_1 - \sin\varnothing_2 = \pm m\frac{\lambda}{d} \quad (13.4)$$

This implies that at a periodicity of $\frac{\lambda}{d}$, the array factor will repeat [8], which is shown in Figure 13.2 as per our proposed microstrip array antenna.

13.2 RELATED LITERATURE

The authors have presented a high-performance rectangular shape microstrip patch antenna with a resonant frequency of 26 GHz, which is suitable13 for high-quality online education and a variety of 5G applications in the millimetre-wave band between 24 and 30 GHz [9]. At a resonant frequency of 28 GHz, Leeladhar and Ajay (2020) proposed a MIMO microstrip patch antenna array with a size of 2 * 3 for 5G applications [10]. Rohde and Schwarz (2016) demonstrated a straightforward theory of beamforming antennas, techniques for calculating radiation patterns, and a few findings regarding linear arrays for real-world measurement [11]. The authors designed, analysed, and reviewed the performance of a variety of ultra-wideband antennas and their application to 5G communications [12]. The authors described the modelling and implementation of a millimetre-waveband 28 GHz transceiver with a beam-steered patch antenna array using a four-channel transmit chain [13]. The authors have presented and demonstrated multiband microstrip patch antennas with rectangular slots of appropriate size for wireless communication in the frequency range of 5.70–12.60 GHz [14]. The authors demonstrated a small ultra-wideband MIMO microstrip square patch antenna in the frequency range of 3.1–11 GHz with a single column EBG (electromagnetic bandgap) structure and ground stub, resulting in extremely low mutual couplings [15]. The authors demonstrated a two-slot element MIMO patch antenna operating at 2.45 GHz on a Taconic RF-35 substrate [16]. The authors proposed a MIMO patch antenna array with an FR-4 substrate and a 50-Ω transmission line feeding in the frequency range of 24.22–28.557 GHz for 5G applications [17].

13.3 METHODOLOGY

The authors developed and studied a bio-inspired shape-based microstrip patch antenna at a resonant frequency of 30 GHz using this methodology,

Table 13.1 Shows the parameter details used for the design

Name of parameters	Dimensions of parameters in mm
Length of the major axis	1.9
Length of the minor axis	0.6
Length of microstrip line feed	1.9
Width of microstrip line feed	0.35
Length of substrate	8
Width of substrate	8

starting with a single elliptical patch (single leaf) fed with a 50 Ω transmission line (branch) and progressing to multiple elliptical patches up to three and combining them to form a bio-inspired shape-based microstrip patch antenna. Additionally, the same patch antenna is used to create a 3 mm × 2 mm MIMO sensor array. The authors used a Rogers RT/Duroid 5880 substrate in this proposed structure, which has a dielectric constant of 0.0010, making it ideal for 5G high-frequency applications. The proposed substrate has a height of 0.6 mm. The design is investigated in the millimetre-wave band between 26 and 40 GHz. The length of the semi-major axis is determined using Damiano's resonant frequency formula [18].

$$f_{m,n} = \frac{c_o K_{mn} c}{2\pi a e \sqrt{\varepsilon_r}} \tag{13.5}$$

where c_o is the velocity of light, e is the eccentricity of the ellipse, a is the length of the semi-major axis, c is the focal point of the ellipse, and K_{mn} is the nth parametric zero of the even and odd modes of the boundary equations (Table 13.1). The focal point of the ellipse 'c' is given by

$$c = \sqrt{a^2 - b^2} \tag{13.6}$$

where a and b are the radii of circular patches.

The eccentricity of the ellipse 'e' is given by

$$e = \frac{c}{a} \tag{13.7}$$

13.3.1 Design of single elliptical patch (antenna I)

The top view of the proposed single elliptical patch feed with a 50 Ω microstrip transmission line is shown in Figure 13.4.

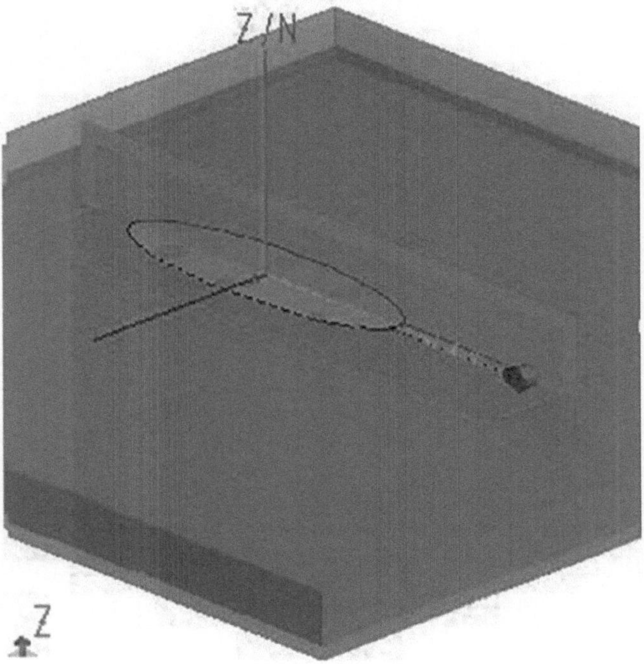

Figure 13.4 Geometrical 3D view of the single elliptical patch.

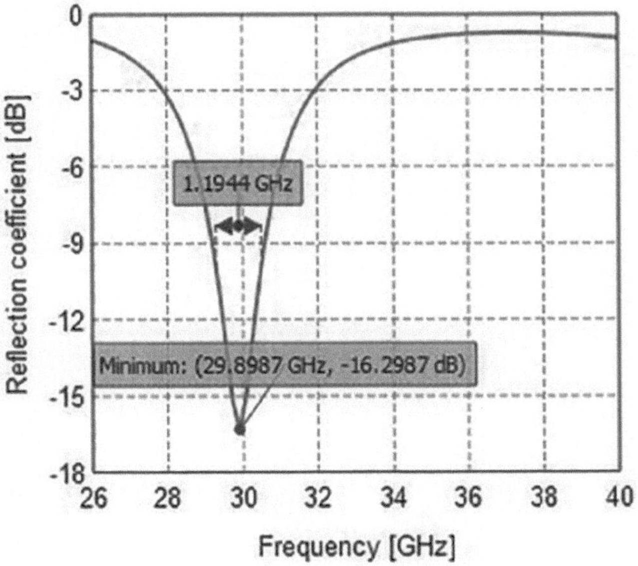

Figure 13.5 Return loss and bandwidth at −10 dB of a single elliptical patch.

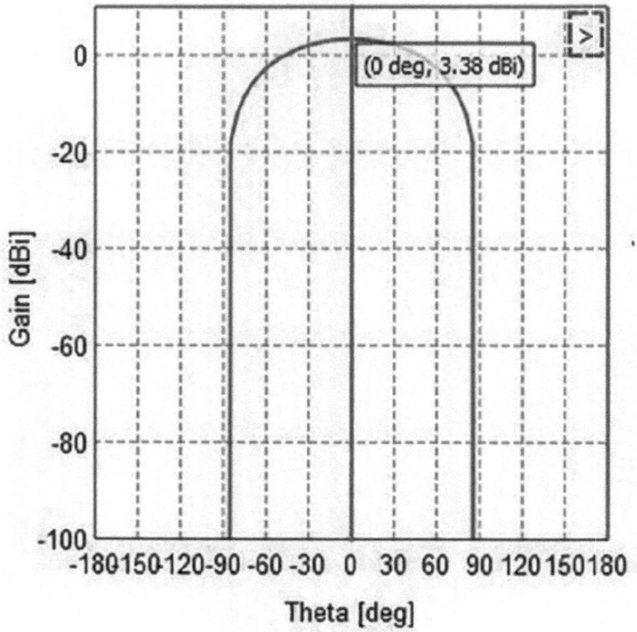

Figure 13.6 Gain w.r.t theta of the single elliptical patch.

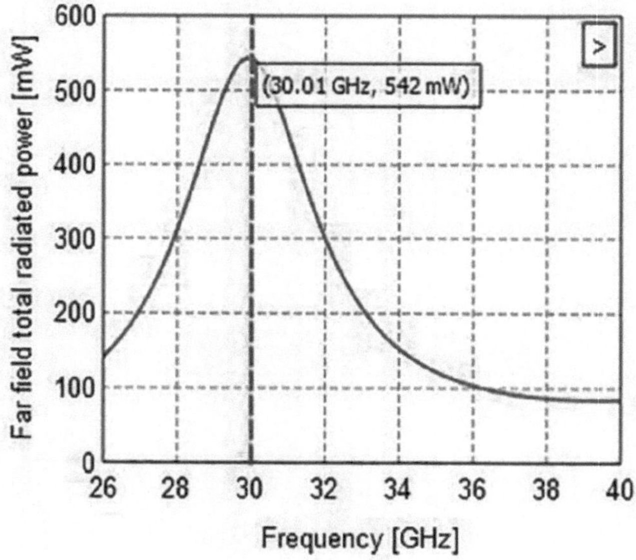

Figure 13.7 Total radiated power of single elliptical patch.

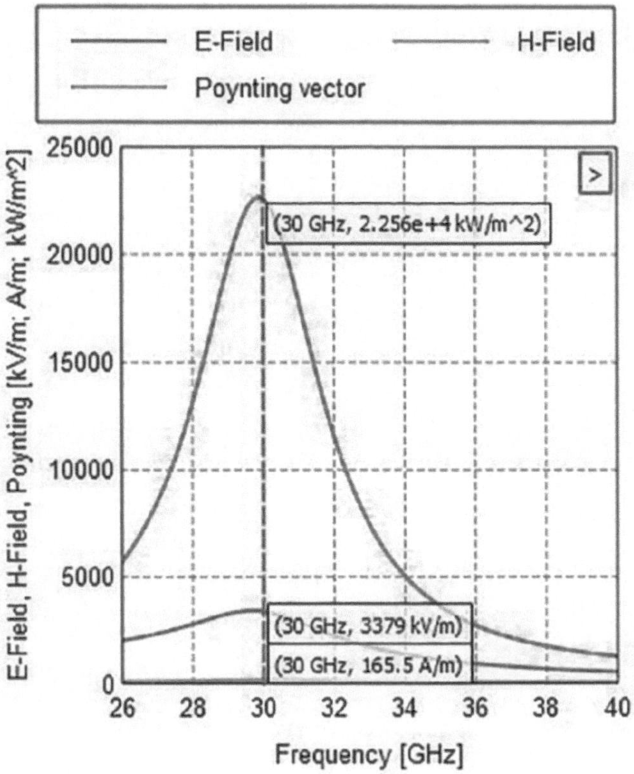

Figure 13.8 Radiation density of single elliptical patch.

13.3.2 Design of double elliptical patch (antenna II)

The proposed double (two leaves) elliptical patch is depicted in Figure 13.10.

13.3.3 Design of triple elliptical patch (antenna III)

The proposed triple elliptical patch antenna is shown in its top view in Figure 13.16 (three leaves).

13.3.4 Characteristics analysis of proposed bio-inspired microstrip antennas I, II, and III

Return Loss and Bandwidth: This property is a positive quantity that is the inverse of the reflection coefficient. It denotes that when any power is applied to the microstrip antenna via the transmission line, say 50 Ω, it measures how much energy is reflected from the antenna and how much energy is delivered to the antenna in order for it to radiate the delivered

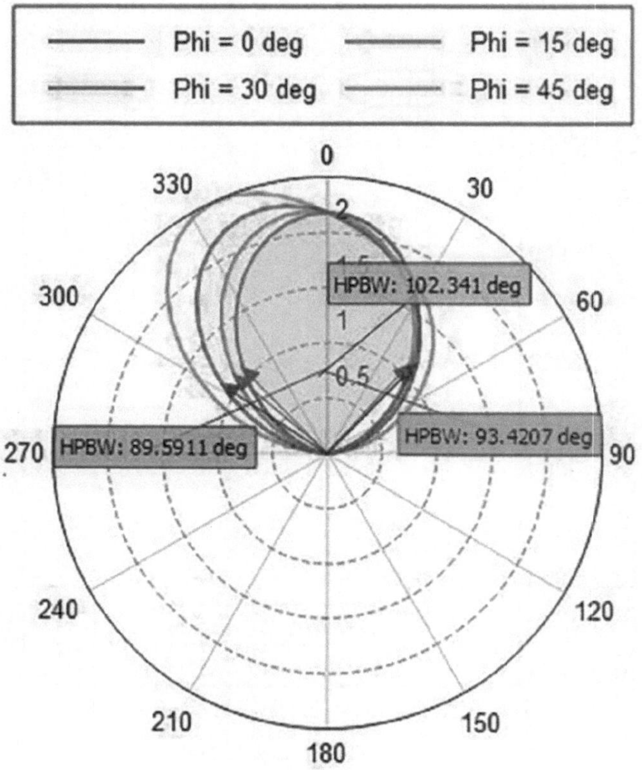

Figure 13.9 Beamwidth with w.r.t theta of the single elliptical patch.

energy. This gives the amount of voltage standing wave ratio, which is used to measure impedance matching using the maximum power transfer theorem. As a result, the relationship between return loss and voltage standing wave ratio is given as [19]

$$\text{Return loss} = -20\log_{10}\left(\frac{1+\text{Voltage standing wave ratio}}{\text{Voltage standing wave ratio}-1}\right)\text{dB} \qquad (13.8)$$

The voltage standing wave ratio for 5G communication must be between 1 and 2 to achieve a minimum of –10 dB bandwidth, referred to as impedance matching bandwidth, or better to radiate at least 90% of delivered power. Figures 13.5, 13.11, and 13.18 illustrate the return loss of –16.30, –18.77, and –19.77 dB, respectively, and the bandwidth of 1.194, 1.56, and 1.78 GHz, respectively, at an impedance bandwidth of –10 dB, indicating that the resonant frequency antenna performs well at 30 GHz, radiating more than 95% of incident power with high-efficiency data transmission

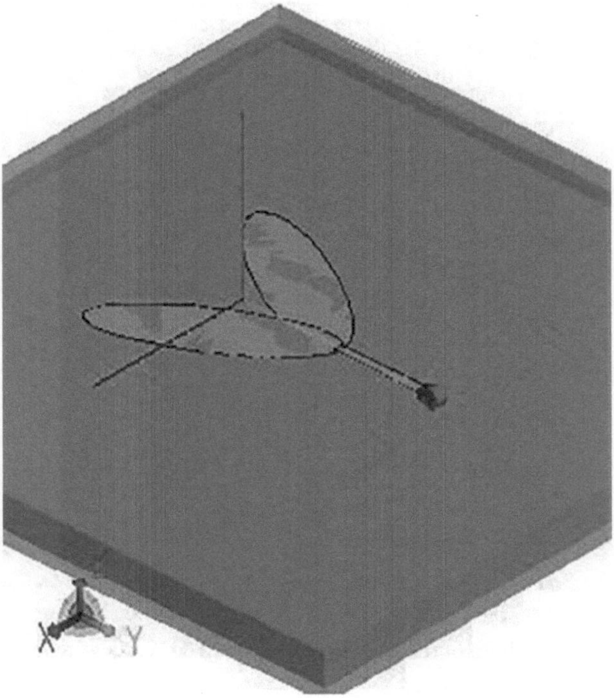

Figure 13.10 Geometrical 3D view of the double elliptical patch.

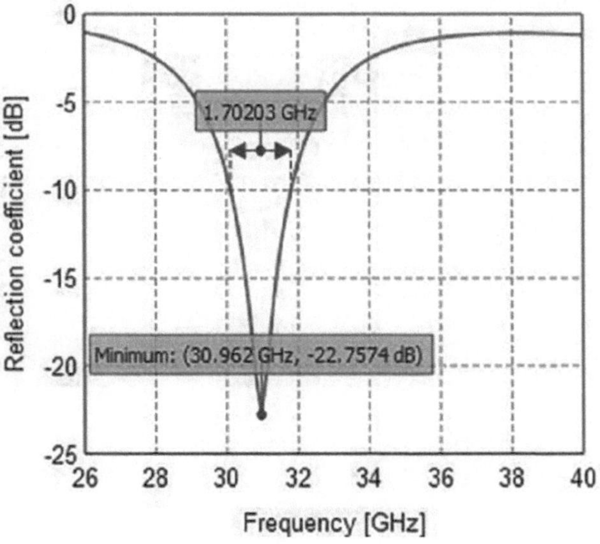

Figure 13.11 Return loss and bandwidth at −10 dB of the double elliptical patch.

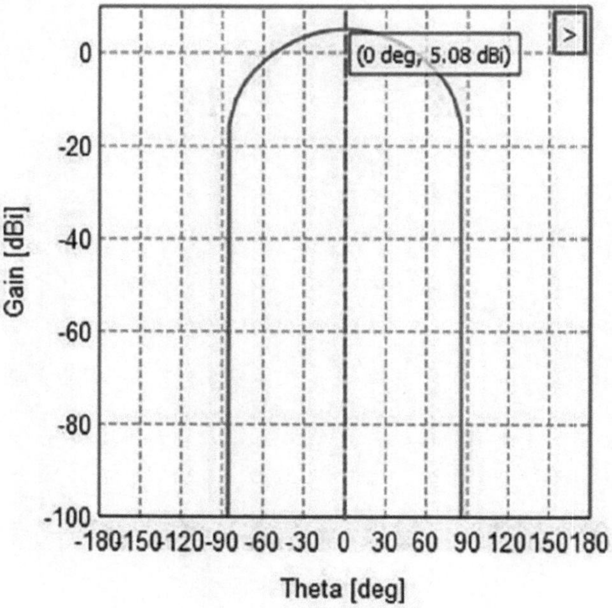

Figure 13.12 Gain w.r.t theta of the double elliptical patch.

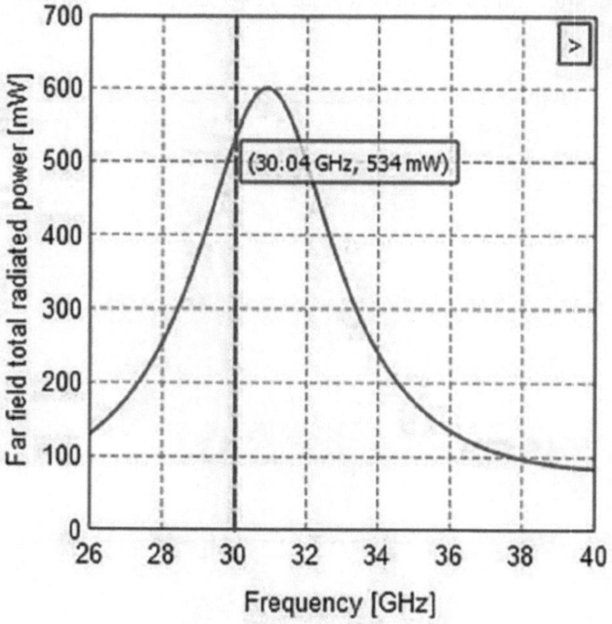

Figure 13.13 Total radiated power of double elliptical patch.

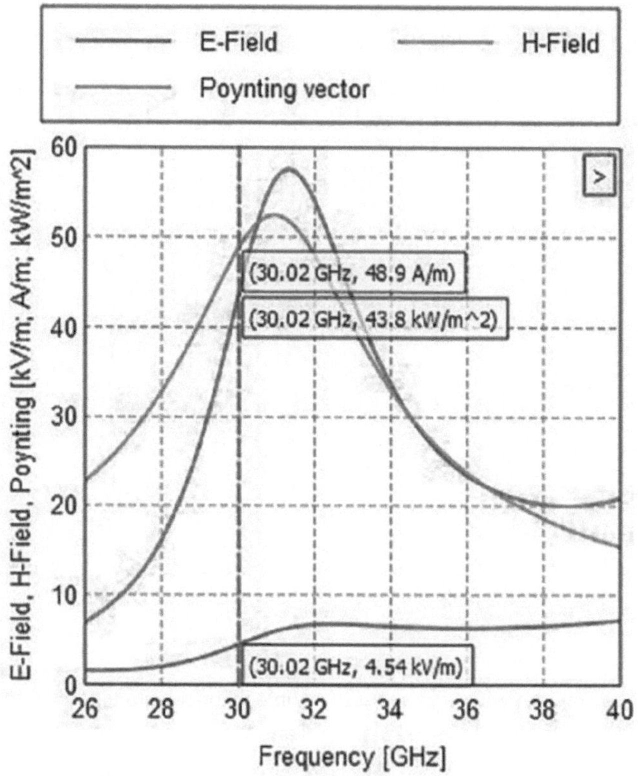

Figure 13.14 Radiation density of double elliptical patch.

rates, indicating constant imped. Thus, as illustrated in Figure 13.17, the proposed antenna satisfies the standard requirement for 5G communication. This increases the reliability and efficiency of 5G-enabled healthcare systems and applications.

Gain: The gain of 3.38, 5.08, and 5.67 dBi radiating isotropically in all directions at phi = 0° is depicted in Figures 13.6, 13.12 and 13.19, indicating that the proposed antenna has good directional capabilities and will provide sufficient signal strength for 5G-based systems and applications. This directional capability is the measure of the ratio of radiation intensity $U(\theta,\varphi)$ corresponding to isotropically radiated power to that of the total inputted power P_{in} divided by 4π (steradians) [19]. Mathematically, it is given as [18]

$$\text{Gain} = 4\pi \frac{U(\theta,\varphi)}{P_{in}} (\text{dimensionless}) \qquad (13.9)$$

Total Power Radiated: The proposed antenna's power output is depicted in Figures 13.7, 13.13, and 13.20, which is obtained by integrating the

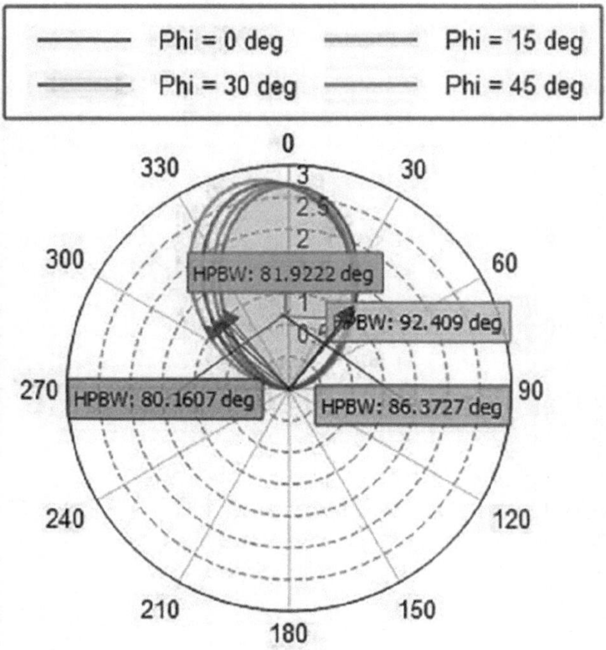

Figure 13.15 Beamwidth with w.r.t theta of double elliptical patch.

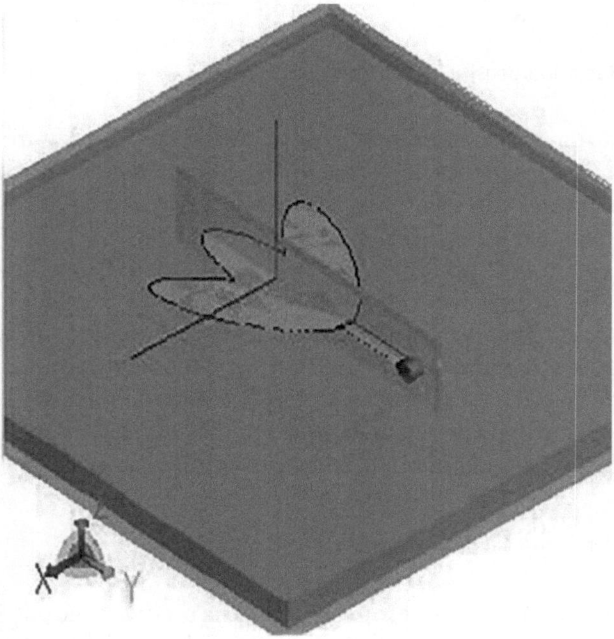

Figure 13.16 Geometrical 3D view of the triple elliptical patch.

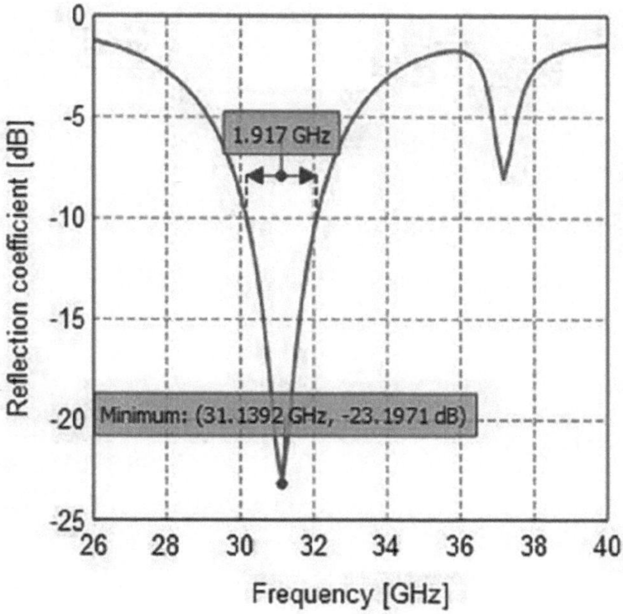

Figure 13.17 Return loss and bandwidth at −10 dB of triple elliptical patch.

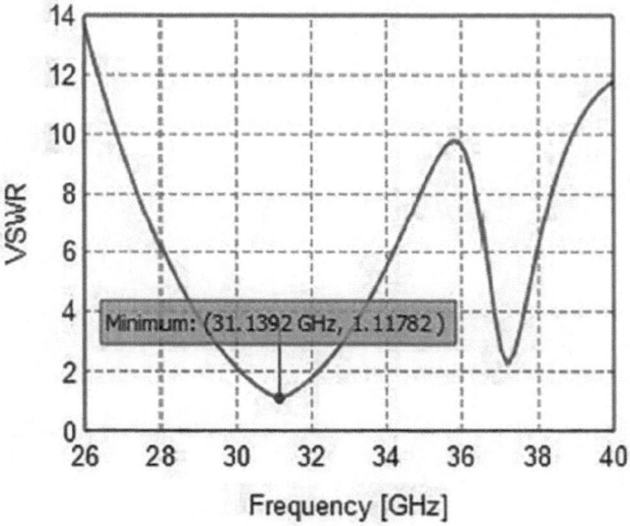

Figure 13.18 Voltage standing wave ratio of triple elliptical patch.

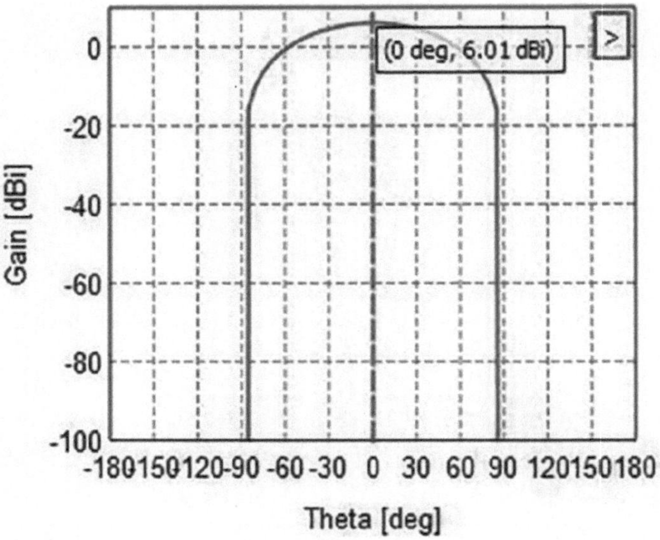

Figure 13.19 Gain w.r.t theta of triple elliptical patch.

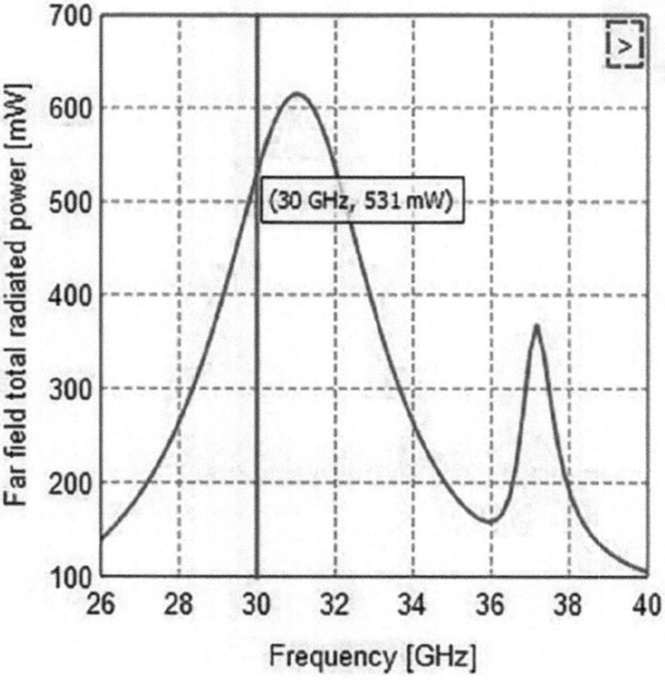

Figure 13.20 Total radiated power of triple elliptical patch.

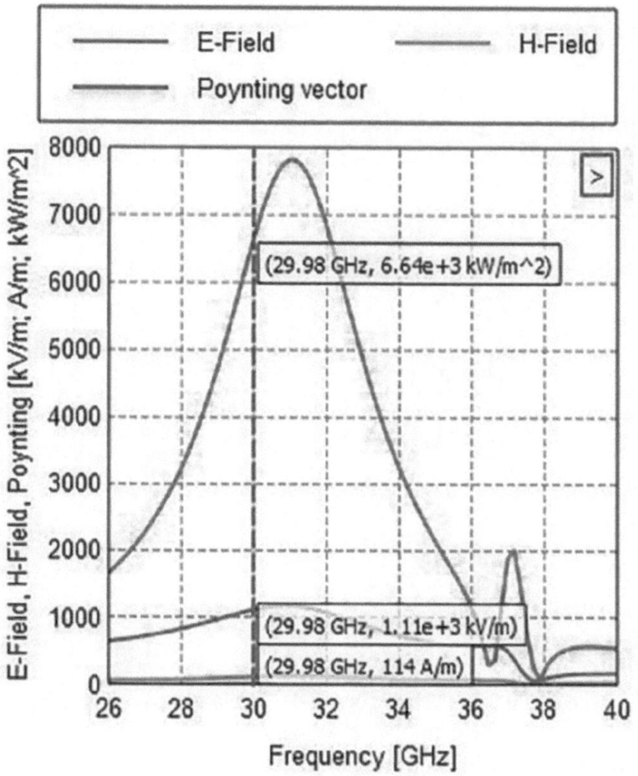

Figure 13.21 Radiation density of triple elliptical patch.

radiation intensity $U(\theta,\varphi)$ over a solid angle of 4π steradians measured in watts and mathematically given as [19]

$$P_{rad} = \oiint_{4\pi} U(\theta,\varphi)ds = 4\pi U_0 \tag{13.10}$$

where U_0 = radiation intensity of the isotropic source.

Radiation Density: Figures 13.8, 13.14 and 13.21 illustrate the radiation density corresponding to the involvement of power via electromagnetic waves in the far-field region of the proposed antenna sui Table 13.for 5G communication. The Poynting vector is defined as [19] and is measured in W/m^2.

$$W = E \times H \tag{13.11}$$

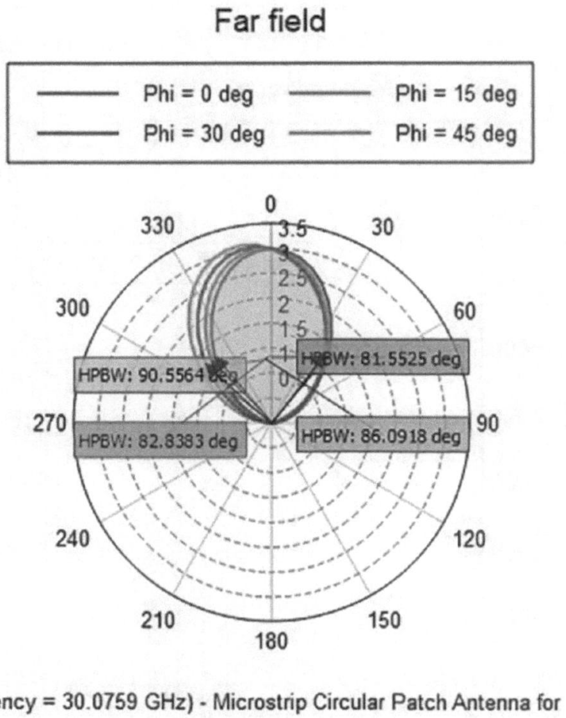

Figure 13.22 Beamwidth with w.r.t theta of double elliptical patch.

Where E is the instantaneous electric field intensity measured in V/m and H is the instantaneous magnetic field intensity measured in A/m. This radiation density measures the power radiations per unit volume in all directions.

Half-power Beamwidth: Figures 13.9, 13.15 and 13.22 illustrate the radiation pattern associated with a half-power beamwidth of 0°, 15°, 30°, and 45°. At phi = 0°, the maximum and narrowest beamwidth are obtained. This half-power beamwidth value indicates the angle of separation between the main lobes at 3 dB points and serves as a trade-off between gain and beamwidth angles [19]. This narrow beamwidth enables stronger connectivity for 5G mobile users, 5G-based IoT, and biomedical electronic systems.

13.3.5 Design of microstrip sensor array (antenna IV)

In this section, the above-mentioned design (antenna III) of the bio-inspired flower-shaped based microstrip patch antenna is converted into a linear array with a size of 3 mm × 2 mm resonating at a 30 GHz millimetre-wave band. To make the array suiTable 13.for advanced 5G applications such as satellite communication, remote sensing and military applications, the transmission line is loaded with a dielectric resonator made of silicon. Additionally, the patch is loaded with circular split-ring-based (SRR) metamaterial composed of copper material with a thickness of 0.035 mm to improve sensitivity and other properties. The array consists of six elements separated by an equal distance *d*, which is half the wavelength. This significantly mitigates the effects of multipath fading.

The top view of the proposed patch array without SRR is shown in Figure 13.23 and 13.24

After conversion to a MIMO microstrip patch antenna linear array of size 3 mm × 2 mm, the designed bio-inspired microstrip patch antenna resonates at dual-band frequencies of 30 and 34 GHz. This dual-frequency band is caused by the structure's design (bio-inspired). This is an excellent achievement as it results in a better return loss and a wide bandwidth at both frequencies. As a result, the return loss and bandwidth are increased. At 30 GHz, the return loss for each element varies between –25 and –36 dB, while at 34 GHz, it varies between –13 and –18 dB. This return loss contributes to the antenna's radiation capacity being maximised. Additionally, the impedance matching bandwidth at 30 GHz is between 1.2 and 1.6 GHz, while at 34 GHz, it is between 400 and 600 MHz. This additional bandwidth will enable 5G MIMO to maximise channel capacity and spectral efficiency in comparison to 4G.

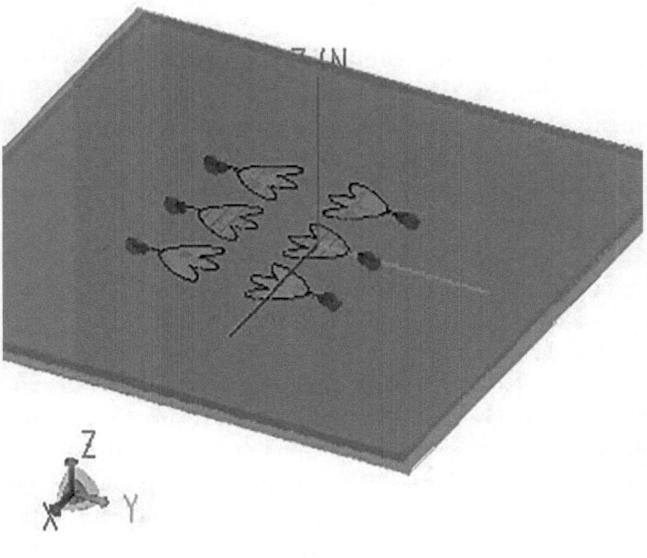

Figure 13.23 Geometrical 3D view of six elements elliptical patch array.

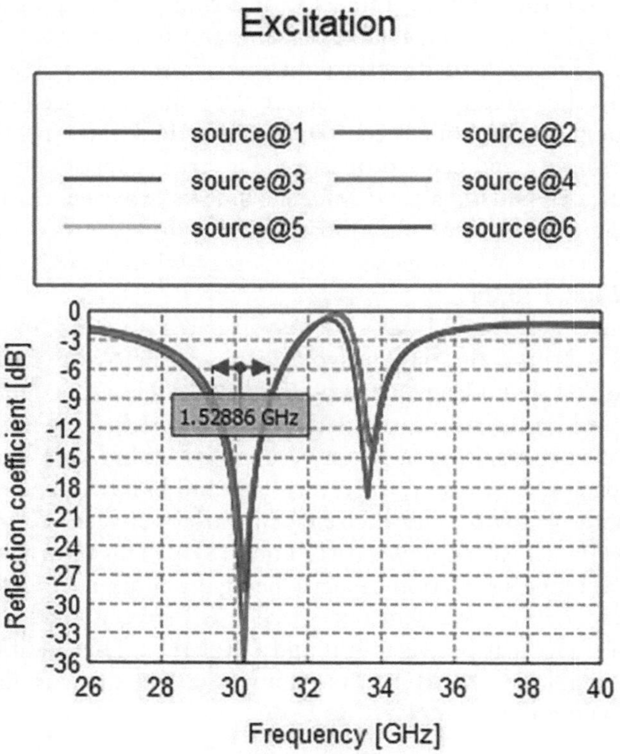

Figure 13.24 Return loss and bandwidth at −10 dB of six elements elliptical patch array.

Figure 13.28 illustrates the beamforming process with respect to various phi angles. This is because of the phased array concept, in which the antenna's six elements combine to form the primary radiation pattern referred to as the main lobe. This results in the energy radiating in the desired direction with a beamwidth of as little as 27.4° at phi = 0°. However, the side lobes generate unwanted radiation and signal interference [20].

The gain of this array antenna, as shown in Figure 13.25, is up to 14.2 dBi, as calculated using the equations antenna gain, $Gt = 10 \log N + Ge$, where N is the element count and Ge is the element gain [21]. Due to the narrow beams used to achieve the high gains, this antenna is well suited for remote sensing via satellite communication [22]. As a result of this trade-off, the microstrip array antenna's physical function becomes one of increased gain and narrow beamwidth. Figure 13.26 illustrates the total radiated power of 870 mW at 30 GHz and 710 mW at 34 GHz which is shown in Figure 13.27 illustrates the increased radiation intensity after converting to a MIMO array antenna, indicating an increased radiation efficiency suitable for any 5G application.

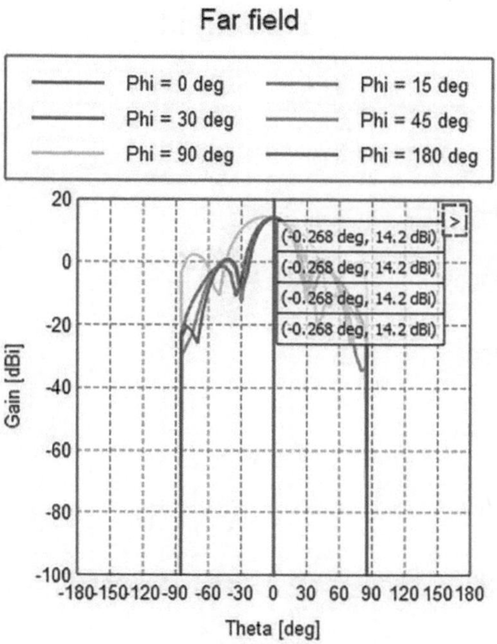

Figure 13.25 Gain w.r.t theta of six elements elliptical patch array.

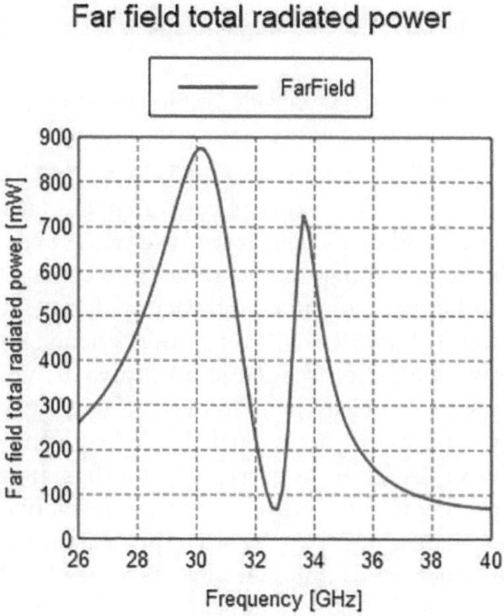

Figure 13.26 Total radiated power of six elements elliptical patch array.

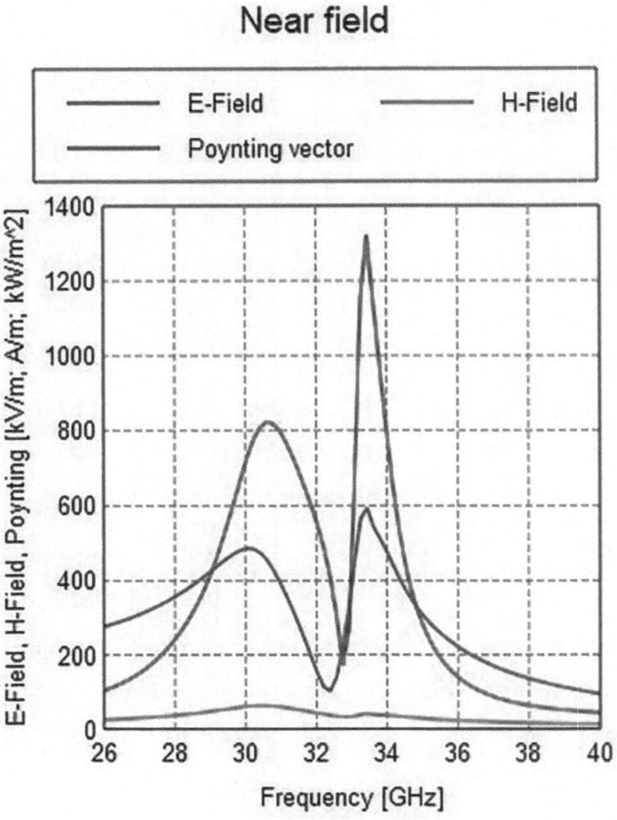

Figure 13.27 Radiation density of six elements elliptical patch array.

The proposed unit cell for metamaterial-based split-ring resonators (SRR) is depicted in Figure 13.29 along with an equivalent circuit containing equivalent inductance (L_{eq}) and equivalent resistance (R_{eq}) forming circular rings coupled with inductance having magnetic flux and capacitance having electric flux and a gap between them as a capacitor (C) and thus forming the RLC equivalent circuit. This SRR will act as an artificial magnetic dipole and will give negative permeability (magnetic property, μ) when a plane wave is an incident on a double negative medium (DNG) as seen in Figure 13.30. Likewise, the thin wire made of copper is loaded into the substrate to give negative permittivity (electrical property, ε), and thus, in combination with SRR and thin wire medium will give a negative refractive index using the equation $n = -\sqrt{\varepsilon\mu}$ indicating that the wave propagation is in a backward direction and the structure of unit cell SRR becomes left-handed metamaterial-based [23]. These phenomena indicate that SRR-based metamaterials are purposefully engineered to possess unusual properties that enhance the reliability of wireless communication technology [24–26].

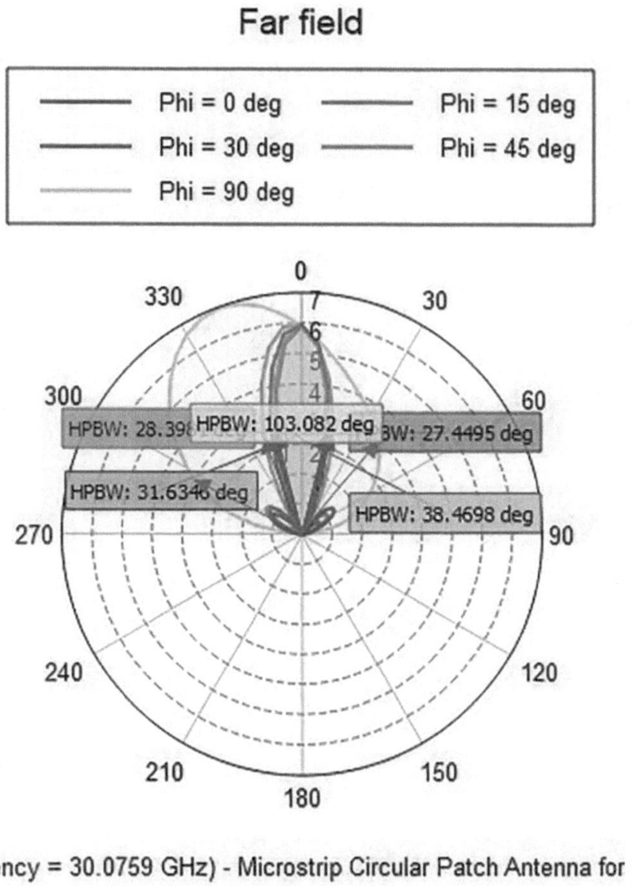

Figure 13.28 Beamwidth with w.r.t theta of six elements elliptical patch array.

The effect of the proposed metamaterial-based split-ring resonator's negative refractive index is depicted in Figure 13.30. This property enhances the performance of the proposed microstrip patch array antenna once it is loaded onto the patch.

The dual-band absorptivity of 84% and 58% shown in Figure 13.31 is calculated using the magnitude of the transmission and reflection coefficients value given by the formula [27].

$$1-|S_{11}|-|S_{21}| \tag{13.12}$$

This absorptivity enhances the microstrip patch array antenna's sensing capability by absorbing electromagnetic waves.

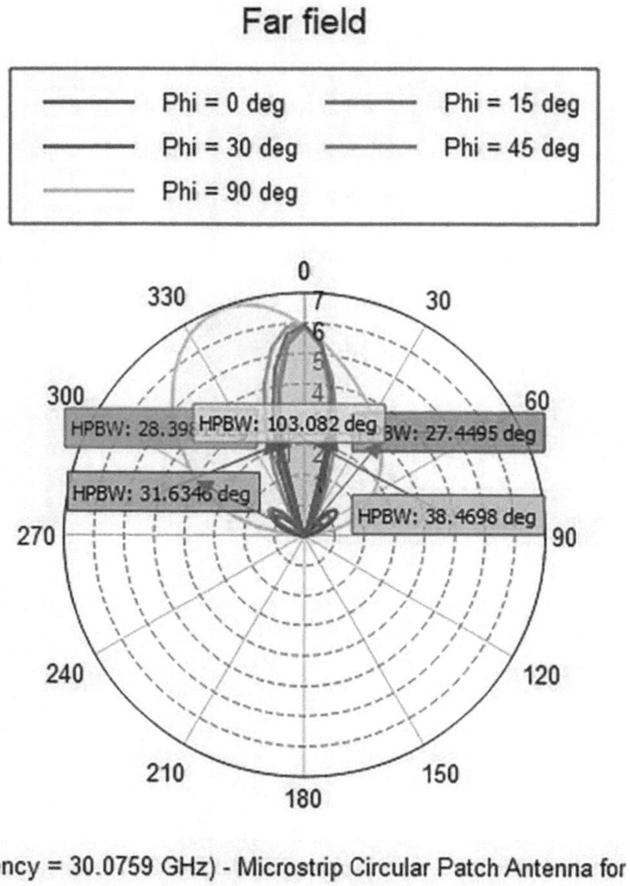

Figure 13.29 Proposed metamaterial-based split-ring resonator with an equivalent circuit.

The top view of the microstrip patch antenna array shows the patch loaded with a metamaterial-based split-ring resonator and the microstrip feed transmission line loaded with an artificial dielectric resonator.

Figure 13.32 illustrates the improvement in return loss and bandwidth at –10 dB following patch loading with a metamaterial-based split-ring resonator. Thus, the proposed microstrip patch array antenna's performance is enhanced and the patch's sensing capability is increased, making it suitable for major 5G applications in satellite communications, healthcare, industry 4.0, and education 2.0, among others.

The improved gain at various theta angles after loading a metamaterial-based split-ring resonator is shown in Figures 13.33 and 13.34. This increases the signal strength for each user and enables faster and more efficient data communication.

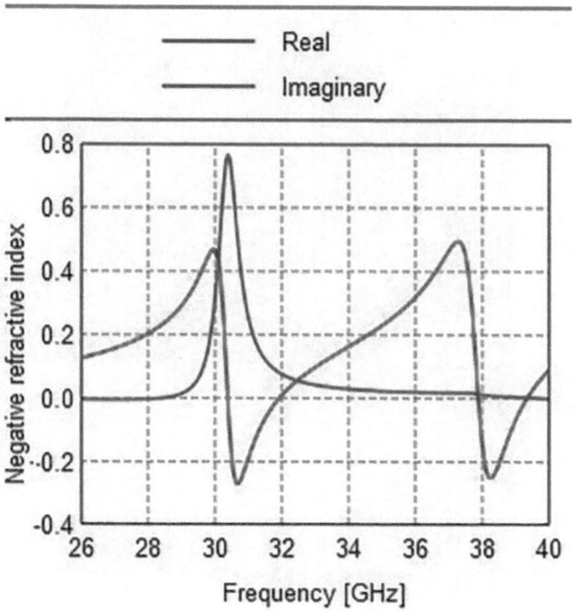

Figure 13.30 Proposed metamaterial-based split-ring resonator with a negative refractive index.

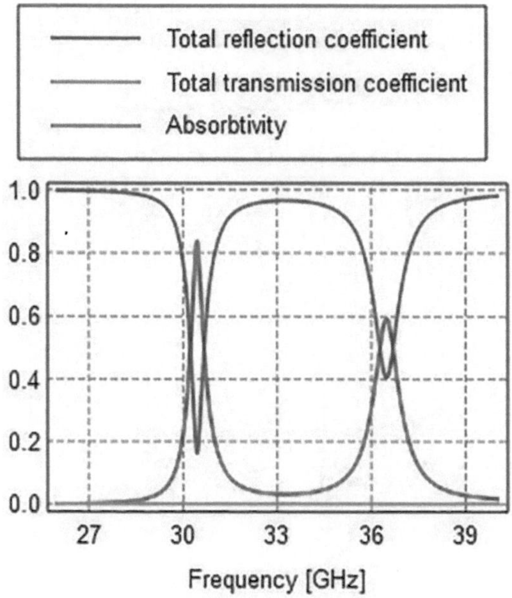

Figure 13.31 Proposed metamaterial-based split-ring resonator absorptivity.

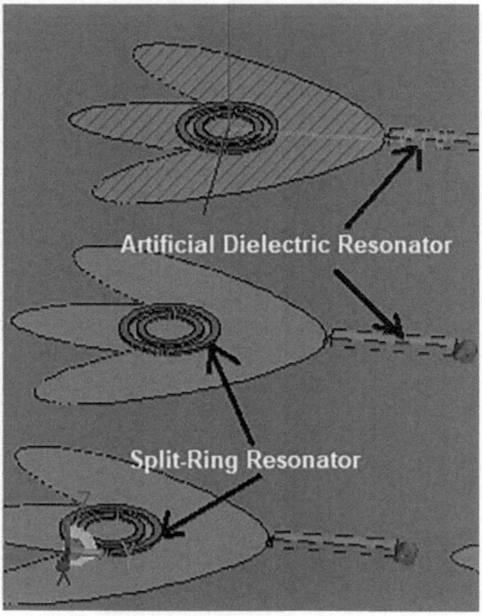

Figure 13.32 Geometrical 3D view of metamaterial loaded six elements elliptical patch array.

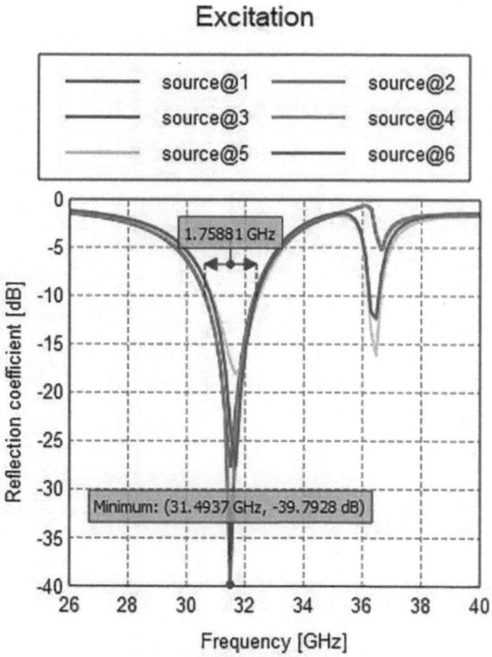

Figure 13.33 Return loss and bandwidth at −10 dB of metamaterial loaded six elements elliptical patch array.

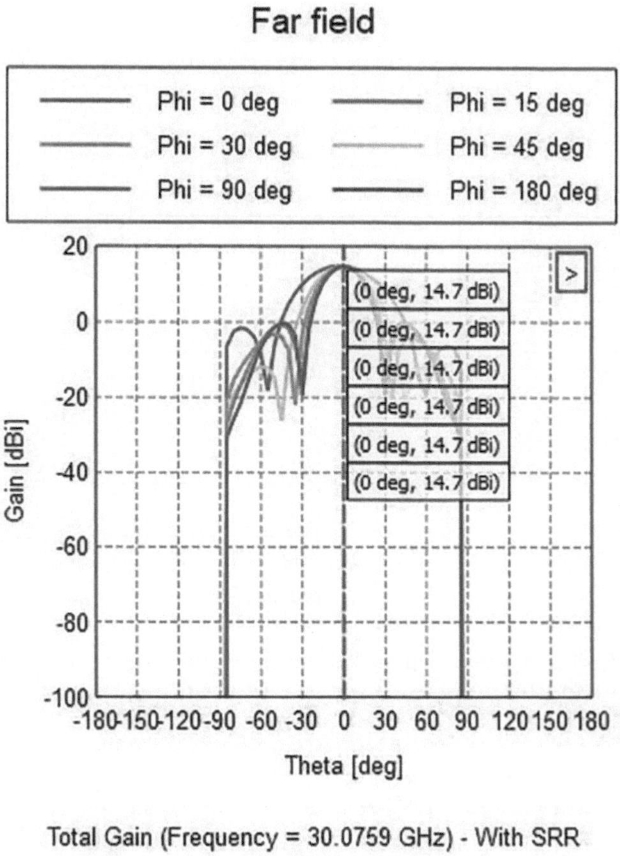

Figure 13.34 Gain w.r.t theta of metamaterial loaded six elements elliptical patch array.

Figure 13.35 illustrates the increased radiation density after loading a metamaterial-based split-ring resonator, indicating that 5G-based wireless machine-machine communication will be significantly more reliable and efficient, particularly in terms of wireless power transmission, which will have a significant benefit in the future for measuring wireless charging of health care and environmental-based systems.

Figure 13.36 illustrates a significant reduction in the minor and side lobes after loading a metamaterial-based split-ring resonator onto a microstrip patch array, thereby, reducing undesirable signal radiations and avoiding extreme signal interference. As a result, the primary beam receives increased focus. Additionally, the beam is steered to the left at phi = 90° and to the right at phi = 270°, with a beamwidth of 44.87°.

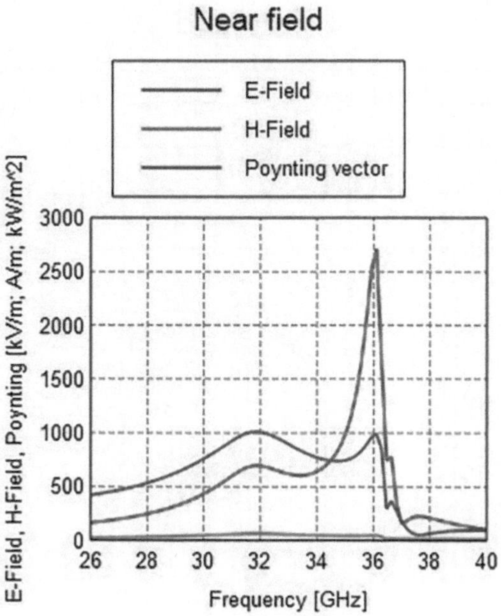

Figure 13.35 Radiation intensity power of metamaterial loaded six elements elliptical patch array.

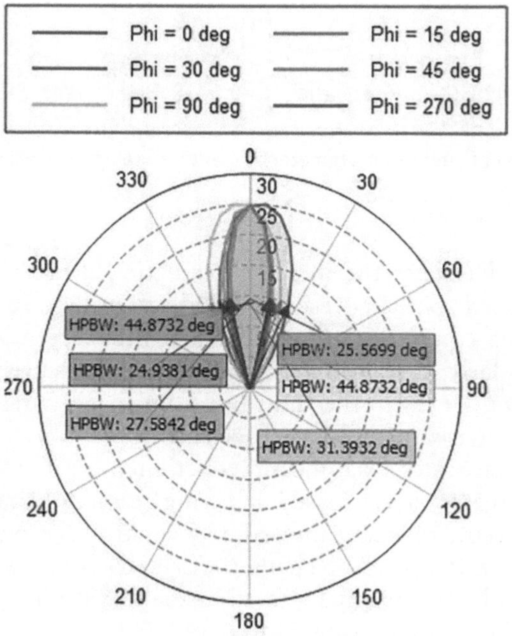

Figure 13.36 Beamforming and beam-steering with w.r.t phi of metamaterial loaded six elements elliptical patch array.

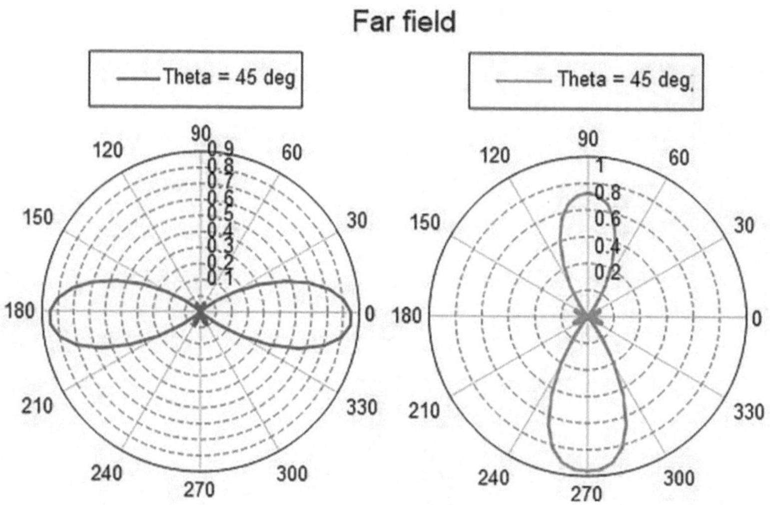

Figure 13.37 Boresight pattern w.r.t phi and theta at $\theta = 45°$ of metamaterial loaded six elements elliptical patch array.

The boresight antenna patterns in Figure 13.37 indicate that the phase shift system has been adjusted to both beams a signal in the expected directions and to cancel out the transmission of an interfering signal from the direction of theta [28].

The phase and frequency shifts caused by loading a metamaterial-based dielectric magnetic resonator made of iron ferromagnetic material on the transmission feed line to alter the phase of the input signal are depicted in Figure 13.38. In this case, unlike with the conventional phase shifters, the phase shift caused by the metamaterial-based dielectric resonator can be both positive and negative (Table 13.2).

13.4 CONCLUSION AND FUTURE SCOPE

The FEKO EM software was used to design and analyse the structure of the rose leaf as a single leaf as well as multiple leaves. In addition, a uniform, linear microstrip sensor array that uses split-ring resonator-based metamaterial and artificial dielectric resonator is built from the patch to the transmission line. The following conclusions were reached because of this analysis:

- The characteristics of the proposed bio-inspired microstrip patch antenna are improving from the design of a single elliptical patch (single leaf) to the combination of multiple elliptical patches (leaves), indicating an improvement in the performance of the proposed microstrip patch antenna with respect to the presence of several patches (leaves).

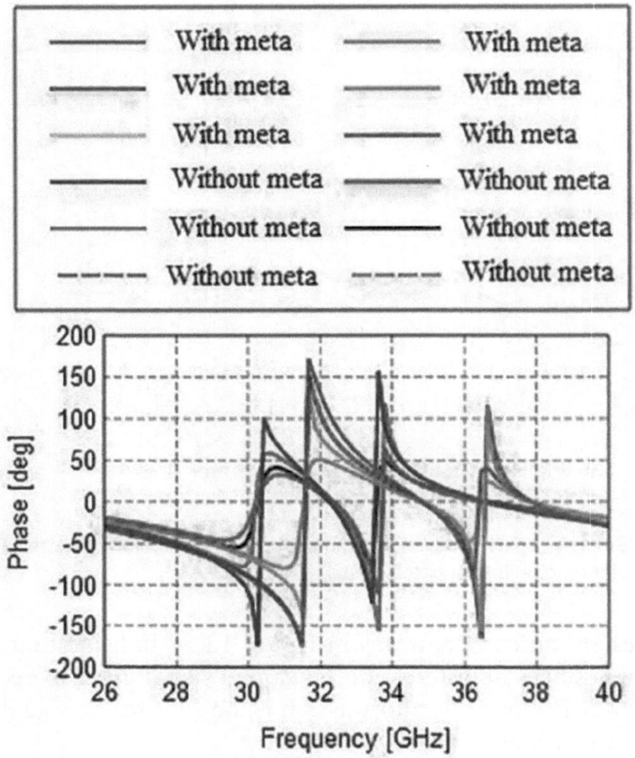

Figure 13.38 Phase shift due to metamaterial loaded six elements elliptical patch array.

Table 13.2 Comparative analysis of all proposed antennas

Characteristics	*Proposed antenna I*	*Proposed antenna II*	*Proposed antenna III*	*Proposed antenna IV (without SRR)*	*Proposed antenna IV (with SRR)*
Return loss (dB)	−16.2	−22.66	−23.20	−25 to −36	−26 to −40
Bandwidth (GHz)	1.19	1.70	1.92	1.2–1.6	1.2–1.8
Gain (dBi)	3.38	5.08	6.01	14.2	14.7
Beam width at Phi = 0 (°)	89.59	80	81.55	27.44	24.93
Radiated power (mW)	542	534	531	851	850
Power density (W/m2)	22,560	50	6640	1350	2550

- Because of the smart beam-steering and smart phase-shifting mechanisms, the proposed bio-inspired microstrip patch array is a smart sensor array.
- The proposed metamaterial-based bio-inspired microstrip patch array antenna performs significantly better with the addition of the metamaterial signifying that the proposed antenna is highly efficient and reliable for 5G-based healthcare systems, IoT-based systems, environmental care-based systems, and other diverse 5G applications. However, the drawback will be mutual coupling effects and isolation between the elements.

The proposed work's future scope includes incorporating electromagnetic bandgap structures between the antenna elements to improve performance, reduce mutual coupling, and improve isolation.

REFERENCES

1. 5G & healthcare: The dawn of 5G technology is here. [online] Available at: https://www.business.att.com/content/dam/attbusiness/briefs/5g-healthcare-ebook-brief.pdf [Accessed 23 Dec 2021].
2. J. Colaco and R. Lohani, Design and implementation of microstrip circular patch antenna for 5G applications, *2020 International Conference on Electrical, Communication, and Computer Engineering (ICECCE)*, pp. 1–4, 2020.
3. J. Gielis, A generic geometric transformation that unifies a wide range of natural and abstract shapes, *American Journal of Botany*, 2003; 90:333–338.
4. P. K. Choudhury and M. Abou El-Nasr, Massive MIMO toward 5G, *Journal of Electromagnetic Waves and Applications*; 34(9):1091–1094.
5. D. A. Hall, Understanding benefits of MIMO technology, *Microwaves and RF*, 2009. Available at: https://www.mwrf.com/markets/article/21846554/understanding-benefits-of-mimo-technology [Accessed 1 Jan 2022]
6. J. Colaco and R. Lohani, High performance and efficient microstrip square patch antenna design for 5G wireless network technology useful for smart TV applications, *IEEE Bangalore Humanitarian Technology Conference (B-HTC)*, India, pp. 1–5, 2020.
7. M. Garg, P. Bhartia, et al., *Microstrip Antenna Design Handbook*, Artech House, London.
8. 5G Americas white paper: Advanced antenna systems for 5G, 2019. [online] Available at: https://www.5gamericas.org/advanced-antenna-systems-for-5g/ [Accessed 1 Jan 2022].
9. J. Colaco and R. Lohani, Design and implementation of microstrip patch antenna for 5G applications, *2020 5th International Conference on Communication and Electronics Systems (ICCES)*, pp. 682–685, 2020.
10. M. Leeladhar and P. Ajay, Highly isolated inset-feed 28 GHz MIMO-antenna array for 5G wireless application, *3rd International Conference on Computing and Network Communications*, pp. 1286–1292, 2020.

11. Rohde and Schwarz, Millimeter-wave beamforming: Antenna array design choices & characterization, white paper, 2016. [online] Avaialble at: https://mtt.org/app/uploads/2019/01/beamform_mmw_antarr.pdf.
12. P. Tiwari and P. K. Malik, Design of UWB antenna for the 5G mobile communication applications: A review, *2020 International Conference on Computation, Automation and Knowledge Management (ICCAKM)*, pp. 24–30, 2020.
13. J. Sifri, Design and simulation of 5G 28-GHz phased-array transceiver, *Webcast, Keysight Technology*, 2017.
14. P. K. Malik and M. Singh, Multiple bandwidth design of microstrip antenna for future wireless communication, *International Journal of Recent Technology and Engineering*, July 2019'; 8(2):5135–5138, ISSN: 2277-3878.
15. A. Khan, S. Bashir, S. Ghafoor and K. K. Qureshi, Mutual coupling reduction using ground stub and EBG in a compact wideband MIMO-antenna, *IEEE Access*, 2021; 9: 40972–40979.
16. T. Prabhu and S. C. Pandian, Design and development of planar antenna array for MIMO application, *Wireless Network*, 2021; 27:939–946.
17. A. Abdulkareem and M. J. Farhan, A novel MIMO patch antenna for 5G applications, *IOP Conference Series: Materials Science and Engineering*; 870(1):012040.
18. P. Mythili and A. Das, Simple approach to determine resonant frequencies of an elliptical microstrip antenna, *Indian Journal of Radio and Space Physics*, 1997; 26:204–207.
19. C. A. Balanis, *Antenna Theory*, John Wiley & Sons, 2012.
20. T-H. Chang, *Ferrite Materials and Applications, Electromagnetic Materials and Devices*, ed. M-G. Han, IntechOpen.
21. K. Benson, Phased array beamforming ICs simplify antenna design, *Analog Dialogue*, 2019; 53(01).
22. 4G Americas MIMO and smart antennas for mobile systems, 2012.
23. K. Gangwar, Paras and R. P. S. Gangwar, Metamaterials: Characteristics, process and applications, *Advance in Electronic and Electric Engineering*, 2014; 4(1):97–106, ISSN: 2231-1297.
24. *Planar Antenna: Design, Fabrication, Testing, and Application*, Nova Science Publishers Inc, New York, 2021, ISBN: 9781536198980.
25. Malik, P., Lu, J., Madhav, B. T. P., Kalkhambkar, G. and Amit, S. (Eds.), *Smart Antennas: Latest Trends in Design and Application*, Springer, doi: 10.1007/978-3-030-76636-8, ISBN: 9783030766368.
26. P. K. Malik, S. Padmanaban and J. B. Holm-Nielsen, *Microstrip Antenna Design for Wireless Applications*, Taylor & Francis, 2021, ISBN: 9780367554385.
27. Z. Weiren, *Electromagnetic Metamaterial Absorbers: From Narrowband to Broadband, Metamaterials and Metasurfaces*, Josep Canet-Ferrer, IntechOpen, 2018.
28. A. Bensky, Chapter 2: Radio propagation, In: *Short-range Wireless Communication*, Third edition, Newnes, 2019, pp. 11–41, ISBN: 9780128154052.

Chapter 14

Development of machine health monitoring and fire alarm using Internet of Things

Ramesh C., Yogeshwaran K., and Udayakumar E.
KIT-Kalaignarkarunanidhi Institute of Technology

R. Gowrishankar
KIT-Kalaignarkarunanidhi Institute of Technology

Madan Singh
National University of Lesotho

CONTENTS

14.1 INTRODUCTION

Electric engines are electromechanical gadgets utilized for converting electric energy into mechanical energy. Engines are the fundamental segment of almost all electromechanical frameworks and have a wide scope of modern applications. The coherence of administration with a significant level of unwavering quality is a significant attribute of the modern framework that requires constant observing of the framework and its segment [1–3]. This energized numerous researchers and designers to investigate mechanical

DOI: 10.1201/9781003347057-14

machines with the end goal to upgrade dependability with the use of issue recognition procedures [4–6]. The extent of model-based deficiency identification strategy is restricted because of the difficult explicit plan structure. The operation of machines under appraised conditions ensures their long service life. However, if they are exposed to excessive heat, over-vibration, high temperatures, or sudden fire due to electrical segments shorting out, their life is significantly reduced, leading to shocking disappointments and the destruction of creation [7–9]. Machine anomalies are associated with various boundaries like motor temperature, coolant oil temperature, motor vibration, and fire [10–12]. High temperature, over-vibration, overheating, and surprising fire are the significant reasons for disappointment in engines [13–15]. As the machine is a blend of numerous parts, all parts should be checked consistently to keep the machine in an ideal working condition. The checking gadgets or frameworks currently utilized for observing machines have a few issues or shortcomings.

As indicated by the above prerequisites, we need machine well-being observing framework to screen all the fundamental boundary's activity, and gathered information's are ship off the checking focus as expected. It prompts web-based checking of the primary utilitarian boundary of the machine which will give fundamental data about the well-being of the machine [16–18] This will help and guide the utilizers to in a perfect world use the machine and save this in movement for a period. An electronic checking system is used to accumulate and examine the different limits of the machine and transported off the noticing center by methods for Wi-Fi.

Fire and smoke murder a greater number of people reliably than various structures consistently produce fake alarms. The proportion of bogus alerts is higher in traditional caution frameworks contrasted with addressable, yet addressable caution fire frameworks are more costly. The most probable reason for a bogus notice is diverse for particular sorts of recognition frameworks, for example, a smoke sensor regularly being enacted erroneously because of an ecological impact [19]. In this way, there is a requirement for a savvy multi-sensors master alert framework that is misleadingly prepared and helps the FDWS (fire location and cautioning situation) to settle on the correct choices and to decrease the number of bogus cautions. Bogus alert admonitions are basic to the point that the London fire unit alone goes almost every 10 min to attend fake alerts causing them an insufficiency to maintain a strategic distance from bogus alerts. The FDWS is prepared with a neuro-fluffy creator. The motivation behind this canny alarm framework is to detect genuine events of fire, alert legitimate specialists, and inform the tenants using GSM to make a fundamental move right away.

A bogus alert can trouble the fire detachment and end up being an exorbitant occasion; countless examinations led to diminishing them. Past investigations proposed various strategies, for example, self-sufficient firefighting robots, alarm frameworks with notice machines, and remote admonition frameworks. Alarm frameworks with notice apparatuses can be exorbitant because they utilize obvious and discernible boosts to tell occupants. The essential goal of this paper is to build up a reproducible and affordable arrangement with the least bogus cautions and a framework that alarms through GSM (worldwide framework for versatile correspondence). The inventive thought is to utilize a neuro-fluffy rationale to plan a keen alert framework.

With the quick improvement of China's monetary turn of events, raised growth of metropolitan people, and the suffering advancement of city building thickness, underground planning, raised structures, and colossal public improvements become to a steadily expanding degree, unbiasedly it progresses more outrageous test to metropolitan fire security. To change as per the current city and the public prosperity of social improvement, the ceaseless removed checking course of action of metropolitan fire affirmation considering IoT is proposed in this paper as an answer to address the above issue. The plan and execution of checking items for the fire assurance framework dependent on IoT are presented in detail in this paper. The IoT terminal is used for fire quenching, and assurance framework. For eg. fire offices caution, hydrant pipe stream, ecological temperature, variable detecting, solid transmission and proficient administration of the fire framework can be accessible. The created observing terminal module can gather the data of real-time caution of fire control bureau, current and voltage of electrical hardware, pipeline stream and weight, surrounding temperature and stickiness, valves switch and transfer activity with the set recurrence. The gathered information including the hub number and time stamp will be pressed as per the Wi-Fi or CAN transport organization. Simultaneously, the camera inside the structure may transfer the gathered realistic data to the client data transmission module through the neighborhood network. RFID fire gear stock activity and upkeep subsystem are remembered for the framework and the stock or upkeep data can be sent to the client data transmission module through the versatile RFID handheld terminal or the USB association port. In the framework, the client data transmission module is identical to the passage which sums up all the checking data in the discernment layer.

In the vehicle layer, we built up the C/S information transmission programming dependent on TCP/IP and windows stage. To guarantee the well-being and dependability of the information transmission, RSA+AES twofold encryption is applied in information transmission [20–22]. Simultaneously, the customer has the capacity for the reconnection of broken organization and the retransmission of impermanent information. The sensor esteem, the video picture just as the RFID data as per the TCP/IP are stuffed, stacked with the GPS stamp, and afterward are transferred to the cloud worker through the switch and web from the client data transmission module. The productive ORACLE away layer is filled in as a capacity information base.

The cloud worker gets the fire-checking data of the structures through web transmission and gives information backing to the accompanying savvy choice and application layer. In the interim, we embrace the innovation of develop load balance, twofold hot reserve, and different innovations to guarantee the well-being and unwavering quality of information stockpiling and virtual capacity innovation to guarantee the adaptable extension of information limit.

In the wise dynamic layer, the fire-checking caution and fire well-being assessment concerning the gathered information are accomplished. By alarm, it implies the procurement to the utilization of electrical gear spillage flow, working voltage, working current, and link temperature esteems as a contribution to producing the fire likelihood by utilizing BP neural organization calculation. Fire well-being assessment is about cosmology displaying, cosmology examination, and thinking dependent ablaze data gathered by the structure to give fire well-being assessment of the checking building [11]. In the application layer, the connection between the observing framework and the client is mostly figured out. The executive staff can see the structure's constant area, pipe stream and weight, temperature and moistness of the climate, current more, voltage of the electric hardware, valve switch, transfer, fire control bureau's continuous alert data, activity and support data, chronicled information, the structure alarm more, fire security assessment data through the site as well as different information reports. Furthermore, observing the site too covers two modules: the client, the executives and the site the executives [23].

The previous is advantageous for the board faculty to add and erase clients, alter client data, oversee client rights, and so forth, while the last is primarily about the executives of each checking site, including essential data about the site of augmentations and cancellations, threshold settings of site cautions just as the expansion and inquiry of site upkeep logs. IoT has a serious level of insight for keeping up with numerous item classifications, amounts, complex fire risk variables, and huge scope of supplies for fire checking and battling. IoT has high versatility and high asset-sharing abilities for dealing with different complex business data. IoT joined with WSN assumes a significant part in the fire alert, fire control office checking, and fire hardware of the board. IoT innovation is joined with fire battling for danger source observing, fire checking, putting out fires salvage, fire early admonition, avoidance, and early removal. It is utilized adequately to upgrade the fire unit's fire terrifying and crisis salvage capacities.

Flame mishaps are turning out to be more arrangement due to greater structure thickness and higher metropolitan structures. Unintentional flames caused 6% of all unnatural passings in India. Detonating cooking gas chambers and ovens represented almost one-sixth of all passings from inadvertent flames somewhere in the range between 2010 and 2014, with an aggregate of 19,491 passings. Electrical short circuits murdered 7,743 individuals over a similar period. Fire mishaps execute 54 individuals every day in India [1] and direct property misfortunes are obscure. To ensure individuals and secure properties from fire, it is important to plan a great genuine time-high solid fire-checking framework. There is a parcel of hindrances in the accessible fire identification, checking, and alert framework. The meager few inconveniences are little observation limit, basic human PC interface framework, poor dependability in location, slow reaction time, and non-adaptable organization interface framework. The customary fire observing framework has bogus negative reactions and bogus positive reactions are high in number. The pace of event of breakdowns in these frameworks are enormous and the time delay in location is intense. It is important to plan a framework to beat these issues and fulfill the application client's necessities.

14.2 RELATED WORKS

The remote framework execution and cost are definitely not great contrasted and the wired framework. Even though the points of interest of remote have many. Some of them are quick speed of establishment, little harms to the structures, remote sensor cost drops down, application places are more extensive, and less force utilization. The remote framework is a basic, adaptable organizational structure, with minimal effort and short delay. It could meet all the prerequisites of the client. The estimating capacity, cautioning of fire, and improvement in the unwavering quality are likewise the upsides of the remote fire caution framework. Lately, fire recognition has become an exceptionally large issue as it has caused genuine mischief tallying the insufficiency of living spirits. Now and again, these scenes are more ruinous when the fire spreads to the ecological variables. The early ID of a fire occasion is a possible method to save lives and lessen property hurt. To move away from a blasting spot and to drench the fire source, the fire should be seen at its beginning stage. The establishment of an alarm structure is the most obliging approach to manage to separate a fire early and evade disasters. Cautions include various contraptions working together that can recognize fire and arranged individuals through visual and sound mechanical gatherings. The affirmation gadgets (i.e., heat, smoke, and gas locators) perceive occasions and start the caution typically, or infrequently, the alarms are impelled genuinely [33].

The alert may incorporate rings, mountable sounders, or horns. By far, most of the ready structures use the improvement of a distant sensor alliance (WSN). WSNs have got conspicuous since they have a strategy of occupations in different applications, for instance, target following [1,2], hindrance, clinical associations, smart transpiration, standard checking, and mechanical robotization. WSN is similarly used in gathering data and checking, both self-rulingly and with the help of customers. WSN applications additionally help humans and animals and are likewise used for mechanical purposes, for example, underground pipeline observing. In a WSN, sensor contraptions every now and then are little, battery-energized, and thickly populated with the advantage of seeing a few imperatives of the climate. The perceived data are passed on off the central social gathering unit (i.e., the sink, bunch head, etc.) for overseeing [8]. WSNs used for fire in an area structure in like manner have indistinct utilitarian properties. Each sensor sees rising warmth, smoke, or gas in unequivocal spots in a home and makes an alert in its brain to place a point in an association. The head place gathers reports from various sensors, and what's more, sees the presence of a fire [34].

Next, different heads figure out the data sources and comprehend with a distant war space to plan a response that may contain in a chief prepared age or then again in muddled clearing systems. Different types of progress reliant on WSN have starting late been proposed to see fire. Some of them are free with WSN and some have cross combination degrees of progress. There are diverse event-perceiving confirmation structures, which help to see warmth, gas, and smoke. Today, bewildering houses and sharp metropolitan associations are furnished with different kinds of WSNs [12]. In WSNs, more energy may be eaten up because of the correspondence overhead. Appropriately, if all else fails, a sensor's battery is drained fast and it may cause the mistake of the sensor or the breakdown of the whole relationship, as houses have gathered sub-packs and each piece is outfitted with one sensor with a specific cutoff, which, if there should be an event of bewilderment, causes a system disfigurement. In the current circumstance, if an event occurs in a particular part and the sensor fails to see the incident, then there is no substitute framework to perceive the event at its hidden stage. As unifunctional sensors are essentially have the choice to recall one event, there is another discernable issue concerning the opportunity of fake alerts. For example, a radiance finder sees temperature in the climate and produces an alert if the temperature increases past an edge [35].

In any case, the growth in warmth may be a quick delayed consequence of standard changes or human activity in the room. Because of smoke alarms, the smoke may come from outside or various sources. Today, sensors are unbelievably simple and insignificant in size. In this manner, to address the starting late alluded to difficulties, we propose a capable, IoT-based sharp home fire presumption system using various sensors. The total of the sensors uses its section for ID. Our strategy confines fire productively and decreases fake positives by using the Global System for Mobile Communications. The commitment of this article is confounded [36].

The proposed need revelation and isolation plot contains three stages: data getting, unite extraction, and multiclass keep up vector machine classifier. This paper looks at single and changed disfigurements in single-stage selection motors including bearing deficiency, load issue, and their blend.

The demonstrating ground incorporates ½ hp, 220V squirrel cutoff enrollment motor with load, vibration sensor, current sensor, data picking up structure, and controller. Two features, standard deviation and regular worth, are figured for each sensor data. Vector machine classifier is executed using an inconsequential exertion Arduino controller for deficiency divulgence and partition. The presentation assessment of the classifier with persevering sensor data is presented which shows pervasive limitations of made framework.

Fire is a catastrophe that isn't compelled by man and is achieved by consuming [7]. The three essential segments of fire are combustible, comburent, and start source. The burnable material is gas, solid, and liquid; "comburent" essentially insinuates oxygen. For burnable gas ignition, as per the combination of flammable gas and air, it very well may be isolated in two unique ways. If it is in the ignition before the air has been blended in with the gas, it is called premixed ignition; if the air and flammable gas don't enter the ignition state simultaneously but are blended and consumed, this is called dispersion burning. Fluid and strong substances are consolidated issues, which are hard to blend in with air. The fundamental cycle of burning is as per the following: when it gets enough energy from an external perspective, the dense issue dissipates into steam or breaks down, and the flammable gas atoms, cinders, and unburned matter particles are suspended noticeable all around called mist concentrates. Typically, vaporized particles are moderately little. During the creation of mist concentrates, huge atoms of strong or fluid particles are delivered simultaneously, known as smoke. Regardless, when consuming, warmth will be created, causing a temperature rise while producing a ton of smoke; with the temperature, weight, and smoke dust boundaries, it very well may be resolved whether the fire happened [8].

By and large, combustibles while consuming produce the accompanying a few types of articulation as appeared in Figure 14.1: for fluid and strong ignitable materials, the principal created is a flammable gas, trailed by smoke; in the event of adequate ignition, the gas must be completely singed, delivering a ton of warmth to advance the current surrounding temperature. During the time spent on fire, the underlying stage delivers an enormous measure of smoke, however, the temperature isn't extremely high. On the off chance that the finder at this stage starts to test, you can limit the misfortune brought about by the fire. After the fire begins, the fire will immediately spread and produce a great deal of warmth to the current climate, expanding the temperature and consuming oxygen, so the gaseous tension is decreased. On the off chance that as of now the current temperature and weight can be viably distinguished, the fire can be controlled. At present, some new smoke alerts use LoRa to accomplish remote correspondence, and they don't have to introduce wires. This takes care of the issue of troublesome establishment of customary smoke cautions and empowers individuals to see the status of the alert distantly to guarantee the idealness of the caution. In any case, there was no adjustment in the manner in which the fire was judged, and the exactness of the caution was not improved.

For the principal question, we accept that utilizing the AI characterization calculation and an assortment of sensors to screen the most extreme genuine climate can reestablish the genuine scene of the climate roughly and extraordinarily upgrade the precision of fire notice while lessening bogus, exclusion, and late cautions' recurrence. When arranging the correspondence structure, we need it to have the alternative to help multipoint data transmission with low flightiness, insignificant exertion, and high faithful quality. Along these lines, we choose to use ZigBee. Also, we arranged and executed data portrayal on the web to ensure that the customers can screen their homes remotely. Examinations show that the insightful smoke-ready system has a high faithful quality in data transmission and caution, can screen various scenes at the same time, and has high practicability. In this paper, the start cycle is bankrupt down, and a variety of WLAN advances and ML computations are examined. The sensor sort of the structure is given, and the accommodation and steady nature of the system are attempted through reenactment tests. Finally, the end is given and the improvement plot is progressed.

A large portion of the disappointments in the modern frameworks are because of engine flaws which can be calamitous and cause significant personal times. Consequently, ceaseless well-being observing, exact issue location, and advance disappointment cautioning for engines are vital and financially savvy. The distinguishing proof of engine flaws requires complex sign handling strategies for speedy issue location and detachment. This paper presents a constant well-being checking strategy for enlistment engines utilizing the design acknowledgment technique. The irregularity has appeared as of late referred to as predefined headings. This adaptable structure will assist the utilities with ideally using transformers and perceiving issues before any ruinous dissatisfaction happens. This structure will be a high-level improvement to mechanization by decreasing human reliance as it is a faraway presentation.

14.3 SYSTEM DESIGN

The machine is the most fundamental resource in any modern organization and it needs exceptional consideration. Along these lines, in the new years, machine well-being checking happens in the enterprises. This framework utilizes numerous sensors to screen the machine's boundary and contrast the current qualities and the predefined limit esteem and if any of the boundaries surpass the worth, it will demonstrate through LED on the machine itself. The LED will sparkle constantly until the supervisor comes and perspectives it. Along these lines, it takes a long effort to realize that

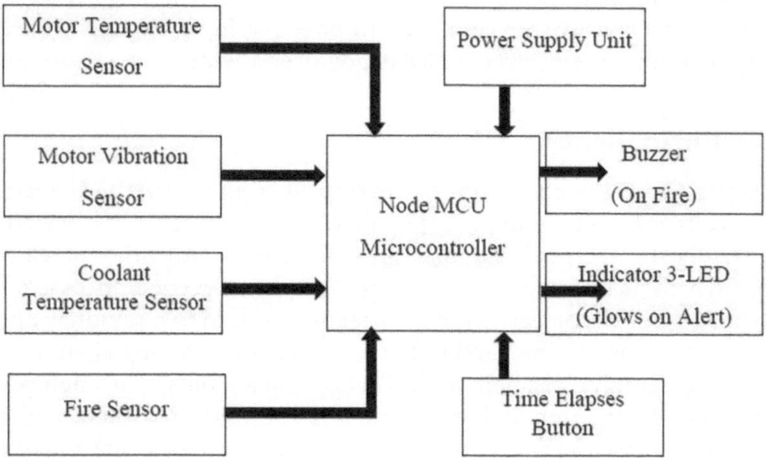

Figure 14.1 Block diagram of machine health monitoring and fire alarm system.

the issue has occurred. Because of the previously mentioned reason, the machine's lifetime gets diminished.

Figure 14.1 shows the machine well-being observing framework. We have utilized a miniature control unit hub which controls the signs from the sensors. The signs are communicated to check the focus through the Wi-Fi module, in which signs are put away in a cloud-based IoT framework. The engine vibration sensor gauges the vibration level of the engine in the machine. The engine temperature sensor gauges the temperature of the engine. The coolant temperature sensor quantifies the temperature of the coolant oil utilized in the machine. The fire sensor is utilized to identify the fire in the business. The framework has its limit esteem independently for various boundaries if any of this sensor esteem surpasses the particular predefined esteem. The framework demonstrates it on the LED. Also, the fire is being recognized by the sensor and the signal jumps ON and makes a caution to the individuals in the business. This framework likewise consistently screens the machine and sends the incentive to the observing focus through Wi-Fi. This encourages the concerned individual to keep the machine from significant disappointment.

14.3.1 Temperature sensor

A thermistor is a resistor used to gauge temperature changes, depending on the adjustment in its obstruction with evolving temperature. A thermistor is a mix of the words warm and resistor. Thermistors can be characterized into two kinds relying upon the indication of *k*. If *k* is positive, the obstruction increments with expanding temperature, and the gadget is known as a positive temperature coefficient (PTC) thermistor, posistor. If *k* is negative, the opposition diminishes with expanding temperature, and the gadget is known as a negative temperature coefficient (NTC) thermistor. Resistors that are not thermistors are intended to have the lowest conceivable *k*, so their opposition remains practically steady over a wide temperature range.

14.3.2 Piezo electric effect

Piezoelectricity is the constraint of pearls and certain tasteful materials to make a voltage thinking about applied mechanical weight. Piezoelectricity was invented by Pierre Curie and the word comes from the Greek word *piezein*, which means to pound or press. The piezoelectric impact is reversible meaning the piezoelectric jewels when introduced to a remotely applied voltage, can change shape barely. (For example, the bending is about 0.1% of the fundamental assessment in PZT.) The impact finds huge applications, for example, the creation and affirmation of sound, the season of high voltages, electronic recurrent age, microbalance, and too fine centering of optical gatherings. Arduino Ide.

Loader induces draws start following force on, rapidly, and you have some additional impact memory open to your assignments. Utilizing the (discretionary) even more moderate inner clock choices recommends you can get a decent arrangement on parts, yet likewise on effect. The Internet of Things (IoT) is the relationship between genuine contraptions, vehicles, home mechanical gatherings, and various things gave equipment, programming, sensors, actuators, and connection transparency which engage these things to accessory and exchange data. Everything is amazingly evident through its introduced figure structure at any rate can between work inside the stream Internet establishment. Experts see that the IoT will fuse around 30 billion things by 2020.

The IoT grants objects to be seen or controlled remotely across existing connection structures, making open entryways for more direct association of this present reality into PC-based systems and achieving improved limit, precision, and a monetary bit of breathing space notwithstanding diminished human intervention. Correctly when IoT is loosened up with sensors and actuators, the progression changes into an event of the wider class of mechanized certified structures, which other than sets propels, for instance, shrewd lattices, virtual power plants, equipped homes, quick transportation, and noteworthy metropolitan associations.

"Things," in the IoT sense, can propose a wide combination of contraptions, for instance, heart-checking embeds, biochip transponders on animals, cameras streaming live feeds of wild animals in coastline waters, vehicles with worked in sensors, DNA evaluation devices for trademark/food/microorganism seeing, or field movement devices that help firefighters in pursuit and rescue errands. Authentic specialists propose concerning "things" as an "ambiguous blend of hardware, programming, data, and affiliation." These contraptions accumulate huge data with the help of various existing advances and thus self-rulingly stream the data between various devices.

14.3.3 Wi-Fi wireless networking

Wi-Fi has emerged as irrefutably the most amazing unavailable connection show of the twenty-first century. While other inaccessible shows work better in unequivocal conditions, Wi-Fi progress controls most home affiliations, distinctive business locale, and public hotspot affiliations. A few people erroneously name a wide degree of eliminated structures relationship as "Wi-Fi." When in doubt, Wi-Fi is just one of the various far away kinds of progress.

14.3.4 Techniques for Wi-Fi network operation

Wi-Fi can be designed in one of the two modes called establishment mode Wi-Fi and phenomenally named mode Wi-Fi. Generally, all Wi-Fi techniques use the establishment mode, where customer contraptions inside show up at all cooperate with and present through a focal distant area. Unrehearsed Wi-Fi licenses customers to relate directly to one another without the utilization of a passage.

14.3.5 Wi-Fi hardware

Distant broadband switches commonly utilized in home affiliations serve (near to their different cutoff points) as Wi-Fi ways. Besides, public Wi-Fi hotspots use at any rate one different ways introduced inside the consolidation zone. Little Wi-Fi radios and receiving wires are presented inside cell phones, PCs, printers, and different customer contraptions drawing in them to round in as sort out customers. Sections are configured with network names that customers can find while checking the zone for open affiliations.

14.3.6 Wi-Fi hotspots

Hotspots are structure mode networks proposed for public or metered consent to the Internet. Different hotspot ways use remarkable programming bunches for managing client enlistments and keeping Internet access also.

14.3.7 Wi-Fi network protocols

Wi-Fi includes an information interface layer show that runs over any of the two or three specific certified later (PHY) joins. The information layer underpins a remarkable Media Access Control (MAC) show that utilizes influence shirking frameworks (in reality considered Carrier Sense Multiple Access with Collision Avoidance or CSMA/CA) to help handle different customers on the affiliation passing on right away. Wi-Fi keeps up stations like those of TVs. Every Wi-Fi channel uses a particular recurrent reach inside the more prominent sign of social events (2.4 or 5 GHz). This awards near relationship in close real closeness to give without meddling with one another. Wi-Fi shows besides testing the possibility of the sign between two contraptions and changing the alliance's information rate down if essential to develop relentlessness. The fundamental show thinking is inserted in express gadget firmware pre-introduced by the maker.

14.4 RESULTS

Figure 14.2 setup consists of a motor temperature sensor, coolant temperature sensor, vibration sensor, and time elapses button interfaced with a Wi-Fi module. All sensors perform different functions which are involved in machine health monitoring and fire alarm system. The time elapses button helps us to check the levels of temperature, vibration, and fire per hour. These all will be stored in a server so that the monitoring center can check them at any time.

Figure 14.3 represents the login page of the machine health monitoring and fire alarm system. This page consists of the user name and password features. It helps to prevent the data from unauthorized persons.

Figure 14.4 represents the output for machine health monitoring and fire alarm system. It consists of parameter values like motor temperature, motor

Figure 14.2 Hardware output.

Figure 14.3 Login page.

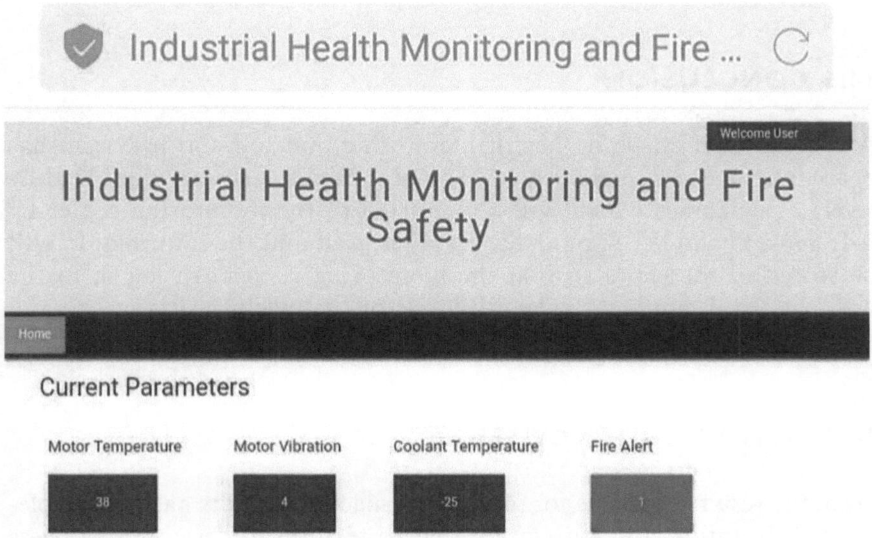

Figure 14.4 Web page for output.

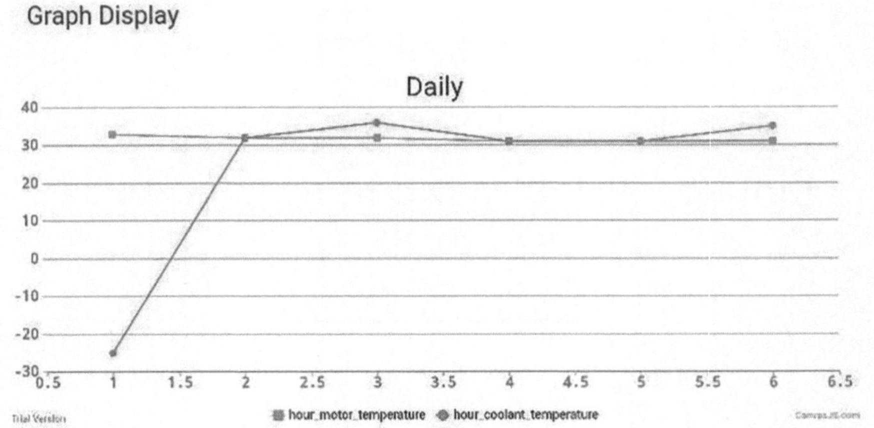

Figure 14.5 Software output graph.

vibration, coolant temperature, and fire detection. This project has the predefined threshold values separately for different parameters. If any one of the parameters exceeds the predefined value, it automatically indicates a red color alert.

Figure 14.5 shows the hour-based reading of the machine's motor temperature and coolant oil temperature. The monitoring center can easily monitor the machines' health conditions through this graph. The graph gives the accurate levels.

14.5 CONCLUSION

A cost-effective machine health monitoring and fire alarm system has been designed and implemented. The proposed system provides healthy levels of machines which will be reported to the monitoring center by IoT. Sensors and Wi-Fi modules are interfaced and the information will be stored in the cloud so that the monitoring center can log in to the web page and monitor the health levels of the machine at anytime and anywhere.

14.6 FUTURE ENHANCEMENTS

From the result, a module for monitoring machines' faults has been implemented. In future, this system updated its features to design the wireless module which will helps in easy monitoring.

REFERENCES

1. A. Abid, M. T. Khan, A. Ullah, M. Alam and M. Sohail, Real time health monitoring of industrial machine using multiclass support vector machine, in IEEE 2017 2nd International Conference on Control and Robotics Engineering.
2. D-H. Kang, M-S. Park, H-S. Kim, D-y. Kim, S-H. Kim, H-J. Son and S-G. Lee, Room temperature control and fire alarm/suppression IoT service using MQTT on AWS, *IEEE*, 2017, ISBN: 978-1-5090–5140–3/17.
3. C. Gajanayake, Bhangu B. S., Foo G., Zhang X., Tseng K. J. and Don Vilathgamuwa M., Sensor fault detection, isolation and system reconFigureuration based on extended Kalman filter for induction motor drives, *IET Electric Power Applications*, vol. 7, no. 7, pp. 607–617, Aug. 2013.
4. H. Ahmed and A. Nandi, Three-stage method for rotating machine health condition monitoring using vibration signals, *IEEE*, 2018, doi: 10.1109/PHM-Chongqing.2018.00055.
5. K. M. Reichard, et al., Application of sensor fusion and signal classification techniques in a distributed machinery condition monitoring system, sensor fusion: Architectures, algorithms, and applications IV, *Proceedings of SPIE*, vol. 4051, pp. 329–336, 2000.
6. S. Rahman, S. K. Dey, B. K. Bhawmick and N. K. Das, Design and implementation of real time transformer health monitoring system using GSM technology, in IEEE 2017 International Conference on Electrical, Computer and Communication Engineering (ECCE), February 16–18, 2017, Cox's Bazar, Bangladesh.
7. H. A. Talebi and K. Khorasani, A neural networl-based multiplicative actuator fault detection and isolation of nonlinear systems, *IEEE Transactions on Control Systems Technology*, vol. 21, no. 3, pp. 842–851, 2013.
8. J. M. Werzer, et al., Diagnostic and condition assessment techniques for condition based maintenance, in 2000 Conference on Electrical Insulation and Dielectric Phenomena, pp. 47–51.
9. J. Yuan, G. Liu, S. Member and B. Wu, Power efficiency estimation-based health monitoring and fault detection of modular and reconfigureurable robot, *IEEE Transactions on Industrial Electronics*, vol. 58, no. 10, pp. 4880–4887, 2011.
10. Barera Sarwa, et al., An intelligent fire warning application using IoT and an adaptive neuro-fuzzy inference system, *Sensors (Basel) Mdpi*, vol. 19, no. 14, p. 3150, 2019.
11. Y. Li, et al., Developing a fire monitoring and control system based on IoT, developing a fire monitoring and control system based on IoT, vol. 133, pp. 174–178.
12. S. R. Vijayalakshmi and Muruganand, Internet of things technology for fire monitoring system, *International Research Journal of Engineering and Technology*, vol. 04, no. 06, June 2017.
13. F. Saeed, et al., IoT-based intelligent modeling of smart home environment for fire prevention and safety, *Journal of Sensor and Actuator Networks*, vol. 7, p. 11, 2018.
14. https://www.hindawi.com/journals/wcmc/2018/8235127/.

15. S. Farhana, M. M. Billah, Z. M. Yusof, K. Kadir, A. M. M. Ali, A. A. Razak, M. Izani, Intelligent, low cost, real time flame alarm system, in 2018 IEEE 5th International Conference on Smart Instrumentation, Measurement and Application (ICSIMA), 2018.
16. N. Savitha and S. Malathi, A survey on fire safety measures for industry safety using IOT, in 2018 3rd International Conference on Communication and Electronics Systems (ICCES), 2018.
17. H. Ahmed and A. Nandi, Three-stage method for rotating machine health condition monitoring using vibration signals, in 2018 Prognostics and System Health Management Conference (PHM-Chongqing), 2018.
18. K. Srihari, et al., Automatic battery replacement of robot, *Advances in Natural and Applied Sciences*, vol. 9, pp. 33–38, 2015.
19. P. Vetrivelan, et al., A neural network based automatic crop monitoring robot for agriculture. The IoT and the next revolutions automating the world, *IGI Global*, pp. 203–212, 2019.
20. P. K. Malik, P. Kumar, S. Kumar and D. K. Singh, *Smart Antennas: Recent Trends in Design and Applications*, Bentham Science, Sharjah, UAE, Aug 2021, ISSN: 2717-5421 (Print), ISSN: 2717-543X (Online), ISBN: 9781681088600.
21. P. K. Malik, Chapter 4. Mathematical modeling and principle of wireless communication, *Energy Harvesting Technologies for Powering WPAN and IoT Devices for Industry 4.0 Up-Gradation*, Nova Science Publishers, Inc., Hauppauge, NY, April 2020, ISBN: 9781536169430.
22. R. Roges and P. K. Malik, Planar and printed antennas for internet of things-enabled environment: Opportunities and challenges, *International Journal of Communication Systems*, vol. 34, no. 15, p. e4940, 2021, https://doi.org/10.1002/dac.4940, ISSN: 1099-1131.
23. A. Rahim and P. K. Malik, Analysis and design of fractal antenna for efficient communication network in vehicular model, *Sustainable Computing: Informatics and Systems*, vol. 31, p. 100586, 2021, Elsevier, https://doi.org/10.1016/j.suscom.2021.100586, ISSN 2210-5379.

Chapter 15

Textile-based wearable antenna for wireless applications

Mehaboob Mujawar
Annamalai University

Subuh Pramono
Universitas Sebelas Maret

CONTENTS

15.1 INTRODUCTION TO TEXTILE AND NON-TEXTILE ANTENNAS

Two Types of Flexible Antennas: Flexible antennas can be broadly classified into two types: antennas that are made from textile materials and antennas that are made from non-textile materials. For antennas to be a part of accessories such as T-shirts, garments, dresses, etc., we need antennas that are designed or fabricated using textile materials. When the antenna has to be

DOI: 10.1201/9781003347057-15

integrated with such type of clothes, the antenna size is not a major issue. Non-textile antennas are integrated with small accessories like buttons, parts of glass, pieces of jewelry, integrated with shoes, belts, etc. In these scenarios, the antenna size should be very small. So, the size of the antenna is very important for antennas that are going to be part of non-textile solutions.

Antenna Fabrication Requirements: From the fabrication point of view, we require at least two materials. One is the non-conductive substrate layer and the second is a conductive material for the conductive layer. We already know that in microstrip patch antenna, we have top and bottom conductive layers and they are separated by a substrate layer. There can be multilayer approaches depending upon the antenna to be designed for specific applications. Non-textile antennas are antennas that use materials other than fabric for the substrate and/or conductive path, so one of the materials is not fabric. In a fully textile antenna, both the substrate and conductive parts are fabricated using textile materials or fabric materials.

15.1.1 Non-textile antennas

In non-textile antennas, there are three broad categories, namely flexible copper antennas, ornamental customized antennas, and plaster-like antennas.

15.1.1.1 Copper antennas

These antennas have a conductive part made of copper. The flexible fabrication techniques similar to semiconductor IC manufacturing technology can be used in flexible copper antennas. The semiconductor IC manufacturing technology needs a clean room. It also needs a different type of controlled environment. All these things add up to the cost. This leads to the high-cost fabrication technique. In the case of copper antennas, there is an oxidation of the materials due to which the performance degrades. The performance degradation of the copper antennas for different types of fabrication techniques is also a major issue. In addition to the semiconductor IC manufacturing type of technology, certain low-cost alternatives can be used for fabricating these flexible copper antennas. Adhesive copper tapes can be used to design the wearable antenna. This is a very popular and low-cost fabrication technique, but in this case, there are flexibility issues. Washing and ironing are other issues which need to be addressed when we are using flexible copper tapes [1]. Copper antennas can also be designed on flexible substrates using copper-clad laminates and a solid ink printer. Adhesive copper sheets also help in fabricating low-cost copper antennas. With adhesive copper sheets, fabrication is very fast but their use is limited.

15.1.1.2 Ornamental antennas

Ornamental antennas are highly embeddable solutions. When we want to embed an antenna in a particular button or any other type of clothing object, we can go for the ornamental antennas. These antennas are specifically designed to copy small clothing objects. In an ornamental antenna, the antenna shape is copied from the clothing object, so that the antenna can be easily integrated over the clothing object while fabrication clothing items are modified, so that they contain the antenna and that particular antenna becomes the part of that particular clothing object.

15.1.1.3 Metal-button antenna

Metal-button antenna is a very common antenna. Generally, these antennas are mounted on the textile jacket or textile clothes. These types of antennas can be fabricated by placing the antenna on the brass button. These antennas can be designed using two copper disks with a thickness of 3 mm and separated by a layer of polytetrafluoroethylene and finally integrated into the brass button. These antennas are designed for the wearable application in the WLAN band, so 2.4 and 5.2 GHz of the WLAN bands were targeted for this typical design.

15.1.1.4 Transparent and flexible monopole antenna

This is another example of an ornamental antenna, where a transparent and flexible monopole antenna is embedded into a glass. This antenna can be made a part of the glass as it is transparent and not easily visible to the user or to the person who is interacting with the user. These antennas are mostly designed to work with a 2.45 GHz frequency. These antennas are fabricated using a transparent and conductive film, which is composed of 100 nm thick indium zinc tin oxide. This is the material which can be used for implementing transparent antennas. The substrate used to design this type of antenna is polyimide.

15.1.1.5 3D-printed antennas

3D-printed antennas are also an important part of ornamental antennas. The implementation of very compact antennas is also possible with the 3D printed antennas because of which it is proving to be a promising fabrication technique. It can be used for a wide range of electronic devices. We can easily fabricate various customized types of wearable antennas using this 3D printing technique. It has a special feature of fast prototyping. We can have reasonable accuracy with the availability of highly accurate high-resolution printers. This technique is a fully 3D topology, where we can print in all three dimensions and all three layers. Also, the cost of fabrication is less in the case of 3D-printed antennas.

15.1.1.6 Plaster-like antennas

These antennas are generally implemented on polymeric substrates, which are used for wearing directly on the skin. Most of the plaster-like antennas are designed to be directly worn on the human skin. Polymeric substrates, especially polydimethylsiloxane (PDMS), have attracted a large interest in the field of plaster-like antennas, which helps in implementing flexible and stretchable antennas. Since these antennas are directly placed on the skin, they are also called epidermal antennas. These antennas can be used for real-time and continuous wireless measurement, therefore, these antennas are specially used in sports and health monitoring applications. Therefore, when they are placed directly on the skin, we can very easily have real-time and continuous measurements of different parameters from the user.

Fabrication Techniques Used in Plaster-Like Antennas: A thin gold film can be used on a PDMS substrate to implement a plaster-like antenna. Conductive silver inks are also used on PDMS substrates. Adhesive copper tapes are integrated with a particular substrate [2–4].

15.1.2 Fully textile antennas

In fully textile antennas, both the substrate and conductive parts are implemented using textile fiber or fabrics. When we have these parts implemented using the fabric or fibers, it is very easy to integrate with the clothes. These antennas are also resultant antennas, which are flexible and lightweight. Both the parts in fully textile antennas are implemented using fabrics, so we need accurate EM characterization of the fabric which is being used for the substrate or the conducting part. In general, electric permittivity and loss tangent are very small for the fabrics. The implementation of antennas with large bandwidth and controlling the antenna radiation characteristics are not possible with the low electric permittivity and loss tangent. Also, the range of permittivity is very narrow for different types of fabrics that are available for the antennas. The thickness of the fabric plays a very important role. The multiple layers of the fabric can be integrated to get the desired thickness for the substrate to implement large bandwidth solutions.

Major Challenges in Fully Textile Antennas: There is a continuous exchange of moisture from the environment with the use of fabrics. Due to the continuous exchange of moisture from the environment, there will be variations in the dielectric properties of that particular fabric. Dealing with the variations in dielectric properties is a major challenge. These fabrics are porous materials, even with the application of a small or low value of pressure, there may be a change in the thickness and density of the particular fabric which is being used for implementing the antenna. There will be variations in the performance of the antenna due to these environmental effects. To determine the behavior of the antenna, there is a need for accurate sensitivity analysis during the designing of the antenna before integrating that antenna with the wearable device. There are also issues with integrating different components of the wearable system into the clothes. Two approaches are suggested for easy integration of components with the clothes: the use of tin soldering and SIW technology. Power consumption is also a major issue when an antenna or wearable device is used. Most of the time, these devices are battery-powered. If large power is absorbed or large power is consumed, then replacing or recharging the battery is a major issue. In this direction, energy harvesting techniques have been proposed to charge or power the wireless devices which are integrated with the wearable devices. Fully textile antennas are classified as embroidered antennas, E-textiles-based antennas, inkjet, and screen-printed antennas.

15.1.2.1 Embroidered antennas

These antennas are implemented in the form of embroidery on a particular cloth. The embroidery is done using conductive wires or conductive threads or conductive yarns. These conductive wires or conductive threads may be multifilament threads or monofilament threads. These are the two major categories of these conductive wires or conductive threads. They differ in electrical properties, wearability, comfort, reliability, and mechanical properties. The important research trend in the field of embroidered antennas is to determine the electrical and mechanical properties of the conductive threads. **Important issues:** In the case of embroidered antennas, determining the electrical and mechanical properties of the antennas is a major issue. The electrical resistance at dc is specified in the datasheet prepared by the manufacturer but we need to determine the resistance values and other electrical parameters at microwave frequencies. Similarly, conductive yarns are preferred over other conductive wires. The direction of the stitches also influences the current flow. Generally, the current tends to follow the stitches' direction rather than jumping between the threads. Large density of the stitches leads to the improvement in the antenna efficiency. **Fabrication process:** These antennas typically use digitally controlled sewing machines for complex geometries. These digitally controlled machines need conductive yarns, which have high flexibility and strength, otherwise, there will be breakage of thread during the fabrication process. Due to digitally controlled sewing machines, the threads or conductive yarns are subjected to high tension. Due to fabrication issues, mass production is difficult in the case of embroidered antennas. But it is feasible to fabricate these antennas using sewing machines. Another important requirement for embroidered antennas is that they should be highly embeddable and washable. To achieve these embeddable types of characteristics, we should have a conductive path and textile substrate as one cloth or there should be a seamless integration. Glues and adhesive layers are used to integrate these two layers. Protective layers on the antennas have also been suggested to avoid deterioration of the threads' conductivity.

15.1.2.2 E-textiles-based antennas

Different conductive fabrics differ in conductivity, thickness, flexibility, and strength. The first step is to select the desired conductive fabric for implementing the antenna. There is also reliability of self-adhesive E-textiles, which can be used for fabricating wearable devices. Self-adhesive E-textiles are easy to integrate with the substrate or with the wearable device and it results in a fast fabrication process. But in terms of E-textiles antennas, there is a specific requirement—the electrical surface resistivity should be low. Typically, it should be less than 1 ohm per square meter. There are two main types of E-textile antennas. The first type is when there is a conductive coating or conductive plating. In this type, non-conductive threads are coated with conductive material. In this type of E-textile antenna, there are issues with the coating being non-uniform and the entire structure is not coated. When the conductive coating is used, there might not be a continuous conductive path and also the sheet resistance is more when using this conductive coating. In-homogeneity of the sheet resistance is another major issue. The second type of E-textile antenna is interweaved conductive fibers. Conductive fibers are interweaved with clothes while implementing the textile. There are different types of fabrication methods used in the implementation of E-textile-based antennas. Hand-cutting is one solution, where the desired antenna pattern is cut using manual cutting methods. These antennas are an impracticable solution for complicated geometries. But just for simple geometries, we can easily cut the conductive fibers and the antenna can be designed. For better accuracy, laser machines are suggested. A laser machine results in better accuracy, and most of the time, it is a computer-controlled machine, so it can easily duplicate the geometries and we can produce the desired antennas at a large scale. Another major issue in E-textile-based antennas is the assembly process.

In the assembly process, the conductive part is designed separately from the dielectric, which is a major issue. There are three main assembly methods used: glues, sewing, and adhesive sheets.

A. **Glue-Based Method:** If we want to assemble the antenna with a very thin film, then the glue method is not practical. After the application of glue on the antenna surface, the surface becomes stiff. When the antenna is implemented on cloth, there will be comfort issues, so will have to compromise on the antenna flexibility as well as wearability.
B. **Sewing-Based Method:** Sewing machines are used for non-conductive threads for stitching. We cannot use those types of non-conductive sewing machines to assemble these antennas. Using non-conductive threads antenna is sewed with the substrate material. This method is not recommended for thick substrate because the stitches will exert pressure on the substrate, and sometimes it could permanently compress the substrate, so the desired substrate parameters may vary. There may be drift in the values of permittivity and the resonance frequency.
C. **Use of Adhesive Sheets:** Conductive fibers, which are self-adhesive, are used, which can be easily pasted on the substrate. But in such cases, there are issues with the accurate alignment of different parts of the

antenna. There are certain advantages of the adhesive sheets method, so we have uniformly deposited a thin layer of the conductive textile which is achieved by the ironing process. When we use these types of adhesive sheets, the sheet resistance and substrate permittivity are not affected. The advantages of the adhesive sheets are that they are faster than easier assembly and we can easily scale these adhesive sheets at the industrial level.

15.1.2.3 Non-woven conducting fibers (NWCFs)-based antennas

NWCFs are basically fabric-like materials which are ranged or bonded together by chemical, mechanical, or heating processes. These fibers have properties similar to that of E-textile. There is a wide variety of the types available for non-woven conductive fibers, so we can have adhesive-backed solutions.

15.1.2.4 Inkjet and screen-printed antennas

In this case, the use of conductive ink is proposed to directly print the antenna on the fabric. The major issues faced by these antennas are resolution and surface roughness. When we want to print the antenna geometries, which have very small size elements or parts, the resolution of the printer matters a lot and also the roughness of the surface after printing. The roughness of the substrate on which the printing is done also affects the properties of the fabricated antenna. There are industrial screen printers available which can be used for getting accuracies in the range of 50 and 200 μm. These techniques are more suitable for wearable applications up to 3 GHz, because achieving very high accuracy is not easy using these inkjet and screen-printed types of fabrication techniques. These techniques are suggested to be more suitable for wearable applications, relatively low-frequency applications and not for millimeter-wave applications. There are washing durability issues for conductive inks. The screen-printed antennas show a more stable behavior even after repeated washing.

15.2 LITERATURE SURVEY

Paper [5] presents the design of an antenna using adhesive flexible copper foil and a laser cutting machine. When we have a particular adhesive copper tape or copper foil, cutting the desired antenna pattern is a major issue. Especially when we want to design the antenna for high frequencies, because for high frequencies, the antenna size becomes very small. One of the solutions is the use of a laser cutting machine. They have printed the desired pattern directly on the substrate. Additional fabrication issues may be avoided using this particular type of fabrication machine. In this paper, they have proposed an accuracy level of approximately 10 μm. This design provides a very high accuracy level, which is a major requirement for such high-frequency applications. This type of fabrication technique can easily result in accurate antennas and when we are using this type of laser cutting machine, we can have high reproducibility. We can produce a large number of antennas with the same accuracy but in such type of fabrication techniques, we need to be careful about the power of the laser. Very high laser power can damage the fabric, which is used as a substrate for the antenna. Paper [6] presents the design of an antenna using solid ink printers and flexible PCBs. The wax layer is printed on the flexible PCB. Flexible PCB has copper all over the region. In this paper, they used the wax layer as the sacrificial layer, so that a solid ink printer deposits a particular wax layer in the form of an antenna shape on the flexible PCB. The next step is the wet-etching of the copper. This approach is low-cost and time-saving. In this paper, they have proposed an accuracy level of approximately 250 μm, which is suitable for low-frequency applications like microwave frequency rays applications. Here, the antenna design is relatively large compared to Ref. [5]. In this paper [7], they have designed phase-graded frequency selective surface. There are two layers of frequency selective surface. One is the top layer and the second is the bottom layer. The bottom layer has an array of circular patches and the top layer has circular patches of different diameters that are called graded. The bottom side is designed with circular patches and the patch is basically in the capacitive form. The top side is in the form of circular apertures which forms the inductive grid. The magnitude and phase of the partially reflecting surfaces have been controlled and it helps to achieve the beam tilt. By varying the dimensions of the capacitive grid or inductive grid reflection magnitude and phase of the beam tilt have been controlled. In this antenna design, inductive grids are generally varied by varying the diameter of the top surface to obtain large tilt angles. In this paper [8], the basic idea is the design of a monopole antenna with some gaps. A bandstop filter has been designed to remove the unwanted interaction between the two antennas. The isolation between the ports has been enhanced by 27 dB. To balance the circuit, stubs have been added to the circuit. Therefore, the capacitance and inductance of the circuit are balanced.. The transmission from port 1 to port 2 is quite high initially. After placing the bandstop filter, the transmission from port 1 to port 2 is low. So, the $S11$ characteristics are matching with the $S21$ characteristics after placing the bandpass filter. Therefore, proper isolation is achieved between the two antennas, which is acceptable in the MIMO antenna design [9]. In this paper [10], a reconfigurable MIMO antenna has been designed. The antenna design consists of a four-port system. Isolation between the antenna elements as well as frequency reconfigurable were obtained by including the PIN diode. This means that at a time, the same structure will be working at 2.4–2.5 and 5.1–5.8 GHz—both bands can be tuned electronically. The main aim was to design a four-port MIMO antenna for WLAN frequencies [11]. The design of the antennas with dual bands increases the size of the antenna. The main aim of this paper was to design a compact antenna. Therefore, dual band antenna design has been avoided and PIN diode as a switching element has been used. Switching elements are used for ON/OFF operation. Switching elements make some structures/stubs ON/OFF for a particular interval of time. Hence, both the bands have been easily managed. *S*21 is –7 dB, which means the isolation offered by the antenna was not good. Meander lines have been removed to improve the S-parameters. Hence, with the introduction and removal of meander lines during the ON/OFF time of the PIN diode, isolation as well frequency re-configularity was obtained. The observed S-parameters exhibit good isolation, that is, 15 dB between the antenna elements. In this paper [12], a reflector has been placed in between the two antennas constituting the reflector-based antenna design. Any radiations coming from the antenna will get reflected to the antenna itself because of the reflector. Therefore, the radiations of one antenna element will not get interfered with other antenna elements. The gain and radiation pattern of the antenna are degraded if the antenna elements are not properly analyzed concerning one-fourth of the wavelength. In this paper [13], adjacent antennas are orthogonally placed, which helps to improve the isolation. When antenna elements are placed away from each other, it will help to obtain isolation. But this is not a preferred technique since a lot of space will be utilized. This antenna resonated at 5.2 and 5.6 GHz. The simulated and measured results were similar. In this paper [14], a wideband antenna was designed for a frequency range of 2.5–5.5 GHz. The 2.5–5.5 GHz bandwidth can be used to fulfill the 5G requirements. The same antenna was placed in a 2*2 array form. With a 2*2 array pattern, high isolation was obtained. Instead of placing the filter for reducing isolation, a reflector was used because placing the filter can create a bulky or complicated system. Placing a reflector in the antenna structure resulted in very good isolation of around –20 dB. The surface current distribution on the two-port antenna is at 3.5 GHz, when excited at port 1. With the reflector, there was no surface current, which indicates that isolation was created successfully on the antenna structure. In this antenna design, multifunctional filter was also developed. The multifunctional filter behaves like a bandpass filter, all-pass filter, and bandstop filter. The filter was tuned to a specific range of frequencies to obtain the desired results. The envelope correlation coefficient is less than 0.5. The isolation between the ports is more than 15 dB. The gain of the antenna is 1.5 dBi, and its efficiency is more than 70%.

15.3 ANTENNA DESIGN

As illustrated in Figure 15.1, the suggested antenna is a patch antenna with square slots. Jeans substrate with a dielectric constant of 1.6 and a thickness of 3.5 mm is used to make the antenna. The proposed antenna structure's total dimension is 24×38 mm^2. The dimensions of the proposed

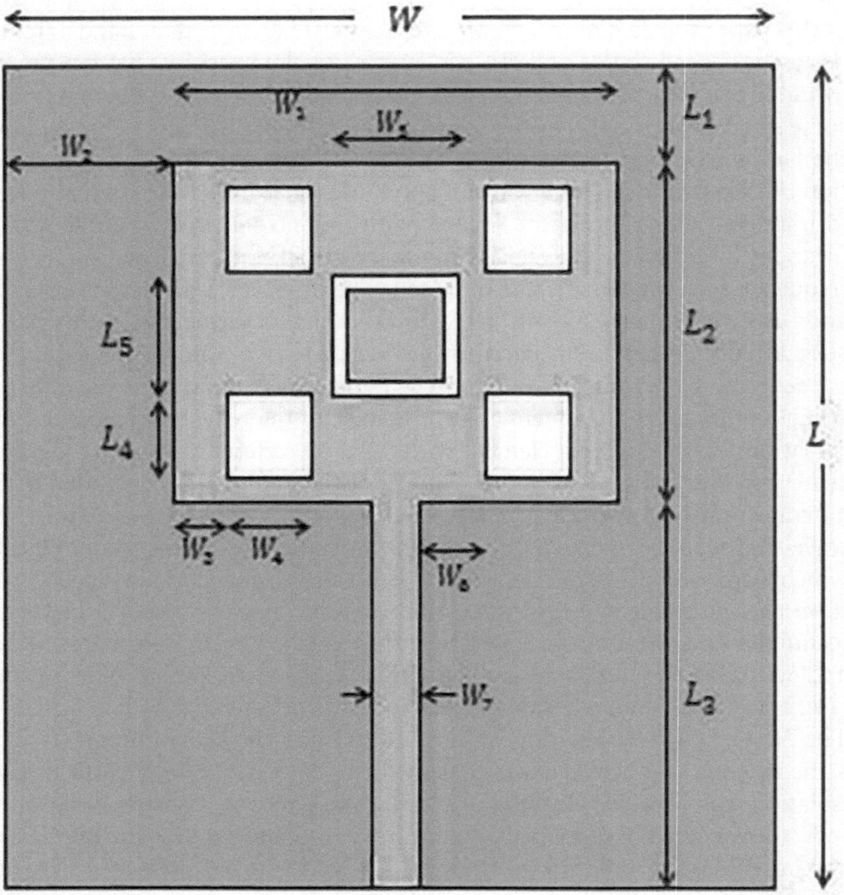

Figure 15.1 Structure of the antenna.

antenna have been listed in Tables 15.1 and 15.2. This antenna resonates at a 2.45 GHz frequency.

15.4 RESULTS

From the simulated results obtained for the proposed antenna, the return loss as shown in Figure 15.2 is obtained to be –29 dB at a resonating frequency of 2.45 GHz. For better performance, the antenna return loss should be less than –10 dB. The simulated results from Figure 15.3 indicate that the gain of the antenna is 27 dB.

Table 15.1 Dimensions of the proposed structure

Dimensions	*Values in mm*	*Dimensions*	*Values in mm*
W	24	W_1	14
L	38	W_2	5
L_1	3	W_3	3
L_2	15	W_4	4
L_3	20	W_5	5
L_4	4	W_6	3
L_5	5	W_7	2

Table 15.2 Dimensions of the proposed structure

Dimensions	*Values in mm*	*Dimensions*	*Values in mm*
W	24	W_1	14
L	38	W_2	5
L_1	3	W_3	3
L_2	15	W_4	4
L_3	20	W_5	5
L_4	4	W_6	3
L_5	5	W_7	2

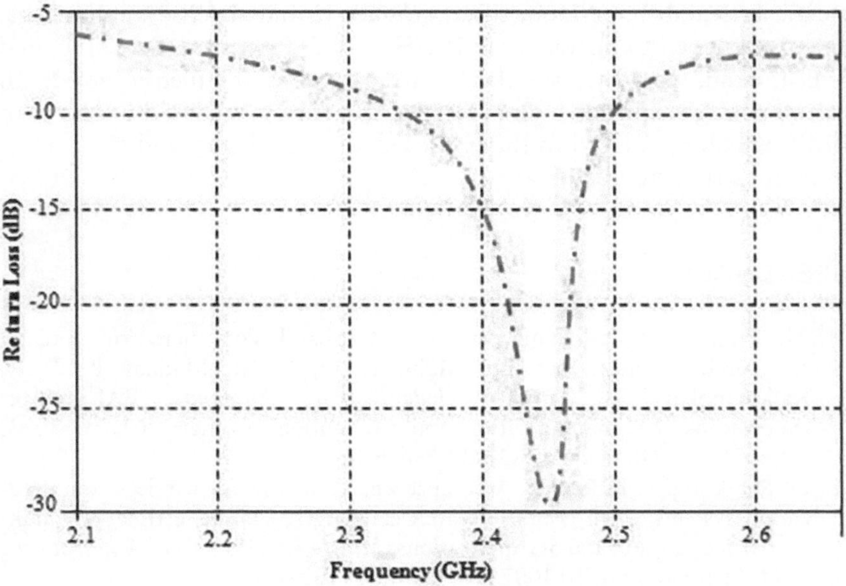

Figure 15.2 Return loss v/s frequency at 2.45 GHz.

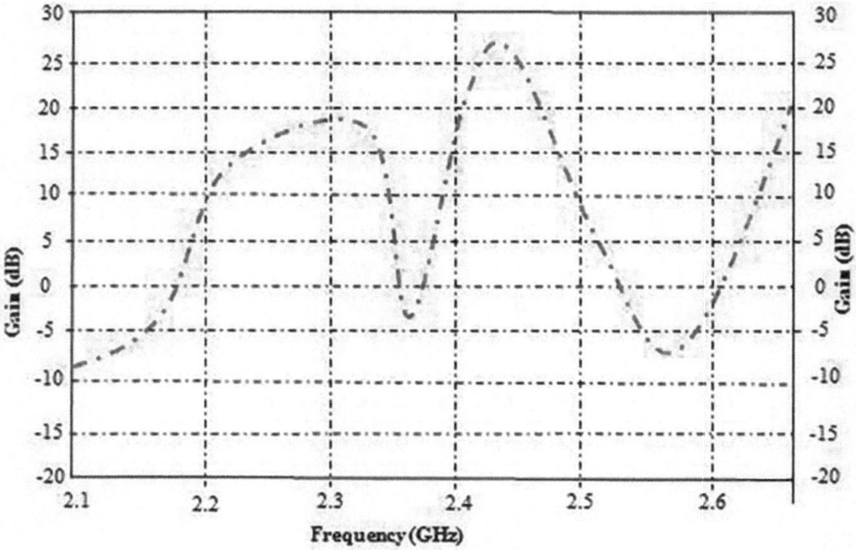

Figure 15.3 Gain v/s frequency for the proposed antenna.

15.5 CONCLUSION

The author offers a new microstrip patch antenna design that is appropriate for wearable applications. The simulated findings show that the proposed antenna is suitable for biomedical, telemetry, and defense applications, which is designed to operate at 2.45 GHz in the Industrial, Scientific, and Medical band, would work well. The return loss is obtained to be –29 dB at a resonating frequency of 2.45 GHz. The return loss should be less than –10 dB, which is satisfied in the proposed antenna design and the gain of the antenna is 27 dB.

REFERENCES

1. M. Mujawar and T. Gunasekaran, Multiband slot microstrip antenna for wireless applications. In: Malik P. K., Lu J., Madhav B. T. P., Kalkhambkar G., Amit S. (eds) *Smart Antennas*. EAI/Springer Innovations in Communication and Computing. Springer, Cham, 2022, https://doi.org/10.1007/978-3-030-76636-8_3.
2. N. Shaik and P. K. Malik, A comprehensive survey 5G wireless communication systems: open issues, research challenges, channel estimation, multi carrier modulation and 5G applications, *Multimedia Tools and Applications*, 2021, https://doi.org/10.1007/s11042-021-11128–z.
3. P. Tiwari and P. K. Malik, Wide band micro-strip antenna design for higher “X” band, *International Journal of e-Collaboration (IJeC)*, vol. 17, no. 4, pp. 60–74, 2021, http://doi.org/10.4018/IJeC.2021100105, ISSN: 1548-3673.

4. D. S. Wadhwa, P. K. Malik and J. S. Khinda, High gain antenna for n260- & n261-bands and augmentation in bandwidth for mm-wave range by patch current diversions, *World Journal of Engineering*, 2021, https://doi.org/10.1108/WJE-03-2021-0133, ISSN: 1708-5284.
5. N. Chahat, M. Zhadobov, S. A. Muhammad, L. Le Coq and R. Sauleau, 60-GHz textile antenna array for body-centric communications, *IEEE Transactions on Antennas and Propagation*, vol. 61, no. 4, pp. 1816–1824, April 2013, doi: 10.1109/TAP.2012.2232633.
6. C. Luca, C. Riccardo and T. Luciano, Smart prototyping techniques for UHF RFID tags: Electromagnetic characterization and comparison with traditional approaches, *Progress in Electromagnetics Research*, vol. 132, pp. 91–111, 2012, doi: 10.2528/PIER12080708.
7. D. Gourab, S. Anand, G. Ravi and S. Mohammad, Performance improvement of multi-band MIMO dielectric resonator antenna system with a partially reflecting surface, *IEEE Antennas and Wireless Propagation Letters*, 2019, doi: 10.1109/LAWP.2019.2938004.
8. T. K. Roshna, U. Deepak, V. R. Sajitha, K. Vasudevan and P. Mohanan, A compact UWB MIMO antenna with reflector to enhance isolation, *IEEE Transactions on Antennas and Propagation*, vol. 63, pp. 1873–1877, 2015, doi: 10.1109/TAP.2015.2398455.
9. M. Mehaboob, Antenna array design for massive MIMO system in 5G application, 2021, doi: 10.1201/9781003175155–18.
10. R. Sreenath, K. Rajkishor and C. Raghvendra, Isolation and frequency reconfigurable compact MIMO antenna for WLAN applications, *IET Microwaves, Antennas & Propagation*, vol. 13, 2019, doi: 10.1049/iet-map.2018.5895.
11. M. Mehaboob, Compact microstrip patch antenna design with three I-, two L-, one E- and one F-shaped patch for wireless applications, 2021. doi: 10.1201/9781003093558–7.
12. R. Sreenath and C. Raghvendra, Mu-negative metamaterial filter based isolation technique for MIMO antennas, *Electronics Letters*, vol. 53, 2017, doi: 10.1049/el.2017.0809.
13. H. Hao-Tao, C. Fu-Chang and C. Qing, A compact directional slot antenna and its application in MIMO array, *IEEE Transactions on Antennas and Propagation*, pp. 1–1, 2016, doi: 10.1109/TAP.2016.2621021.
14. R. Sreenath, A. Mohammad and C. Raghvendra, Four-port MIMO cognitive radio system for midband 5G applications, *IEEE Transactions on Antennas and Propagation*, pp. 1–1, 2019, doi: 10.1109/TAP.2019.2918476.

Chapter 16

Smart antenna for emerging 5G and application

Sandeep Singh Kang, Kiran Deep Singh, and Shalini Kumari
Chandigarh University

CONTENTS

16.1 INTRODUCTION

In optical remote correspondence frameworks, brilliant radio wires, otherwise called different receiving wires, versatile exhibit reception apparatuses, etc. are utilized to boost execution. This works by exploiting the advantages of the variety's impact on the handset of the remote gadget, which is the source and the objective. The word diversity effect refers to the transmission and receipt of intense radio frequencies used during the transmission and reception of data communication to minimize errors and also to improve the

DOI: 10.1201/9781003347057-16

speed of data between the source and the destination. In most wireless communication systems, this type of technology has already found its utility as extraordinary reception apparatus clusters are utilized with calculations for signal preparation that can undoubtedly find and screen different remote targets, such as cell phones. It is also used to calculate the arrival vectors that form the signal's beam and DOA direction [1].

The shrewd reception apparatus framework comprises numerous radio wires or receiving wire exhibits and calculations for advanced sign handling that are in charge of significant capacities, for example, DOA signal assessment. Overall, the phases of development of remote correspondence frameworks can be ordered dependent on the received advancements guided by the difficulties of prerequisites for limited interest and nature of administration (QoS). The stages are summed up as follows:

Omni-directional frameworks: In the base station at the focal point of every cell, with traditional cell structure, recurrence reuse (seven cell reuse designs), omni-directional reception apparatus types.

Cell splitting and sectorized systems: Smaller (micro-cell) cells, cell sectoring at the base station with multiple directional antennas.

Smart antenna systems: Multiple antennas (antenna arrays), creative signal processing algorithms, and beamforming techniques with dynamic cell sectorization (user location-based beam assignment).

The latest developments in telecommunications, such as the Internet of Things (IoT), confirm that people are willing to broaden the existing technologies and hire or build new ones that generate many new criteria and drive connectivity standards beyond the current constraints. In reality, some countries have already completed national IoT networks (such as South Korea and the Netherlands). The practicality (or preparation) of the IoT projects for savvy houses, shrewd urban communities, and vehicles to be presented is along these lines high and can be remembered for the arranged keen receiving wire frameworks examined. For instance, SK Telecom's portable operation operator in South Korea has assembled and finished a long-range wide-region organization (LoRaWAN) zeroed in on long haul development (LTE) foundation (4G or 5G organizations). This IoT network permits public and private savvy gadgets to get and handle information for various and different purposes. These cases may likewise empower and cause extra endeavors and ventures pointed toward upgrading the remote availability of gadgets in such organizations [2].

16.1.1 Difference between conventional antenna and smart antenna

The key difference is related to how both systems deal with the issues created by the propagation of multipath waves. When a wireless signal is transmitted over a wide distance, several obstacles such as high buildings, mountains, utility cables, and so on can have to be passed. The wavefronts of these signals will thus be distributed and will take several paths to reach the receiver. A structure called single info single yield (SISO) is utilized in a customary Wi-Fi correspondence framework, which implies that one receiving wire will be connected to the source and another will be connected to the objective. They can show up blurred, cut-out, and even with basic correspondence issues including picket fencing when the signs show up later than expected at the objective. This is one of the core concerns of the SISO system. So, if in an Internet connection we use the SISO method, the data will arrive late and that is too wrong. With the support of smart antennas, all these issues can be resolved [3].

16.2 SMART ANTENNA – FUNCTIONS

There are fundamentally two essential elements of a brilliant radio wire. They are explained in detail below.

16.2.1 Assessment of direction of appearance (DOA)

Different techniques such as MUSIC (Multiple Signal Classification) and signal parameter estimation through rotational invariance techniques (ESPRIT) algorithms are used for smart antennas to find a signal's DOA. This approach includes a lot of calculations and rhythms. Even the Matrix Pencil approach is widely used to locate the DOA in smart arrays. In real-time systems, the Matrix Pencil form is more widely used as they are highly efficient than the other two. The reception apparatus fills in as a sensor where the spatial scope of the cluster is chosen and the DOA from the pinnacles of that range is found [4].

16.2.2 Beamforming method

By adding the sign stages, the mobiles or focuses at which the signs are to be sent are first looked for and a radiation example of the radio wire cluster is then settled. The mobiles that do not need the signal will be out of the pattern at the same time. While this method can seem a little too complex, it can be easily done with the aid of a delay line filter tapped by FIR. It is also possible to adjust the weight of the FIR filter accordingly to the signal used. To reduce the MMSE between the actual and desired beam designs that are produced, the filters will also be useful in providing optical beam formation [5].

16.2.3 Types of smart antenna

The categorization of smart antennas depends on the type of environment and device specifications. Two kinds of smart antennas are mainly available.

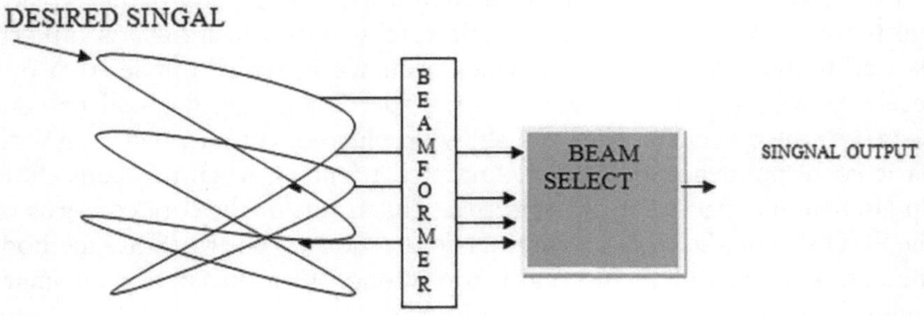

Figure 16.1 Phased array antenna [6].

16.2.3.1 Phased array/beam smart/multi-beam antenna

There will be several fixed beams in this type of array, including one beam that will turn on or be guided toward the desired signal. Only with the aid of phase shift can this be achieved. In other words, the beam will also be steered as the desired target changes. A phased array antenna is shown as a figure-low (Figure 16.1) [6].

16.2.3.2 Adaptive array antenna

In such a radio wire, there will be a change in the shaft plan according to the development of the ideal buyer and the development of the check. The signals obtained will be weighted in addition to the noise and power ratio [S/N] and later combined to optimize the desired interference signal. Since the desired signal will be in the direction of the main beam, balance will also be in the direction of interference. The antenna can quickly redirect the main beam in either direction simultaneously nullifying the interfering signal. It is conceivable to characterize the way the pillar utilizes the DOA strategy. Figure 16.2 of an adaptive array antenna is given below. Capture a sight [7].

The number of inputs and outputs that the system uses is another way of categorizing intelligent antennas. The categories are listed below according to this classification:

1. SIMO (Multiple Output-Single Input)
   One radio wire will be utilized at the source and different receiving wires will be utilized at the destination in this cycle.
2. MISO (Single Output-Multiple Input)
   Numerous reception apparatuses will be utilized at the source in this procedure and just a single receiving wire will be utilized at the beneficiary.

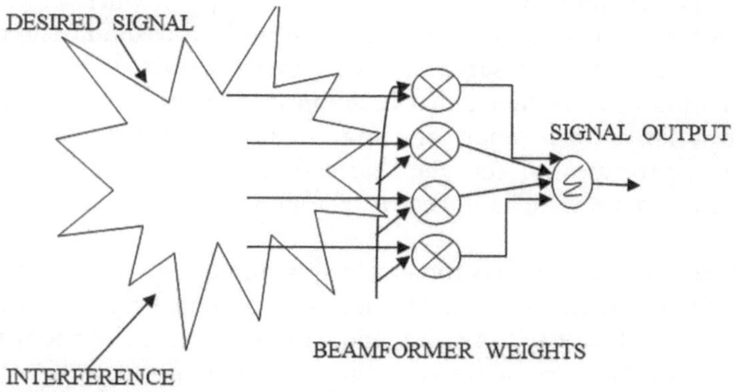

Figure 16.2 Adaptive array antennas [7].

3. Multiple Input-Multiple Output (MIMO)
   Multiple antennas will be used in this phase at both the source and the destination. Among all, this is the most powerful process. This approach explicitly supports spatial information processing [8].

16.2.4 Challenges

This scheme's challenges are as follows:

Restricted Flexibility: The main beam may only be directed in some defined directions and the user doesn't need to be present in the direction of that particular beam, so the overall achievable benefit which not be the gain in the user's actual direction.

Restricted Interference Reduction: Nulls cannot be pointed in random directions in this mechanism, fixing the location and number of nulls. The null can, therefore, not be produced in the direction of the undesirable signal (interferer), i.e., interferer nulling is rather inefficient. Switched-beam antennas seem to be more suitable for CDMA applications for these purposes, where signal enhancement is critical, and not for SDMA applications, where it is important to suppress interference [9].

16.2.5 Advantages and disadvantages

Both beam's intelligent and adaptive arrays provide the desired signal with high efficiency and high power. Beam-smart antennas use narrow pencil beams when a large number of antenna components are used at a greater frequency. High efficiency in the direction of the desired signal is thus obtained. The power gain will be produced at the point when a fixed number of reception apparatus components are utilized a similar number of times with the guide of versatile cluster receiving wires.

The amount of interference that is suppressed offers another gain. With the narrow laser, beam-smart antennas suppress it and versatile exhibit reception apparatuses smother the obstruction by changing the example of the shaft. The main disadvantages are as follows:

Cost:

There would be more costs for such a system, not only in the electronics segment but also in the power field. That is the system, especially if MIMO method is used. It is way too costly and will also reduce the battery life of mobile devices. To decrease the cost, the receiver chains that are used must be decreased. Because of the RF electronics and A/D converters used for each antenna, costs are also rising.

Size:

Large base stations are necessary for this approach to be successful. The size will be increased by this. Apart from this, multiple external antennas on each terminal are needed. This is not realistic. Yet, businesses are seeking to minimize size by using strategies such as dual-polarization.

Diversity:

Where multiple mitigations are needed, diversity becomes a major challenge. There must be many antennas on the terminals and base stations. Three kinds of diversity are primarily present: spatial, polarization, and angle. When used on cell phones, the spatial isolation of the antennas that are used is virtually impossible. In point-to-point systems where a near line-of-sight occurs between the transmitter and receiver, it is also hard to achieve. The above issue can be avoided to a certain extent by using polarized diversity. Without the use of spatial separation, dual-polarization is easy to instigate. The most widely used approach nowadays is angular diversity. From multiple beams, the signals which have the highest signal strength are chosen and used to preserve diversity. But the gain depends on the angular distribution. That is, if the distribution is small, the diversity would be small as well.

- Tracking
- Spatial-temporal processing
- Hooks in international
- Standards to include provisions for smart antennas
- Vertical integration

16.3 SMART ANTENNA TECHNOLOGY OVERVIEW

Because of the advancements and advantages over omni-directional systems, smart (or adaptive) antenna systems are an active research subject today. The IEEE describes a smart antenna system as an antenna system that has circuit elements connected to its radiating elements in such a way that the received signal controls one or more of the antenna properties [10]. Each transmitter located at a certain location has its distinctive pattern in these systems, which is also referred to as a spatial signature [11].

There are many ways of introducing intelligent antennas. A generalized classification is shown in Figure 16.3. Two kinds of intelligent antennas exist—exchanged bar frameworks and versatile cluster frameworks. The exchanged pillar framework includes just basic exchanging between various directional radio wires or cluster predefined radiates while taking into consideration high directivity and gain [12]. It is possible to further split switched-beam systems into two groups: single beam and multi-beam directional antennas. Only one beam at a given time is involved in single beam directional antenna systems. No simultaneous transmissions are permitted because there is only one transceiver in this system, as shown in Figure 16.4a. On the other hand, an example of the Spatial Division Multiple

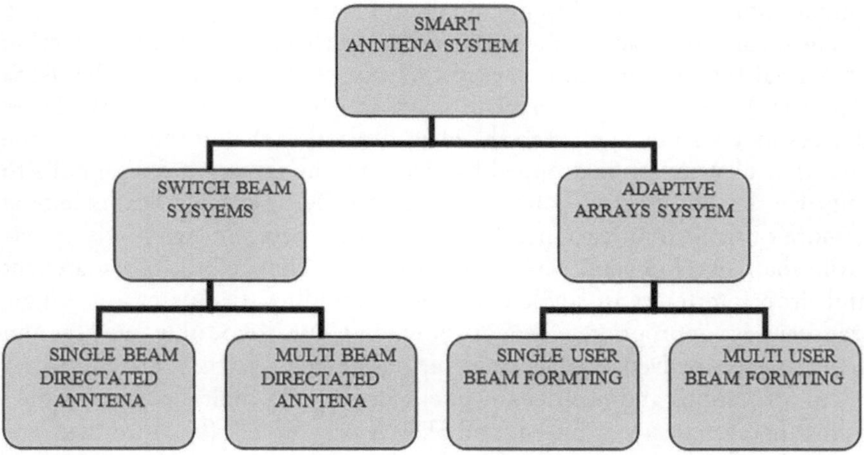

Figure 16.3 Classification of a smart antenna system [11].

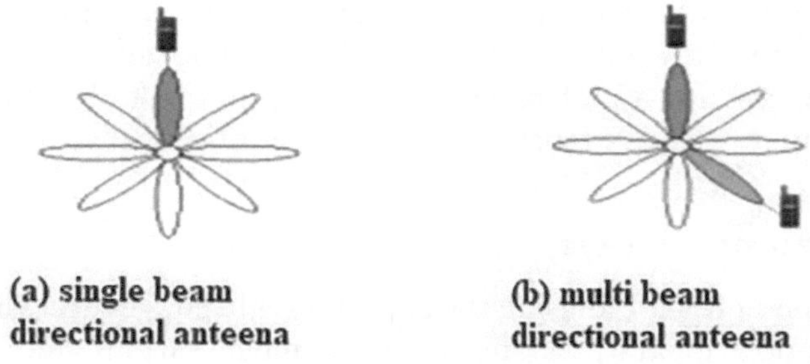

Figure 16.4 Switched-beam-smart antennas [4].

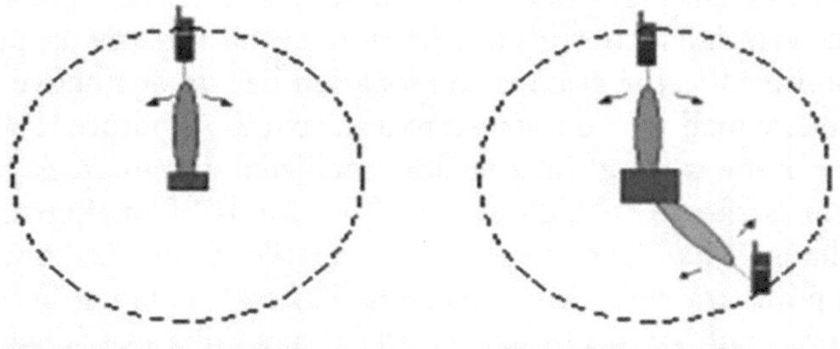

Figure 16.5 Adaptive array smart antennas [14].

Access (SDMA) method is the multiple beam directional antenna system. Here, at the same time and frequency, each directional receiving wire can be utilized and transmissions are permitted. As shown in Figure 16.4b, the number of beams is equal to the number of transceivers [13].

The second approach is the creation of adaptive beams, where a Direction of Arrival (DOA) algorithm is used to evaluate the direction of the user's signal [14]. Continuous control of users can be done in this way. These devices may also be applied to the identification of the interferers with the goal that obstruction is dropped by changing the radiation design nulls to improve the sign to impedance proportion (SIR). Adaptive beam shaping is more difficult than switched-beam systems. There are two kinds of versatile shaft development: The radio wire bar is changed to follow a client and drop interferers in single-client pillar molding. In this case, a single transceiver is appropriate when, as shown in Figure 16.5, only one user at a given time is involved. (a) There are various beam patterns in the multi-user beam formation, and each beam tracks one user. Simultaneous transmissions, therefore, are permitted, and SDMA is achieved. (b) More than one transceiver-beam pair occurs in the creation of multi-user beams, as shown in Figure 16.5b.

The wireless channel's broadcast presence is modified by both kinds of smart antennas. Random MAC access algorithms that rely on link sensing, specifically MAC 802.11, are difficult to implement directly. The real throughput that users will encounter will be determined by the MAC layer. In the ideal case (for a perfect MAC protocol), the N fold would increase the throughput, where N represents the number of spatial channels in use.

16.3.1 Antenna arrays

Effective techniques for abusing the enormous advantages accessible in different reception apparatus systems, for example, various info numerous yield (MIMO) remote frameworks are communicating and getting diversity

[15]. These favorable circumstances incorporate, however, are not restricted to variety acquires (free blurring ways, a decrease of channel changeability), exhibit acquires (normal expansion in the sign to-commotion proportion, beamforming, acquire corresponding to the cluster dimensions), multiplexing acquires (direct expansion in limit or information rate), and impedance acquires (forceful recurrence reuse technique, space-time signal preparing to lessen the obstruction impacts).

The radiation design delivered by a solitary reception apparatus segment is not moderately enormous with low directivity and gain esteems and fewer control capacities over significant boundaries. The augmentation of the receiving wire measurements in mathematical and electrical designs by amassing different emanating radio wire components (clusters) con-recognitions for upgraded order attributes [16–18]. In an exhibit, the receiving wire components in-slowed down might be comparative or extraordinary (same kind of radio wires, for example, dipole, miniature strip, reflector, gap, waveguide, chime, and so forth). Numerous techniques can be utilized to screen and shape the absolute radiation example of the receiving wire cluster, for example:

The mathematical arrangements (straight, organizer, circular, and so forth). The relative distance between the components (area and relocation). The abundance and period of taking care of electrical flow for every radio wire element. The relative radiation example of the individual radio wire component. By design duplication, the absolute radiation example of the reception apparatus cluster of comparable components is gotten where the radiation example of a solitary component set at an audience point is increased by the exhibit factor (AF). The last contention can be well supported and a model can represent it. Consider a direct reception apparatus cluster with a complete number of indistinguishable components equivalent to M with uniform dividing (d) evenly located along with a similar pivot as appearing in Figure 16.1 (outspread distance round directions r, azimuth point φ, and height or polar point φ).

The AF of the linear antenna array presented in Figure 16.6 can be expressed using the following form:

$$\mathrm{AFM} = \Sigma n = 1M/2\omega n\cos\left[(2n-1)\psi n\right] \tag{16.1}$$

where ωn is the amplitude of the feeding electrical current (excitation) for each antenna element and ψn is given by

$$\psi n = \pi d\lambda sin(\theta)sin(\varphi) + \beta n \tag{16.2}$$

where βn is the period of taking care of the electrical flow of the individual component and λ is the frequency (that shows the recurrence connection with the AF definition). In this manner, the absolute radiation design total

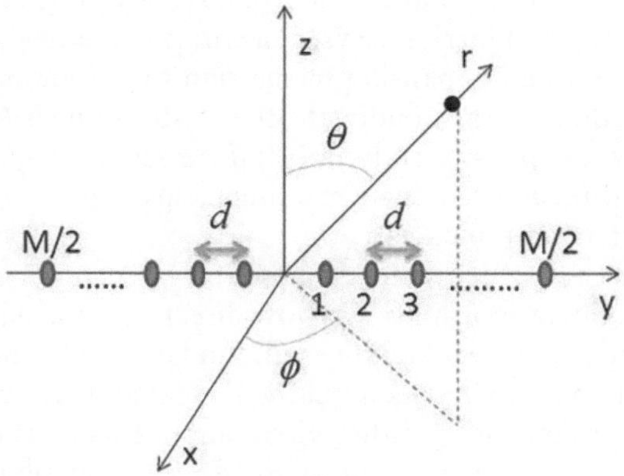

Figure 16.6 Linear antenna array with *M* identical elements [19].

introduced by the adequacy of the electrical field of the straight reception apparatus exhibited in Figure 16.6 is introduced as follows:

$$\text{Etotal} = \text{AFM} \cdot \text{Ese} \tag{16.3}$$

where Ese is situated at the reference purpose of the cluster, the single component radiation pattern. One fascinating perception structure of the last conversation is that by changing the estimations of the AF coefficients ωn and βn, it is conceivable to screen the state of the radiation design in addition to the major to minor flaps level and the examining abilities of the reception apparatus exhibit, individually. Any beamforming procedure can utilize the past control coefficients to shape and divert the radiation flaps or shafts following the buyer area. On account of portable contact, the organizer clusters are supported by the filtering capacities of the three measurements (3D space).

16.4 IMPORTANCE OF 5G NEW RADIO (5GNR)

5GNR can have a range of numerous gigabits every second with much higher information rates. Similarly critical, for some compact or portable applications and administrations, for example, modern computerization, advanced mechanics, haptic Internet, and augmented reality, 5G can support the super low inactivity (under 1-ms delay) required. For a few previously fastened just applications and administrations, for example, 4K video web-based, ongoing controller, haptic interchanges, and that's only the tip of the iceberg, this will permit convey ability/versatility [20].

This will be principally centered around 5GNR's one-of-a-kind abilities that will empower Ultra-Reliable and Low-Latency Communication

(URLLC) for both strategic administrations and shopper benefits that are idleness delicate. Some applications have low dormancy resistance. Video gets nervous and unwatchable if there is an excessive amount of bundle deferral, and basic correspondence, for example, robot tasks, turns out to be completely usable and even risky.

Some URLLC application territories, for example, modern mechanization and mechanical technology, would hence profit essentially from 5G. New use cases and plans of action will emerge, for example, cloud mechanical technology in which a robot can be controlled distantly and without the requirement for restrictive UIs in an untethered manner (for example, uncaged and unconnected by wires). Advancements presently being worked for self-driving vehicles can likewise be utilized to permit 5G-controlled self-sufficient robots. Why is supporting URLLC applications important? For a certain something, by 2023, Mind Commerce sees the worldwide independent 5G-empowered robot market alone as a $14.6 billion opportunity.

16.4.1 5G needs smart antennas

MEC is not sufficient to ensure streamlining, as 5G organizations will also need smart antennas to improve participation, portability, and minimize the 5G to 4G RAN handover requirement. Numerous advanced and demonstrated 5G advances and administrations, for example, extended reality, self-driving cars, wired vehicles, and voice-over 5G are fully essential to providing portability support. Brilliant reception devices are useful for LTE upgrades (Vo5G).

Brilliant Antennas for 5G can improve inclusion and augment power by centering RF signals where they are generally required. Likewise, smart antennas increment 5G gadgets and administration versatility by permitting more touching correspondence, which can turn out to be especially helpful for 5G inclusion creases. Something else, a 5G-driven client experience, would degrade as the handover from 5G to LTE occurs [21].

5G networks are operational and constant voice-over NR (Vo5G) inclusion can be empowered by managing RF where it is required: interchanges. Vo5G calls should be transmitted to LTE without 5GNR for VoNR inclusion, similar to how LTE handover to 3G was portrayed in 4G use. Regardless, there is a large void. Unlike LTE, which is gradually becoming omnipresent (particularly in metro regions), 5GNR will be more constrained, requiring intelligent receiving wires to direct RF signs to improve the quality of VoNR's end-client experience [6].

16.4.2 Working of smart antennas in 5G

To maximize 5G capability and coverage, smart antennas use a few main technologies. Beamforming is one such innovation in which RF energy is moved in a thin bar to exactly where it is required, as opposed to transmitting similar energy in an enormous field. For 5GNR, beamforming is especially valuable as the higher recurrence vmWave RF is liable to remove blurring and deficiency of constriction actuated by ancient rarities hitting (structures, vehicles, foliage, and so on) A more guided RF energy bar guarantees a higher probability of most extreme transfer speed and sign quality. It is imperative to recollect, nonetheless, that view is as yet an issue as beamforming benefits are diminished with lessening [22].

Various reception apparatuses (for example, receiving wire clusters) that utilize Multiple Input/Multiple Output (MIMO) at both the source (transmitter) and the objective (beneficiary) to support the efficiency of the sign are another utilized innovation. This is rather than non-cluster frameworks in which the source and the objective utilize a solitary reception apparatus (and sign way).

As multiple signal paths can compensate for attenuation, MIMO/MIMO is advantageous. While another attenuate is blocked entirely, a given path can experience signal gain. During a given 5G data or voice link, optimal signals will change periodically, meaning that the best signal in an antenna array will change from antenna to antenna.

16.5 5G SMART ANTENNA MARKET SUPPORT APPS AND Vo5G

Furthermore, for voice-over 5G (Vo5G) inclusion, the 5G keen receiving wire market would be basic. If 5G organizations are operational, consistent inclusion of Vo5G can be empowered by coordinating RF where interchanges are generally required. Vo5G calls would be given over to LTE without 5GNR for Voice-over NR (VoNR) inclusion, like how LTE handover to 3G was represented in 4G arrangements. There is a significant gap, however. Unlike LTE, which is becoming increasingly widespread (especially in metro areas), 5GNR will be more restricted to optimize QoE for VoNR, requiring smart antennas to direct RF signals [23].

Adaptive arrays and miniaturized, low-power transmitters provide a few primary technological approaches to developing equipment for the 5G smart antenna industry. Furthermore, keen receiving wires depend on specific methods of RF spread, for example, pillar shaping (shaft directing), as they depend on reception apparatus exhibits to move signals into restricted bars to limit impedance and enhance the sign between the base station and the framework. Brilliant sign preparing calculations permit 5G shrewd reception apparatuses, for example, the heading of appearance, to perceive the spatial sign mark of RF. This is utilized for the estimation of the pillar shaping vectors needed for an objective remote framework to find and screen an RF bar.

RF energy is strengthened in the desired direction with this phased array method, although suppressed in undesired directions. To link simultaneously to two different eNodeB network components, phased array antenna systems may also lock on to two waves (from a given device signal). This offers enhanced signal selection/use management and eventually enhanced compatibility and QoS, which are essential advantages for the 5G smart antenna market.

To boost signal quality, smart antenna arrays use Multiple Input/Multiple Output (MIMO) at both the source (transmitter) and the destination (receiver). This is concerning non-array systems in which the source and the destination are used with a single antenna (and signal path). There are some major variations between the smart antenna market for LTE and 5G. For example, Multi-User MIMO (Mu-MIMO) is leveraged by 5G keen radio wires, while LTE just uses single-client MIMO (Mu-MIMO) (Su-MIMO). As the name suggests, in a multi-client climate, in other words, any 5G versatility environment, Mu-MIMO is utilized to give operational execution. In the examination, some 5G innovations, for example, fixed remote and other single remote arrangements, other single-client use cases approve of just Su-MIMO [24].

Nonetheless, numerous versatile applications should uphold the 5G savvy reception apparatus market, including numerous huge IoT arrangements, for example, the resource following business sector in which many connected business resources are observed at the same time in different areas. As numerous gadgets (cell phones, wearables, and different IoT gadgets) seek a connection, Mu-MIMO permits 5G receiving wires to give upgraded admittance control, wiping out lining delays. This reduces RAN clog and builds throughput, which will give im-demonstrated QoS, particularly, later on, gigantic IoT climate.

It is difficult to grasp the theoretical benefits of MIMO, such as antenna spacing/density, for several reasons. Thus, with cooperative MIMO in mind, radio networks are designed, reflecting collaboration and joint signal processing of multiple base stations with multiple point-to-point radio links to/from wireless devices. To reflect the use of arrays and distributed antenna systems, this radio network planning and engineering approach is often referred to as CO-MIMO or associated with the words Distributed MIMO, Virtual Antenna Arrays, or Virtual MIMO.

Cooperative MIMO is available in three forms: Fixed Arrays, Mobile Relays, and Synchronized Multi-Point (CoMP). For the 5G smart antenna market, the latter has especially high importance as it has tremendous potential for radio network efficiency (capacity and coverage) by exchanging data and channel state information between adjacent cells as a means of coordinating downlink transmission and jointly processing uplink received signals. Performance gains are achieved by converting inter-cell interference into useful signals by CoMP. One downside to CoMP is that data sharing requires high-speed backhaul between base stations [10].

16.5.1 Applications

A processor for space-time ("smart antenna") is capable of forming."Transmit/receive beams of interest toward the smartphone. At the hotel, at the same time, spatial nulls can be located in the direction of undesirable intrusion.

This skill can be used for the improvement of A mobile communication system's efficiency. To enhance most wireless applications, smart antenna technologies can be used, including:

Wi-Fi passage way and customers

In vehicle DBS theater setup, for example,

- Mobile video
- Mobile broadband/gaming
  - Satellite/digital radio
  - GPS
  - 3G wireless
  - Wi-Max
  - RFID
  - UWB

16.6 CONCLUSIONS

This chapter of the book gives a brief idea of the intelligent antenna systems and their forms that we use to escape the multipath and interference from a co-channel. Such antennas have advanced features, such as higher efficiency and higher reliability, than standard antennas.

REFERENCES

1. Kassab, W., and K. A. Darabkh. A–Z survey of Internet of Things: Architectures, protocols, applications, recent advances, future directions and recommendations. *Journal of Network and Computer Applications* 163(2020): 102663.
2. Tyagi, A. K., and S. U. Aswathy. Autonomous intelligent vehicles (AIV): Research statements, open issues, challenges and road for future. *International Journal of Intelligent Networks* 2(2021): 83–102.
3. Teng, K. H., et al. Embedded smart antenna for non-destructive testing and evaluation (NDTandE) of moisture content and deterioration in concrete. *Sensors* 19(3) (2019): 547.
4. Moghadam, G. S., and A. B. Shirazi. Direction of arrival (DOA) estimation with extended optimum co-prime sensor array (EOCSA). *Multidimensional Systems and Signal Processing* (2021): 1–21.
5. Hoshino, K., S. Sudo, and Y. Ohta. A study on antenna beamforming method considering movement of solar plane in HAPS system. *2019 IEEE 90th Vehicular Technology Conference (VTC2019-Fall)*. IEEE, 2019.
6. Choi, Y-S., et al. Power transformer excitation force estimation for load noise reduction using experimental apparatus based on beamforming theory. *Noise Control Engineering Journal* 70(1) (2022): 77–89.

7. Lee, K., and J. Lee. Design and evaluation of symmetric space–time adaptive processing of an array antenna for precise global navigation satellite system receivers. *IET Signal Processing* 11(6) (2017): 758–764.
8. Petrariu, A. I., A. Lavric, and E. Coca. VLC for vehicular communications: A multiple input multiple output (MIMO) approach. *2018 International Conference on Development and Application Systems (DAS)*. IEEE, 2018.
9. Sävje, F., P. M. Aronow, and M. G. Hudgens. Average treatment effects in the presence of unknown interference. *The Annals of Statistics* 49(2) (2021): 673–701.
10. Misra, G., et al. Smart antenna for wireless cellular communication – A technological analysis on architecture, working mechanism, drawbacks and future scope. *2018 2nd International Conference on I-SMAC (IoT in Social, Mobile, Analytics and Cloud) (I-SMAC)*. IEEE, 2018.
11. Ganage, D., and Y. Ravinder. Wavelet-based denoising of direction of arrival estimation signals in smart antenna. *2018 IEEE Global Conference on Wireless Computing and Networking (GCWCN)*. IEEE, 2018.
12. Zhang, J., et al. Prospective multiple antenna technologies for beyond 5G. *IEEE Journal on Selected Areas in Communications* 38(8) (2020): 1637–1660.
13. Chen, Z., D. A. Basnayaka, and H. Haas. Space division multiple access for optical attocell network using angle diversity transmitters. *Journal of Lightwave Technology* 35(11) (2017): 2118–2131.
14. Bloessl, B., et al. Performance assessment of IEEE 802.11 p with an open source SDR-based prototype. *IEEE Transactions on Mobile Computing* 17(5) (2017): 1162–1175.
15. Björnson, E., et al. Massive MIMO is a reality—What is next?: Five promising research directions for antenna arrays. *Digital Signal Processing* 94(2019): 3–20.
16. Amandeep, K., and P. K. Malik. Multiband elliptical patch fractal and defected ground structures microstrip patch antenna for wireless applications. *Progress in Electromagnetics Research B* 91(2021): 157–173, doi: 10.2528/PIERB20102704, ISSN: 1937-6472.
17. Nilofer, S., and P. K. Malik. A retrospection of channel estimation techniques for 5G wireless communications: Opportunities and challenges. *International Journal of Advanced Science and Technology* 29(05) (2020): 8469–8479, ISSN: 2005-4238.
18. Malik, P. K., and M. Singh. Multiple bandwidth design of micro strip antenna for future wireless communication. *International Journal of Recent Technology and Engineering* 8(2) (2019): 5135–5138, doi: 10.35940/ijrte.B2871.078219, ISSN: 2277-3878.
19. Singh, U., and R. Salgotra. Synthesis of linear antenna array using flower pollination algorithm. *Neural Computing and Applications* 29(2) (2018): 435–445.
20. Wang, Z., Y. Dong, and T. Itoh. Miniaturized wideband CP antenna based on metaresonator and CRLH-TLs for 5G new radio applications. *IEEE Transactions on Antennas and Propagation* 69(1) (2020): 74–83.
21. Ziegler, V., et al. Stratification of 5G evolution and Beyond 5G. *2019 IEEE 2nd 5G World Forum (5GWF)*. IEEE, 2019.

22. Wane, S., et al. Millimeter-wave beamformer chips with smart-antennas for 5G: Toward holistic RFSOI technology solutions including RF-ADCs. *2019 IEEE Texas Symposium on Wireless and Microwave Circuits and Systems (WMCS)*. IEEE, 2019.
23. Samanta, D., and A. Banerjee. Transformation of Intelligent IoT in the energy sector. In: *Computationally Intensive Statistics for Intelligent IoT*. Springer, Singapore, 2021, pp. 133–164.
24. Basar, E. Reconfigurable intelligent surface-based index modulation: A new beyond MIMO paradigm for 6G. *IEEE Transactions on Communications* 68(5) (2020): 3187–3196.

Chapter 17

Application of approximation theory in antenna design

Sangeeta Garg
Mewar Institute of Management

CONTENTS

17.1 INTRODUCTION

In this chapter, we consider the original Space Mapping (SM) in order to get a better comprehension of the technique. It concentrates on combative space mapping and its applications in antenna design as in Rahim et al. [13]. Antenna Theory and Microstrip Antennas offer a uniquely balanced analysis of antenna fundamentals Malik et al. [11] and microstrip antennas. Eliminating redundancy, it provides the theoretical background, application materials, and details of recent progress. Exploring several effective design approaches, this chapter covers a broad scope and makes it an ideal hands-on resource for professionals seeking a refresher in the fundamentals. Our primary pivot in this chapter is to introduce specific techniques that enable users to use strong mercantile software packages and computational electromagnetics used in full wave analysis, plane circular disk, and antenna design. It can be noted in Malik et al. [12]. Beyond specific numerical computations for wide concepts, this chapter systematically presents the all-important approximation methods using Taylor's expansion and approach to analyzing micro strip structures including antennas.

Srivastava–Gupta [1] proposed to obtain a family of linear positive operators for $x \in [0, \infty)$ and $f \in C^{\infty}[a, b)$, defined by

$$C_n(f, x) = n\sum_{k=1}^{\infty} p_{n;k}(x, c)\int_0^{\infty} p_{n+c;k-1}(t, c) f(t)\,dt + p_{n;0}(x) f(0) \tag{17.1}$$

where $p_{n,k}(x, c) = (-1)^k x^k k! \phi_{n;c}^{(k)}(x)$ for

- $c = 0 \Rightarrow \phi_{n;c}(x) = e^{-nx}$. Then we obtained Phillips operators.

DOI: 10.1201/9781003347057-17

- $c \in \mathrm{N} \Rightarrow \phi_{n;c}(x) = (1+cx)^{-\frac{n}{c}}$. Then we got Baskakov–Durrmeyer operators.

The sequence of the function $\{\phi_n\}_{n\in N}$ defined as an interval $[0, b], b > 0$ satisfies the following properties for every $n \in N,\ k \in N \cup \{0\}$:

- $\phi_{n;c} \in C^{\infty}[a, b)$
- $\phi_{n;c}(0) = 1$
- $\phi_{n;c}$ is completely monotone i.e. $(-1)^k \phi_{n;c}^{(k)} \geq 0,$
- There exist an integer c such that $\phi_{n;c}^{(k+1)} = -n\phi_{n;c}^{(k)}(x)$

Remark 1. *Functions $\phi_{n;c}$ have many applications in different fields like potential theory, probability theory, physics, numerical analysis, etc. Some interesting properties of such a function are read in* Ref. [2].

Kajla [3] has described the Bezier variant for these operators and obtained the rate of convergence of the functions with bounded variations. Motivated by the sequence C_n, Acu-Gupta [4] also defined a mixed family of summation integral operators with different weight functions. Along with the approximation theory, the genuineness of operators is more important as they are defined implicitly with the value of functions at the endpoints of the interval in which the operators are defined. In the year 2018, Tikhonov-Nguyen [5] introduced such operators and discussed these operators in different forms. Sharma-Garg [6] also worked in this direction and discussed different approximation properties of various operators. Based on two parameters α, β satisfying the condition $0 \leq \alpha \leq \beta$, Stancu-type generalization for many operators has been obtained in the last two decades. Sharma-Garg [7] also worked on this generalization and got many interesting results. In the same way, we have considered Stancu-type generalization of operators (17.1) as

$$V_{n;c}^{\alpha,\beta}(f, x) = n\sum_{\upsilon=1}^{\infty} p_{n;\upsilon}(x, c)\int_0^{\infty} p_{n+c;\upsilon-1}(t, c) f\left(\frac{nt+\alpha}{n+\beta}\right)dt + p_{n;0}(x, c) f\left(\frac{\alpha}{n+\beta}\right) \tag{17.2}$$

where $p_{n;\upsilon}(x, c)$ is defined above. This chapter includes the study of simultaneous approximation for the case $c = 1$ for the operators (17.2). We establish the asymptotic formula of Voronovskaya-type and error estimates for the operators. Garg [8] has, in the same manner, found error estimates for mixed Baskakov-Szasz operators. Motivated by Gupta-Yadav [9], we also obtained moments by using the hypergeometric series function.

17.2 ALTERNATE FORMS

The operators $V_{n;1}^{\alpha,\beta}(f, x)$ for the case $c = 1$ is taken as follows:

$$V_{n;1}^{\alpha,\beta}(f,x)=n\sum_{\upsilon=1}^{\infty}p_{n;\upsilon}(x)\int_{0}^{\infty}p_{n+1;\upsilon-1}(t)f\left(\frac{nt+\alpha}{n+\beta}\right)dt+(1+x)^{-n}f\left(\frac{\alpha}{n+\beta}\right)$$

$$=\int_{0}^{\infty}K_n(x,t)f\left(\frac{nt+\alpha}{n+\beta}\right)dt \tag{17.3}$$

where

$$K_n(x,t)=n\sum_{\upsilon=0}^{\infty}p_{n;\upsilon}(x)p_{n+1;\upsilon-1}(t)+(1+x)^{-n}\delta(t)$$

is the kernel of the operators $V_{n;1}^{\alpha,\beta}$ and $\delta(t)$ is the Dirac Delta function. $p_{n,\upsilon}(x)$ is the Baskakov basis function defined by

$$p_{n;\upsilon}(x)=\binom{n+\upsilon-1}{\upsilon}\frac{x^{\upsilon}}{(1+x)^{n+\upsilon}}=\frac{(n)_{\upsilon}}{\upsilon!}\frac{x^{\upsilon}}{(1+x)^{n+\upsilon}}.$$

$(n)_{\upsilon}$ represents the Pochhammer symbol and is defined as

$$(n)_{\upsilon}=n(n+1)(n+2)(n+2)(n+3)\cdots(n+\upsilon-1)$$

Therefore, the operators can be written as

$$V_{n;1}^{\alpha,\beta}(f,x)$$

$$=\sum_{\upsilon=1}^{\infty}\frac{(n)_{\upsilon}}{\upsilon!}\frac{x^{\upsilon}}{(1+x)^{n+\upsilon}}\int_{0}^{\infty}\frac{(n)_{\upsilon}}{(\upsilon-1)!}\frac{t^{\upsilon-1}}{(1+t)^{n+\upsilon}}f\left(\frac{nt+\alpha}{n+\beta}\right)dt+(1+x)^{-n}f\left(\frac{\alpha}{n+\beta}\right)$$

$$=\int_{0}^{\infty}\frac{xf\left(\frac{nt+\alpha}{n+\beta}\right)}{\left[(1+x)(1+t)\right]^{n+1}}\sum_{\upsilon=1}^{\infty}\frac{(n)_{\upsilon}(n)_{\upsilon}}{(\upsilon-1)!\upsilon!}\frac{(xt)^{\upsilon-1}}{\left[(1+x)(1+t)\right]^{\upsilon-1}}dt+(1+x)^{-n}f\left(\frac{\alpha}{n+\beta}\right)$$

$$=n^2\int_{0}^{\infty}\frac{xf\left(\frac{nt+\alpha}{n+\beta}\right)}{\left[(1+x)(1+t)\right]^{n+1}}\sum_{\upsilon=1}^{\infty}\frac{(n+1)_{\upsilon}(n+1)_{\upsilon}}{(2)_{\upsilon}\upsilon!}\frac{(xt)^{\upsilon-1}}{\left[(1+x)(1+t)\right]^{\upsilon-1}}dt$$

$$+(1+x)^{-n}f\left(\frac{\alpha}{n+\beta}\right).$$

Many authors have studied the hypergeometric series function. First of all, this study was collected by Bailey [9]. Using the hypergeometric series as

$$ {}_2F_1\,(a, b; c\,;x) = \sum_{v=1}^{\infty} \frac{(a)_v\,(b)_v}{(c)_v} \frac{(x)^v}{v!} $$

we have

$$ V_{n;1}^{\alpha,\beta}(f, x) = n^2 \int_0^{\infty} \frac{xf\left(\dfrac{nt+\alpha}{n+\beta}\right)}{\left[(1+x)(1+t)\right]^{n+1}}\, {}_2F_1\left(n+1, n+1; 2; \frac{xt}{(1+x)(1+t)}\right) dt $$

$$ +(1+x)^{-n} f\left(\frac{\alpha}{n+\beta}\right) $$

Applying the Pfaff-Kummer transformation

$$ {}_2F_1\,(a, b, c, x) = (1-x)^{-a}\, {}_2F_1\left(a, c-b; c; \frac{x}{x-1}\right) $$

we get

$$ V_{n;1}^{\alpha,\beta}(f, x) = n^2 \int_0^{\infty} \frac{xf\left(\dfrac{nt+\alpha}{n+\beta}\right)}{(1+x+t)^{n+1}}\, {}_2F_1\left(n+1, 1-n; 2; \frac{-xt}{1+x+t}\right) dt $$

$$ +(1+x)^{-n} f\left(\frac{\alpha}{n+\beta}\right) \tag{17.4} $$

This is the required alternative form of the operators (17.3) in terms of the hypergeometric function.

17.3 AUXILIARY RESULTS

Here, we present some Lemmas used in the proof of the main theorems.
Lemma 1. *For* $\alpha = \beta = 0,\ n > 1$ and $r \geq 0$, *we have*

$$ V_{n;1}(t^r, x) = \frac{(n-r-1)!\,r!}{(n-1)!} nx(1+x)^{r-1}\, {}_2F_1\left(1-n, 1-r; 2; \frac{x}{1+x}\right) \tag{17.5} $$

and also

$$ V_{n;1}(t^r, x) = \frac{(n-r-1)!(n+r-1)!}{(n-1)!^2} x^r + r(r-1)\frac{(n-r-1)!(n+r-2)!}{(n-1)!^2} x^{r-1} $$

$$ +O\left(\frac{1}{n^2}\right) $$

Proof: Taking $f(t) = t^r$ such that $t = (1+x)z$, we get

$$V_{n;1}(t^r, x) = n^2 \int_0^\infty \frac{x(1+x)^r z^r}{[(1+x)(1+z)]^{n+1}} \sum_{v=0}^\infty \frac{(n+1)_v (1-n)_v}{(2)_v v!} \frac{(-x(1+x)z)^v}{[(1+x)(1+z)]^v} dz$$

$$= n^2 x \sum_{v=0}^\infty \frac{(n+1)_v (1-n)_v}{(2)_v v!} (-x)^v (1+x)^{r-n} \int_0^\infty \frac{z^{r+v}}{[(1+x)]^{n+v+1}} dz$$

$$= n^2 x \sum_{v=0}^\infty \frac{(n+1)_v (1-n)_v}{(2)_v v!} (-x)^v (1+x)^{r-n} B(r+v+1, n-r)$$

$$= n^2 x \sum_{v=0}^\infty \frac{(n+1)_v (1-n)_v}{(2)_v v!} (-x)^v (1+x)^{r-n} \frac{(r+v)!(n-r-1)!}{(n+v)!}$$

Using $(n+v)! = n!(n+1)_v$, we get

$$V_{n;1}(t^r, x) = n^2 x \sum_{v=0}^\infty \frac{(n+1)_v (1-n)_v}{(2)_v v!} (-x)^v (1+x)^{r-n} \frac{(r+1)_v r!(n-r-1)!}{n!(n+1)_v}$$

$$= n^2 x (1+x)^{r-n} \frac{(n-r-1)!r!}{n!} \sum_{v=0}^\infty \frac{(r+1)_v (1-n)_v}{(2)_v v!} (-x)^v$$

$$= nx(1+x)^{r-n} \frac{(n-r-1)!r!}{(n-1)!} {}_2F_1(1-n, r+1; 2; -x)$$

$$= \frac{(n-r-1)!r!}{(n-1)!} nx(1+x)^{r-1} {}_2F_1\left(1-n, 1-r; 2; \frac{x}{1+x}\right)$$

using the Pfaff-Kummer transformation. Hence, the first part (17.5) has been proved. Using this result, we can easily find the result (17.6).

Lemma 2. For $0 \le \alpha \le \beta$, we have

$$V_{n;1}^{\alpha,\beta}(t^r, x) = x^r \frac{n^r}{(n+\beta)^r} \frac{(n-r-1)!(n+r-1)!}{(n-1)!^2}$$

$$+x^{r-1} \left\{ \begin{array}{l} r(r-1) \dfrac{n^r}{(n+\beta)^r} \dfrac{(n-r-1)!(n+r-2)!}{(n-1)!^2} \\ +r\alpha \dfrac{n^{r-1}}{(n+\beta)^r} \dfrac{(n-r)!(n+r-2)!}{(n-1)!^2} \end{array} \right\}$$

$$+x^{r-2}\left\{\begin{array}{l} r(r-1)(r-2)\alpha\dfrac{n^{r-1}}{(n+\beta)^r}\dfrac{(n-r)!(n+r-3)!}{(n-1)!^2} \\ +r(r-1)\alpha^2\dfrac{n^{r-2}}{(n+\beta)^r}\dfrac{(n-r+1)!(n+r-3)!}{(n-1)!^2} \end{array}\right\}+O\left(n^{-2}\right)$$

Proof: The relation between operators $V_{n;1}^{\alpha,\beta}(f, x)$ and operators (17.3) can be defined as

$$V_{n;1}^{\alpha,\beta}\left(t^r, x\right)=\sum_{j=0}^{r}\binom{r}{j}\frac{n^j\alpha^{r-j}}{(n+\beta)^r}V_{n;1}\left(t^j, x\right)$$

$$=\frac{n^r}{(n+\beta)^r}V_{n;1}\left(t^r, x\right)+r\alpha\frac{n^{r-1}}{(n+\beta)^r}V_{n;1}\left(t^{r-1}, x\right)$$

$$+\frac{r(r-1)}{2}\alpha^2\frac{n^{r-2}}{(n+\beta)^r}V_{n;1}\left(t^{r-2}, x\right)+\cdots+\frac{\alpha^r}{(n+\beta)^r}V_{n;1}(1, x)$$

and applying (17.6), we get the required result.

Lemma 3. [10] Let $m \in N \cup \{0\}$ and

$$T_{n;m}(x)=\sum_{v=0}^{\infty}p_{n;v}(x)\left(\frac{v}{n}-x\right)^m.$$

Then $T_{n;0}(x)=1$, $T_{n;1}(x)=0$ and there holds recurrence formula for $m \geq 2$ as

$$nT_{n;m+1}(x)=x(1+x)\left[T'_{n;m}(x)+mT_{n;m-1}(x)\right]$$

Consequently, $T_{n;m}(x)=O\left(n^{\frac{-[m+1]}{2}}\right)$, where $[\alpha]$ being an integral part of α.

Lemma 4. For $m \in N \cup \{0\}$, we define the central moments as

$$\mu_{n,m}^{\alpha,\beta}(x)=V_{n;1}^{\alpha,\beta}\left((t-x)^m, x\right)$$

$$=n\sum_{v=1}^{\infty}p_{n;v}(x)\int_0^{\infty}p_{n+1;v-1}(t)\left(\frac{nt+\alpha}{n+\beta}-x\right)^m dt+(1+x)^{-n}\left(\frac{\alpha}{n+\beta}-x\right)^m$$

for $n \geq (m+1)$, following recurrence relation holds

$$(n-m-1)\left(\frac{n+\beta}{n}\right)\mu_{n;m+1}(x) = x(1+x)\left[\mu'_{n;m}(x) + m\mu_{n;m-1}(x)\right] + m\left(\frac{\alpha}{n+\beta} - x\right)$$

$$\times\left[\left(\frac{\alpha}{n+\beta} - x\right)\left(\frac{n+\beta}{n}\right) - 1\right]\mu_{n;m-1}(x) + \left\{\begin{matrix}(nx-1)+(n-2m-1)\times \\ \left(\frac{\alpha}{n+\beta} - x\right)\left(\frac{n+\beta}{n}\right) + (m+1)\end{matrix}\right\}\mu_{n;m}(x),$$

where

$$\mu_{n,0}(x) = 1,\ \ \mu_{n,1}(x) = \frac{nx + (n-1)(\alpha - \beta x)}{(n-1)(n+\beta)} \text{ for } n \neq 1, \text{ and so on}$$

Consequently, for each $x \in [0,\infty)$, we have from the above relation that $\mu_{n;m}(x) = O\left(n^{\frac{-[m+1]}{2}}\right)$.

Proof: The values of $\mu_{n,0}(x)$ and $\mu_{n,1}(x)$ easily follow the definition of operators. To prove the recurrence relation, we use identity

$$x(1+x)p'_{n;v}(x) = (v - nx)p_{n;v}(x).$$

Differentiating the given moment relation and multiplying by $x(1+x)$ on both sides, we get

$$x(1+x)\mu'_{n,m}(x) = n\sum_{v=1}^{\infty}(v - nx)p_{n;v}(x)\int_0^{\infty}p_{n+1;v-1}(t)\left(\frac{nt+\alpha}{n+\beta} - x\right)^m dt$$

$$-nx(1+x)^{-n}\left(\frac{\alpha}{n+\beta} - x\right)^m - mx(1+x)\mu_{n;m-1}(x).$$

Therefore,

$$x(1+x)\left[\mu'_{n,m}(x) + m\mu_{n;m-1}(x)\right]$$

$$= n\sum_{\nu=1}^{\infty} p_{n;\nu}(x)\int_0^{\infty}(\nu - nx)\,p_{n+1;\nu-1}(t)\left(\frac{nt+\alpha}{n+\beta}-x\right)^m dt - nx(1+x)^{-n}\left(\frac{\alpha}{n+\beta}-x\right)^m$$

$$= n\sum_{\nu=1}^{\infty} p_{n;\nu}(x)\int_0^{\infty}\left[\{(\nu-1)-(n+1)t\}+(n-1)t+(1-nx)\right]p_{n+1;\nu-1}(t)\times$$

$$\left(\frac{nt+\alpha}{n+\beta}-x\right)^m dt - nx(1+x)^{-n}\left(\frac{\alpha}{n+\beta}-x\right)^m$$

$$= n\sum_{\nu=1}^{\infty} p_{n;\nu}(x)\int_0^{\infty}t(1+t)\,p'_{n+1;\nu-1}(t)\left(\frac{nt+\alpha}{n+\beta}-x\right)^m dt + (n-1)\,n\sum_{\nu=1}^{\infty} p_{n;\nu}(x)$$

$$\times\int_0^{\infty}t\,p_{n+1;\nu-1}(t)\left(\frac{nt+\alpha}{n+\beta}-x\right)^m dt + (1-nx)\,n\sum_{\nu=1}^{\infty} p_{n;\nu}(x)\int_0^{\infty}p_{n+1;\nu-1}(t)$$

$$\times\left(\frac{nt+\alpha}{n+\beta}-x\right)^m dt - nx(1+x)^{-n}\left(\frac{\alpha}{n+\beta}-x\right)^m.$$

Taking $t = \frac{n+\beta}{n}\left[\left(\frac{nt+\alpha}{n+\beta}-x\right)-\left(\frac{\alpha}{n+\beta}-x\right)\right]$, we get

$$x(1+x)\left[\mu'_{n,m}(x) + m\mu_{n;m-1}(x)\right]$$

$$= \left(\frac{n+\beta}{n}\right)\left[\begin{array}{c} n\sum_{\nu=1}^{\infty} p_{n;\nu}(x)\int_0^{\infty}p'_{n+1;\nu-1}(t)\left(\frac{nt+\alpha}{n+\beta}-x\right)^{m+1} dt - \left(\frac{\alpha}{n+\beta}-x\right) \\ n\sum_{\nu=1}^{\infty} p_{n;\nu}(x)\int_0^{\infty}p'_{n+1;\nu-1}(t)\left(\frac{nt+\alpha}{n+\beta}-x\right)^{m} dt \end{array}\right]$$

$$+\left(\frac{n+\beta}{n}\right)^2\left[\begin{array}{c} n\sum_{\nu=1}^{\infty} p_{n;\nu}(x)\int_0^{\infty}p'_{n+1;\nu-1}(t)\left(\frac{nt+\alpha}{n+\beta}-x\right)^{m+2} dt + \left(\frac{\alpha}{n+\beta}-x\right)^2 \\ \times n\sum_{\nu=1}^{\infty} p_{n;\nu}(x)\int_0^{\infty}p'_{n+1;\nu-1}(t)\left(\frac{nt+\alpha}{n+\beta}-x\right)^{m} dt - 2\left(\frac{\alpha}{n+\beta}-x\right) \\ \times n\sum_{\nu=1}^{\infty} p_{n;\nu}(x)\int_0^{\infty}p'_{n+1;\nu-1}(t)\left(\frac{nt+\alpha}{n+\beta}-x\right)^{m+1} dt \end{array}\right]$$

$$+(n+1)\left(\frac{n+\beta}{n}\right)\left[\begin{array}{c} n\sum_{v=1}^{\infty}p_{n;v}(x)\int_0^{\infty}p_{n+1;v-1}(t)\left(\frac{nt+\alpha}{n+\beta}-x\right)^{m+1}dt- \\ \left(\frac{\alpha}{n+\beta}-x\right)n\sum_{v=1}^{\infty}p_{n;v}(x)\int_0^{\infty}p_{n+1;v-1}(t)\left(\frac{nt+\alpha}{n+\beta}-x\right)^{m}dt \end{array}\right]$$

$$-nx(1+x)^{-n}\left(\frac{\alpha}{n+\beta}-x\right)^{m}$$

Integrating by parts and rearranging the terms, we get the required result.

Lemma 5. There exist the polynomials $Q_{i;j;r}(x)$, independent of n and v such that

$$x^r(1+x)^r D^r p_{n,v}(x)=\sum_{\substack{2i+j\le r,\\ i,j\ge 0}} n^i(v-nx)^j Q_{i,j,r}(x)p_{n,v}(x), \qquad D\equiv\frac{d}{dx}$$

17.4 MAIN RESULTS

Here, we prove some desired approximation results including the Voronovskaya asymptotic formula and error estimates.

Definition 1. $C_\gamma[0,\infty)$ is defined as

$$C_\gamma[0,\infty)=\left\{f\in C[0,\infty): f(t)=O\left(t^\gamma\right), \gamma>0\right\}.$$

Then the operators (17.3) are said to be well defined for $f\in C_\gamma[0,\infty)$.

Definition 2. A function f is said to have the Modulus of Continuity represented by $\omega_f(\delta)$ on closed interval $[a,b]$ if

$$\omega_f(\delta)=\sup_{|t-x|\le\delta}\ \sup_{x,t\in[a,b]}\left|f(t)-f(x)\right|$$

In particular, if $f\in C_{x^2}[0,\infty)$ then $\omega_f(\delta)\to 0$.

Theorem 1. If the function $f\in C_\gamma[0,\infty)$ is bounded on every finite subinterval of the interval $[0,\infty)$ having the derivatives of $(r+2)^{\text{th}}$ order at fixed $0\le x<\infty$. Again if $f(t)=O\left(t^\gamma\right)$ as $t\to\infty$, for some $\gamma>0$, then

$$\lim_{n\to\infty} n\left[\left(V_{n;1}^{\alpha,\beta}\right)^{(r)}(f,x)-f^{(r)}(x)\right]$$

$$=r(r-2\beta)f^{(r)}(x)+\left[(2r+\alpha)+x(1+r-\beta)\right]f^{(r)+1}(x)+x(1+x)f^{(r+2)}(x).$$

Proof: We have Taylor's expansion for a function f as

$$f(t)=\sum_{i=0}^{r+2}\frac{f^{(i)}(x)}{i!}(t-x)^i+\epsilon(t,x)(t-x)^{r+2},$$

where $\epsilon(t, x)\to 0$ as $t\to x$ and $\epsilon(t, x)\to O(t-x)^{\delta}$ as $t\to\infty$, for some $\delta>0$.

By using Taylor's expansion theorem, we have

$$n\left[\left(V_{n;1}^{\alpha,\beta}\right)^{(r)}(f, x)-f^{(r)}(x)\right]=n\left[\sum_{i=0}^{r+2}\frac{f^{(i)}(x)}{i!}\left(V_{n;1}^{\alpha,\beta}\right)^{(r)}\left((t-x)^i, x\right)-f^{(r)}(x)\right]$$

$$+n\left(V_{n;1}^{\alpha,\beta}\right)^{(r)}\left(\epsilon(t,x)(t-x)^{r+2}, x\right)$$

$$:= I_1+I_2 \text{ (say)}.$$

Applying Lemma 2, we get

$$I_1=n\left\{\sum_{i=0}^{\infty}\frac{f^{(i)}(x)}{i!}\sum_{j=0}^{i}\binom{i}{j}(-x)^{i-j}\left(V_{n;1}^{\alpha,\beta}\right)^{(r)}(t^j, x)-f^{(r)}(x)\right\}$$

$$=\frac{f^{(r)}(x)}{r!}n\left\{\left(V_{n;1}^{\alpha,\beta}\right)^{(r)}(t^r, x)-r!\right\}+\frac{f^{(r+1)}(x)}{(r+1)!}n\left\{\begin{array}{c}(r+1)(-x)\left(V_{n;1}^{\alpha,\beta}\right)^{(r)}(t^r, x)\\+\left(V_{n;1}^{\alpha,\beta}\right)^{(r)}(t^{r+1}, x)\end{array}\right\}$$

$$+\frac{f^{(r+2)}(x)}{(r+2)!}n\left\{\begin{array}{c}\frac{(r+2)(r+1)}{2}x^2\left(V_{n;1}^{\alpha,\beta}\right)^{(r)}(t^r, x)\\+(r+2)(-x)\left(V_{n;1}^{\alpha,\beta}\right)^{(r)}(t^{r+1}, x)\\+\left(V_{n;1}^{\alpha,\beta}\right)^{(r)}(t^{r+2}, x)\end{array}\right\}$$

$$= nf^{(r)}(x)\left\{\frac{n^r}{(n+\beta)^r}\frac{(n-r-1)!(n+r-1)!}{(n-1)!^2}-1\right\}$$

$$+\frac{f^{(r+1)}(x)}{(r+1)!}n\left[\begin{array}{l}(r+1)(-x)\left\{\frac{n^r}{(n+\beta)^r}\frac{(n-r-1)!(n+r-1)!}{(n-1)!^2}r!\right\}+\frac{n^{r+1}}{(n+\beta)^{r+1}}\times\\ \frac{(n-r-2)!(n+r)!}{(n-1)!^2}r!+(r+1)\alpha\frac{n^r}{(n+\beta)^r}\frac{(n-r-1)!(n+r-1)!}{(n-1)!^2}r!\end{array}\right]$$

$$+\frac{f^{(r+2)}(x)}{(r+2)!}n\left[\begin{array}{l}\frac{(r+2)(r+1)}{2}x^2\frac{n^r}{(n+\beta)^r}\frac{(n-r-1)!(n+r-1)!}{(n-1)!^2}r!+\\ (r+2)(-x)\left\{\begin{array}{l}\frac{n^{r+1}}{(n+\beta)^{r+1}}\frac{(n-r-2)!(n+r)!}{(n-1)!^2}(r+1)!x\\ +(r+1)\alpha\frac{n^r}{(n+\beta)^r}\frac{(n-r-1)!(n+r-1)!}{(n-1)!^2}r!\end{array}\right\}\\ \frac{n^{r+2}}{(n+\beta)^{r+2}}\frac{(n-r-3)!(n+r+1)!}{(n-1)!^2}\frac{(r+2)!}{2}x^2+\\ \left\{\begin{array}{l}(r+2)(r+1)\frac{n^{r+2}}{(n+\beta)^{r+2}}\frac{(n-r-3)!(n+r)!}{(n-1)!^2}\\ +(r+2)\alpha\frac{n^{r+1}}{(n+\beta)^{r+2}}\frac{(n-r-2)!(n+r)!}{(n-1)!^2}\end{array}\right\}(r+1)!x+\\ \left\{\begin{array}{l}(r+2)(r+1)r\alpha\frac{n^{r+1}}{(n+\beta)^{r+2}}\frac{(n-r-2)!(n+r-1)!}{(n-1)!^2}+\\ (r+2)(r+1)\alpha^2\frac{n^r}{(n+\beta)^{r+2}}\frac{(n-r-1)!(n+r-1)!}{(n-1)!^2}\end{array}\right\}r!\end{array}\right]$$

$$+O\left(n^{-2}\right)$$

In the above expression, $r(r-\beta)$, $(2r+\alpha)+x(1+r-\beta)$ and $x(1+x)$ are the coefficients of $f^{(r)}$, $f^{(r+1)}$ and $f^{(r+2)}$, respectively, and easily can be obtained by induction postulation for r and then finding limit as $n \to \infty$. To get complete proof of the above theorem, it is ever sufficient to show that $I_2 \to 0$ as the limit $n \to \infty$. Therefore, using Lemma 5, we get

$$|I_2| \le \sum_{\substack{2i+j\le r,\\ i,j\ge 0}} \frac{n^{i+1}|Q_{i,j,r}(x)|}{x^r(1+x)^r} \sum_{v=1}^{\infty} |v-nx|^j p_{n;v}(x) \int_0^{\infty} p_{n+1;v-1}(t)|\epsilon(t,x)| \left|\frac{nt+\alpha}{n+\beta}-x\right|^{r+2} dt$$

$$+\frac{(-1)^r(n+r-1)!}{n!}|\epsilon(0,x)|\left|\frac{\alpha}{n+\beta}-x\right|^{r+2}$$

$$:= I_3 + I_4.$$

Since $\epsilon(t,x) \to 0$ when $t \to x$. Hence, for a given $\epsilon > 0$, there exists a positive number δ such that $|\epsilon(t,x)| < \epsilon$ whenever $|t-x| < \delta$. Now, for $|t-x| \ge \delta$, if β is any integer such that $\beta \ge \max\{\lambda, r+2\}$ then for a constant $M > 0$, which does not depend on t, we have

$$|\epsilon(t,x)| \left|\frac{nt+\alpha}{n+\beta}-x\right|^{r+2} \le M|\epsilon(t,x)| \left|\frac{nt+\alpha}{n+\beta}-x\right|^{\beta}$$

Hence,

$$|I_3| \le M_1 \sum_{\substack{2i+j\le r,\\ i,j\ge 0}} n^{i+1} \sum_{v=1}^{\infty} |v-nx|^j p_{n;v}(x) \left\{ \begin{array}{l} \displaystyle\int_{|t-x|<\delta} p_{n+1;v-1}(t)\,\varepsilon \left|\frac{nt+\alpha}{n+\beta}-x\right|^{r+2} dt \\ \displaystyle +M\int_{|t-x|\ge\delta} p_{n+1;v-1}(t) \left|\frac{nt+\alpha}{n+\beta}-x\right|^{\beta} dt \end{array} \right\}$$

$$:= I_5 + I_6$$

where

$$M_1 = \sum_{\substack{2i+j\le r,\\ i,j\ge 0}} \frac{|Q_{i,j,r}(x)|}{x^r(1+x)^r}$$

Using the Schwarz inequality and then Lemmas 3 and 4, we have

$$|I_5| \le M_1\varepsilon \sum_{\substack{2i+j\le r,\\ i,j\ge 0}} n^{i+1}\left(\sum_{v=1}^{\infty}|v-nx|^{2j}\,p_{n;v}(x)\right)^{1/2}\left(\frac{1}{n}\int_0^{\infty}p_{n+1;v-1}(t)\,dt\right)^{1/2}$$

$$\times\left(n\int_0^{\infty}p_{n+1;v-1}(t)\left|\frac{nt+\alpha}{n+\beta}-x\right|^{2r+4}dt\right)^{1/2}$$

$$\le M_1 \sum_{\substack{2i+j\le r,\\ i,j\ge 0}} n^{i+1}.O\left(n^{\frac{j}{2}}\right).O\left(n^{-|r+2|/2}\right)$$

$$\le \varepsilon O(1) = o(1),$$

since ε is arbitrary. Again, applying the Schwarz inequality, Lemmas 3 and 4, we have

$$|I_6| \le M_2 \sum_{\substack{2i+j\le r,\\ i,j\ge 0}} n^{i+1}\sum_{v=1}^{\infty}|v-nx|^{j}\,p_{n;v}(x)\int_0^{\infty}p_{n+1;v-1}(t)\left|\frac{nt+\alpha}{n+\beta}-x\right|^{\beta}dt$$

$$\le M_2 \sum_{\substack{2i+j\le r,\\ i,j\ge 0}} n^{i+1}\left(\sum_{v=1}^{\infty}|v-nx|^{2j}\,p_{n;v}(x)\right)^{\frac{1}{2}}\left(\frac{1}{n}\int_0^{\infty}p_{n+1;v-1}(t)\,dt\right)^{\frac{1}{2}}$$

$$\times\left(n\sum_{v=1}^{\infty}p_{n;v}(x)\int_0^{\infty}p_{n+1;v-1}(t)\left|\frac{nt+\alpha}{n+\beta}-x\right|^{2\beta}dt\right)^{1/2}$$

$$\le \sum_{\substack{2i+j\le r,\\ i,j\ge 0}} n^{i+1}.O\left(n^{j/2}\right).\left(n^{-\beta/2}\right)$$

$$= \sum_{\substack{2i+j\le r,\\ i,j\ge 0}} O(n^{[r+2-\beta]/2} = o(1).$$

Therefore, we have $I_3 \to 0$ as $n \to \infty$. From here it is clear that $I_4 \to 0$ as $n \to \infty$ and get $I_2 = o(1)$. Combining the estimates of I_1 and I_2, we get the required result and hence the complete proof of the theorem.

Theorem 2. For $f \in C_\gamma[0,\infty)$, $\gamma > 0$ and $r \le m \le r+2$, if m^{th} derivative of f i.e. $f^{(m)}$ exists and is continuous on $(a-\eta, b+\eta) \subset (0,\infty)$, $\eta > 0$ then for sufficiently large n, we have

$$\left(V_{n;1}^{\alpha,\beta}\right)^{(r)}(f, x) - f^{(r)}(x)_{C[a,b]} \le M_1 \frac{1}{n} \sum_{i=r}^{m} f^{(i)}(x)_{C[a,b]} + M_2 n^{-\frac{1}{2}} \omega\left(f^{(m)}, n^{-\frac{1}{2}}\right)$$

$$+ O\left(n^{-2}\right).$$

Here M_1, M_2 are constants independent of f and n; $\omega(f,\delta) \equiv \omega_f(\delta)$ represents the modulus of continuity of f on $(a-\eta, b+\eta)$ and $C[a,b]$ denotes the sup-norm on the interval $[a, b]$.

Proof: Using Taylor's expansion on function f, we have

$$f(t) = \sum_{i=0}^{k} \frac{f^{(i)}(x)}{i!}(t-x)^i + \frac{f^{(k)}(\eta) - f^{(k)}(x)}{k!}(t-x)^k \chi(t) + h(t,x)(1-\chi(t)),$$

where $x < \eta < t$ and $\chi(t)$ shows the characteristic function on the interval $(a-\eta, b+\eta)$. Therefore,

$$\left(V_{n;1}^{\alpha,\beta}\right)^{(r)}(f, x) - f^{(r)}(x) = \left\{\sum_{i=0}^{k} \frac{f^{(i)}(x)}{i!}\left(V_{n;1}^{\alpha,\beta}\right)^{(r)}\left((t-x)^i, x\right) - f^{(r)}(x)\right\} +$$

$$+\left(V_{n;1}^{\alpha,\beta}\right)^{(r)}\left(\frac{f^{(m)}(\eta) - f^{(k)}(x)}{k!}(t-x)^k \chi(t), x\right)$$

$$+\left(V_{n;1}^{\alpha,\beta}\right)^{(r)}\left(h(t,x)(1-\chi(t)), x\right) := L_1 + L_2 + L_3.$$

Using Lemma 2, we have

$$L_1 = \sum_{i=0}^{k} \frac{f^{(i)}(x)}{i!} \sum_{j=0}^{i} \binom{i}{j} (-x)^{i-j} \left(V_{n;1}^{\alpha,\beta}\right)^{(r)}\left(t^j, x\right) - f^{(r)}(x)$$

$$= \sum_{i=0}^{k} \frac{f^{(i)}(x)}{i!} \sum_{j=0}^{i} \binom{i}{j} (-x)^{i-j} \frac{d^r}{dx^r} V_{n;1}^{\alpha,\beta}\left(t^j, x\right) - f^{(r)}(x)$$

$$= \sum_{i=0}^{k} \frac{f^{(i)}(x)}{i!} \sum_{j=0}^{i} \binom{i}{j} (-x)^{i-j} \times$$

$$\frac{d^r}{dx^r}\left[\left\{\begin{array}{c} j(j-1)(j-2)\dfrac{\alpha n^{j-1}\Gamma(n-j+1)\Gamma(n+j-2)}{(n+\beta)^j(\Gamma n)^2} \\ +\dfrac{j(j-1)}{2}\dfrac{\alpha^2 n^{j-2}\Gamma(n-j)\Gamma(n+j-1)}{(n+\beta)^j(\Gamma n)^2} \end{array}\right\}x^{j-2}\right.$$

$$+O\left(\frac{1}{n^2}\right)\frac{n^j\Gamma(n-j)\Gamma(n+j)}{(n+\beta)^j(\Gamma n)^2}x^j$$

$$\left.+\left\{\begin{array}{c} j(j-1)\dfrac{n^j\Gamma(n-j)\Gamma(n+j-1)}{(n+\beta)^j(\Gamma n)^2} \\ +j\alpha\dfrac{n^{j-1}\Gamma(n-j+1)\Gamma(n+j-1)}{(n+\beta)^j(\Gamma n)^2} \end{array}\right\}x^{j-1}\right]-f^{(r)}(x)$$

Hence,

$$L_{1C[a,b]}\le M_1\frac{1}{n}\sum_{i=r}^{m}f^{(i)}(x)_{C[a,b]}+O\left(\frac{1}{n^2}\right)$$

on $[a,b]$. To estimate L_2, we follow

$$|L_2|\le\int_0^\infty\left|K_n^{(r)}(x,t)\right|\left|\frac{f^{(m)}(\eta)-f^{(k)}(x)}{k!}\right|\left|\frac{nt+\alpha}{n+\beta}-x\right|^k\chi(t)\,dt$$

$$\le\frac{\omega\left(f^{(k)},\delta\right)}{k!}\int_0^\infty\left|K_n^{(r)}(x,t)\right|\left|1+\frac{\left|\frac{nt+\alpha}{n+\beta}-x\right|}{\delta}\right|\left|\frac{nt+\alpha}{n+\beta}-x\right|^k dt$$

$$\le\frac{\omega\left(f^{(k)},\delta\right)}{k!}\left[\begin{array}{l} n\displaystyle\sum_{v=1}^{\infty}\left|p_{n;v}^{(r)}(x)\right|\int_0^\infty p_{n+1;v-1}(t)\left(\left|\frac{nt+\alpha}{n+\beta}-x\right|^k+\frac{1}{\delta}\left|\frac{nt+\alpha}{n+\beta}-x\right|^{k+1}\right)dt \\ \dfrac{\Gamma(n+r)}{\Gamma n}(1+x)^{-(n+r)}\left(\left|\dfrac{nt+\alpha}{n+\beta}-x\right|^k+\dfrac{1}{\delta}\left|\dfrac{nt+\alpha}{n+\beta}-x\right|^{k+1}\right) \end{array}\right.$$

By applying Lemma 5 and the Schwarz inequality for summation and integral, we have

$$\sum_{\upsilon=1}^{\infty}|\upsilon-nx|^{j}\, p_{n;\upsilon}(x)\int_{0}^{\infty}p_{n+1;\upsilon-1}(t)\left|\frac{nt+\alpha}{n+\beta}-x\right|^{k}dt$$

$$\leq\left(\sum_{\upsilon=1}^{\infty}|\upsilon-nx|^{2j}\, p_{n;\upsilon}(x)\right)^{1/2}\left(\frac{1}{n}\int_{0}^{\infty}p_{n+1;\upsilon-1}(t)\,dt\right)^{1/2}$$

$$\leq O\left(n^{\frac{j}{2}}\right).1.O\left(n^{-\frac{k}{2}}\right)=O\left(n^{\frac{j-k}{2}}\right)$$

on the interval $[a,b]$. Therefore, Lemma 5 gives us

$$\sum_{\upsilon=1}^{\infty}\left|p_{n;\upsilon}^{(r)}(x)\right|\int_{0}^{\infty}p_{n+1;\upsilon-1}(t)\left|\frac{nt+\alpha}{n+\beta}-x\right|^{k}dt$$

$$\leq\sum_{\upsilon=1}^{\infty}\sum_{\substack{2i+j\leq r,\\ i,j\geq 0}}\frac{n^{i}\left|Q_{i,j,r}(x)\right|}{x^{r}(1+x)^{r}}|\upsilon-nx|^{j}\, p_{n;\upsilon}(x)\int_{0}^{\infty}p_{n+1;\upsilon-1}(t)\left|\frac{nt+\alpha}{n+\beta}-x\right|^{k}dt$$

$$\leq M\sum_{\substack{2i+j\leq r,\\ i,j\geq 0}}n^{i}\sum_{\upsilon=1}^{\infty}|\upsilon-nx|^{j}\, p_{n;\upsilon}(x)\int_{0}^{\infty}p_{n+1;\upsilon-1}(t)\left|\frac{nt+\alpha}{n+\beta}-x\right|^{k}dt$$

$$\leq M\sum_{\substack{2i+j\leq r,\\ i,j\geq 0}}O\left(n^{i}\right).O\left(n^{\frac{j-k}{2}}\right)=O\left(n^{\frac{r-k}{2}}\right),$$

where

$M=\sup_{\substack{2i+j\leq r,\\ i,j\geq 0}}\ \sup_{x\in[a,b]}\dfrac{\left|Q_{i,j,r}(x)\right|}{x^{r}(1+x)^{r}}$. If we chose $\delta=1/\sqrt{n}$, then from the above,

we have

$$L_{2C[a,b]}\leq\frac{\omega\left(f^{(k)},1/\sqrt{n}\right)}{k!}\left\{O\left(n^{\frac{r-k}{2}}\right)+\sqrt{n}\,O\left(n^{\frac{r-k-1}{2}}\right)+O\left(n^{-m}\right)\right\}$$

$$\leq M_2 \frac{\omega\left(f^{(k)}, 1/\sqrt{n}\right)}{k!} O\left(n^{\frac{r-k}{2}}\right).$$

Since $t \in [0,\infty),(a-\eta, b+\eta)$, and we opt δ such that $|t-x| \geq \delta$ for all $x \in [a,b]$. Hence,

$$|L_3|_{C[a,b]} = \sum_{v=1}^{\infty} \left|p_{n;v}^{(r)}(x)\right| \int_0^{\infty} p_{n+1;v-1}(t)\left(h(t,x)(1-\chi(t)),\, x\right) dt$$

$$\leq M \sum_{\substack{2i+j\leq r,\\ i,j\geq 0}} n^i \sum_{v=1}^{\infty} |v-nx|^j\, p_{n;v}(x) \int_0^{\infty} p_{n+1;v-1}(t)\left(h(t,x)(1-\chi(t)),\, x\right) dt$$

$$+ \frac{(n+r-1)!}{(n-r)!}(1+x)^{-(n+r)}\left|h(0,x)\right|.$$

for $(t-x) \geq \delta$, we have a constant K such that

$$\left|h(t,\, x)\right| \leq K \left|\frac{nt+\alpha}{n+\beta} - x\right|^{\rho},$$

where $\rho \geq \max\{\gamma,\, k\}$. Therefore, by using Schwarz inequality for summation and integration and Lemma 3 as well as Lemma 4 [11–13], we follow then $L_3 = O\left(1/n^s\right)$ for some $s > 0$ uniformly on $[a,b]$.

Gathering the estimates L_1, L_2 and L_3, we get the required result.

REFERENCES

1. Srivastava H.M. and Gupta V., A certain family of summation integral type operators, *Mathematical and Computer Modelling* 37(12–13) (2003), 1307–1315.
2. Widder D.V., *The Laplace transform*, Princeton University Press, Princeton (1941).
3. Kajla A., On the Bezier variant of the Srivastava–Gupta operators, *Constructive Mathematical Analysis* 1(2) (2018), 99–107.
4. Acu A.M. and Gupta V., Direct results for certain summation-integral type Baskakov Szasz operators, *Results in Mathematics* 72(3) (2017), 1161–1180.
5. Tikhonov I.V. and Nguyen S.T.V., The solvability of the inverse problem for the evolution equation with a superstable semigroup, *Discrete and Continuous Models And Applied Computational Science* 26(2) (2018), 103–118.

6. Sharma P.M. and Garg S., On q-Baskakov Durrmeyer Stancu operators, *Journal of Mathematics and Statistics, USA* 1(1) (2016), 760–774.
7. Sharma R.R. and Garg S., Simultaneous approximation for Beta-Baskakov Stancu operators, *International Journal of Innovative Technology and Exploring Engineering (Scopus)* 9(4) (2020), 522–528.
8. Garg S., Convergence of derivatives for certain mixed Baskakov-Szasz operators, *IOSR Journal of Applied Mathematics (IOSR-JM)* 15(2) (2019), 19–23.
9. Bailey W.N. *Generalised Hypergeometric Series*, (1935).
10. Gupta V. and Yadav R., Direct estimates in simultaneous approximation for BBS operators, *Applied Mathematics and Computation* 218(22) (2012), 11290–11296.
11. A. Rahim, P.K. Mallik and V.A. Sankar Ponnapalli, Fractal antenna design for overtaking on highways in 5G vehicular communication ad-hoc Networks environment, *International Journal of Engineering and Advanced Technology (IJEAT)* 9(1S6) (December 2019), 157–160. doi: 10.35940/ijeat.A1031.1291S619, ISSN: 2249-8958.
12. P.K. Malik, H. Parthasarthy and M.P. Tripathi, Axisymmetric excited integral equation using moment method for plane circular disk, *International Journal of Scientific and Engineering Research*, 3(3) (March 2012), 1–3, ISSN 2229-5518.
13. P.K. Malik, H. Parthasarthy and M.P. Tripathi, Analysis and design of Pocklingotn's equation for any arbitrary surface for radiation, *International Journal of Scientific and Engineering Research*, 7(9) (September 2016), 208–213, ISSN 2229-5518.

Chapter 18

Antenna selection criteria and parameters for IoT application

Mihir Narayan Mohanty and Sarmistha Satrusallya
ITER, Siksha 'O' Anusandhan

Takialddin Al smadi
Jerash University

CONTENTS

18.1 INTRODUCTION

The Internet of Things is a system of interrelated computing devices, mechanical and digital machines, objects, animals, or people that are provided with a unique identity and the ability to transfer data over a network without requiring human-to-human interaction. Kevin Ashton introduced the phrase Internet of Things to connect network connecting objects in the physical world to the Internet. As the backbone of IoT is a network of connected devices, antennas of different sizes and shapes are designed for efficient transmission of data.

DOI: 10.1201/9781003347057-18

Wireless communication now-a-days is the trending technique to transmit data in any form from one place to another. RF technology enables wireless communication to use the Internet of Things. Though RF technology is not new, it has grown to include cellular devices and other advances to fulfill the demand for new consumer and industrial applications.

As RF technology is used in different IoT applications, antennas become essential for every IoT device. The antenna is described as a metallic structure used to capture and transmit radio electric waves. The antenna has different shapes and sizes depending on the application. Apart from connectivity and power, the antenna needs equal attention to device performance. The antenna is the medium through which an IoT device receives and sends a signal to the outside world. Every Internet of Things (IoT) device needs an antenna to make or break the communication with the device. A properly designed and integrated antenna is directly linked to the total system performance of the device. It improves the data rates, range, safety, and security and provides an improved user experience. IoT application covers low-frequency range covering the ISM band, Wi-MAX, WLAN, and high-frequency range for 5G applications. The design of the antenna depends on the application of the device in the IoT world.

Wireless devices to be marketed the most comply with the recently introduced Electromagnetic Compliance (EMC) and Radio Equipment Directive (RED). As far as EMC is concerned, high-frequency circuitry may cause problem as it is exposed to external EM radiation. EMC helps to minimize the possibility that radiated or conducted emissions produced by the device will interfere with other electronic products in its vicinity. The interference termed Electromagnetic Interference should be minimum to enhance the performance of the device.

18.2 RELATED WORK FOR ANTENNA DESIGN

Many researchers have proposed antennas for IoT applications with various aspects. The proposed antennas have different shapes of patch, ground plane, substrate material, and feeding to match the device of application. Katcoh et al. proposed a compact antenna for dual-band operation. The antenna has a rectangular microstrip patch with two slotted patches and RT Duroid 5880 as the substrate material to be operated in the frequency range of 2.39–3.15 GHz. The antenna had a gain of 5.01 dB [1]. Another compact monopole antenna consisting of a rectangular radiator with a ground plane having a rectangular slit and L-shaped stub was implemented with a size of $9.45 \times 18.5\,mm^2$ [2]. As the narrow bandwidth and small dimension are necessary for antennas used for IoT devices, antenna considered as one of the specific properties of the antenna. The authors in Ref. [3] proposed an antenna based on a folded inverted F antenna which occupies the top section of the PCB with a dimension of $40 \times 25\,mm^2$. The antenna was useful for application in GPS systems. For RFID application, an antenna with a circular ground plane and circular patch was considered. The feed to the patch was considered as the microstrip line which was placed at the edge of the antenna to have greater efficiency and lower losses. The substrate material of RT Duroid 5870 with a height of 1.17 mm was chosen to operate the antenna at the ISM band. The antenna exhibited an efficiency of 81%, a gain of 6.5 dBi, and a directivity of 7.44 dBi [4]. Real-time monitoring in the aircraft is one of the emerging applications of IoT. A microstrip rectangular patch antenna with different slots was proposed to reduce the resonating frequency and to obtain miniaturization of the antenna. The introduction of slots in the design reduced the operating frequency from 2.4 to 1.5 GHz. The use of the inset-feeding technique and FR4 substrate increased the gain of the antenna [5]. Another requirement in IoT is the transmission of large data on multiple frequency ranges and bands. A multiband antenna is the solution to the problem where different bands are active for various frequency ranges. Multiband in an antenna design is achieved by introducing slots in the patch, defected ground structure, and fractal in the antenna. A fractal antenna has the advantage of compact size, wide band, frequency independence, and low mutual coupling [6]. Also, a reconfigurable antenna has the advantage of operating in a multiband. The multiple frequency band application on a common platform is possible for a reconfigurable antenna. Asdallah et al. proposed a reconfigurable antenna which had the aspects to tune into the operating frequency of 2.4 GHz, 1.58 GHz, and 868 MHz. The proposed antenna was composed of a circular patch with a diameter of 58.8 mm mounted on a Roger RO 5880 substrate with a thickness of 0.787 mm. The reconfiguration in the antenna was possible due to the digitally tunable capacitor (DTC). The DTC (pe64907) was activated to achieve the required capacitance and radiation efficiency of 93% [7]. Singh et al. proposed a five-band antenna suitable 18.for WLAN, Wi-Max, and C band application. The antenna in the work had the effect of fractal stripping and rectangular slotted geometry. The ground plane was considered defective with an inverted U slot and two ring slots. Frequency reconfiguration in the design was proposed by simultaneous switching of three PIN diodes with surrounded parasitic structures [8]. For an electronically pattern reconfigurable antenna, a single low power, low insertion loss SP4T switch was enabled for radiation pattern steering. The antenna proposed has two layers of the FR4 substrate material with an air gap and resonates at 2.44 GHz with a higher front-to-back ratio [9]. The antenna designed with ring resonators having stubs and a circular patch is also used for multiband operation. The antenna was designed over a Roger RO4003 substrate and a defective ground structure to have a multiband [10]. A flexible antenna is also an option for IoT applications due to its high radiation efficiency. The authors in Ref. [11] proposed a U-shaped and triangular-shaped radiator with two tuning stubs. The antenna had a Coplanar waveguide (CPW) transmission feed and a flexible substrate formed by mixing natural rubber with SiO_2. The flexibility of the substrate material had a minor effect on the antenna performance. Photo paper was also used as a flexible substrate for ultra-wideband application of antenna [12]. A combination of flexible ceramic, Arlon 25N, and flexible polypropylene was applied for the design of an ultrathin flexible antenna. The antenna fabrications were completed with two suitable techniques, electro textile and inject printing. Inject printing has the limitation of not providing full transparency in the antenna [13]. Another substrate was also considered for the flexible antenna. The acrylic sheet used makes the antenna suitable for conformal application [14]. CPW feeding was also used on a printed patch of rectangular shape on an FR4 substrate with a compact size of $25 \times 35\,mm^2$. A novel rounded corner technique was used to enhance the bandwidth and gain of the antenna [15]. The rounded corner system was also proposed for a CPW-fed antenna with inverted six-shaped configurations. The antenna had radiation, low back lobes, and low cross-polarization [16]. A folded patch antenna is also a suitable choice for IoT devices. The folded antenna has the advantage of decreasing the dimension and maintaining the same frequency [17]. Wireless sensors are one of the important devices in IoT applications. A transparent antenna is the choice for sensors as it has excellent visibility and a low level of fluctuation in realized gain. The transparency in the proposed antenna was due to the polycarbonate substrate with a thickness of 0.5 mm [18]. Eltresy et al. proposed another transparent antenna for radio frequency energy harvesting. The antenna was useful for application in smart museums. The transparency in the antenna was possible by considering glass as a substrate and Indium Tin Oxide as transparent conductors [19]. For low Power Wide Area Network communication with space using the advantage of IoT, a compact three-element radiating structure was proposed with Right-Handed Circular Polarization (RHCP). Considering FR-4 as the substrate of the antenna and feeding network circuit based on Quasi Lumped Quadrature Coupler, the antenna had a radiation efficiency of 69% and a gain of 3.14 dBic [20]. A suitable microstrip antenna with a U-shaped structure was proposed for IoT application. The optimized microstrip antenna had improved bandwidth and gain [21]. Apart from the conventional patch shape, a bloom-shaped antenna was proposed by researchers to obtain multiband operation. The antenna designed on FR4 substrate had improved parameters for IoT [22].

A single antenna has the limitation of narrow bandwidth, low gain, and the IoT device requires a better gain and radiation efficiency. The antenna array, on the other hand, fulfills the demand for enhanced gain. Dardeer et al. proposed a planer circularly suitable antenna array for RF energy harvesting in IoT system applications. The array had an equal gain in a specified direction which is desirable for energy harvesting applications. Though array gain had the same value as the element gain, the received ambient power of the array had exceeded as compared to the element power [23]. The array antenna with flexibility was also considered for MIMO-based IoT and wireless charging platforms. The position of array elements provides the pattern diversity for high-gain antenna. The flexible material Kapton Polymide and the antenna elements made the array thin and useful for wearable devices in IoT, wireless charging, and UWB applications. The radiation pattern of the array is suitable for beam switching application [24,25]. Miniature patch and slot arrays were considered for ISM band application in Internet of Things having a higher gain of 6.89 and 7.63 dBi. The radiation efficiency of the array was above 90% [26]. For the application in 5G communication, a slot antenna array was proposed with a gain of more than 9.5 dBi. The antenna had a radiation efficiency of 84% which made it suitable to integrate with the sensors for 5G communication [27]. As IoT has several applications in different fields, antenna arrays are designed for all-weather large-scale IoT deployment covering both the LTE and 5G sub 6 GHz [28].

18.3 SELECTION CRITERIA OF ANTENNA FOR IoT

The selection of antenna for the IoT devices depends on the PCB layout, operation in the specified range, performance without interference, and the consumption of power [29]. From Ref. [30], the topology of the antenna determines the efficiency, bandwidth, radiation pattern, and gain. So the smallest antenna is not considered the best choice for IoT devices. The antenna suitable is required to place it in small devices. The selection criteria of the antenna depend on different design parameters. The proper selection of ground plane, patch shape, size, and material of the substrate determines the performance of the antenna for various applications [31–34].

In this section, the parameters of antenna selection are summarized as per the previous work discussed earlier.

18.3.1 Selection of substrate

A suitable dielectric substrate of proper thickness and loss tangent is essential in designing the antenna. The substrate in the antenna is needed to provide the mechanical support of the antenna. That's why a substrate must satisfy the electrical and mechanical requirements. The choice of substrate depends on the application of the antenna in the different frequency ranges. The common substrates used for the designing of the antenna are

FR4 [20,22] and Roger RT Duroid [7,10] as they perform well in most environmental conditions and are lightweight and resistant to moisture. But, for the design of a flexible antenna, a substrate with flexibility is required. Kapton Polymide [24], natural rubber [11], Photo Paper [12], flexible ceramic, Arlon 25N, Polypropylene [13], and acrylic sheet [14] are some of the flexible substrates used in antenna design. The flexibility of the material has a minor effect on the performance of the antenna and provides better radiation efficiency. Polycarbonate [18] and glass [19] are considered transparent substrates for sensors in IoT applications due to excellent visibility and low level of fluctuation in gain.

The thickness of the substrate also determines the performance of the antenna. Though a thicker substrate provides an increase in the radiation power and reduced conductor loss, it also provides an increase in weight, dielectric loss, and surface wave loss. A substrate with a high value of the loss tangent decreases the radiation efficiency [35].

18.3.2 Selection of patch shape

The patch of the antenna is responsible for the radiation of electromagnetic waves. A proper selection of the patch shape determines the efficiency of the antenna. Rectangular and circular patches are the choice for antenna design [36]. As the application of the antenna is varying, different shapes of the patch are considered. A modification in the patch shape increases the gain and radiation efficiency of the antenna.

Slots in the patch create a new resonance frequency and increase the gain and radiation efficiency of the antenna. Slots are also considered as the parameters of the antenna to operate in multiband.

18.3.3 Selections of ground plane

One of the parts of the antenna is the ground plane which is a conducting surface used to reflect the RF waves from other antenna elements. The ground plane's shape and size play a major role in determining the radiation characteristics of the antenna. As the IoT device requires an antenna with better gain, a proper selection of the ground plane is necessary. Defective ground structure (DGS) created by etching a portion of the ground plane helps to suppress the cross suitable and increase the gain. The antenna array used for the IoT device in 5G communication requires DGS to decrease the mutual coupling among the array elements.

18.3.4 Selection of the feed

A proper feeding to the patch enhances the performance of the antenna. Microstrip feed, Coaxial feed, and Coplanar waveguide (CPW) are suitable choice of feed to the patch. Most of the IoT requires the antenna to be printed on PCB circuits. CPW feed is suitable for most of the IoT devices for easy integration on the printed circuit boards.

18.3.5 Electromagnetic interference and electromagnetic compliance

Electromagnetic waves have the trend of causing interference in some electrical and electronics pieces of equipment. The interference due to lighting relays and DC motors is also considered for the communication system. Electromagnetic Interference (EMI) is described as undesirable electromagnetic energy which is coupled into a system. Interference can cause serious degradation and destruction in the system. Any electronic system able to function compatibly with other electronic systems and not susceptible to interference is said to be electromagnetically compatible with the environment [37]. An antenna designed for IoT applications must have compatibility with other devices to perform better.

18.4 AN ANTENNA ARRAY FOR IoT APPLICATION

This section represents the verification of the selection criteria of the antenna through an antenna array. The application of an antenna array is required in 5G communication to enhance gain and radiation efficiency with a miniature design.

An antenna array with four triangular elements is proposed in this work. The substrate used is FR4 having a dielectric constant of 4.4 and loss tangent of 0.002. The antenna dimension is considered as 3.5 mm × 5 mm × 0.8 mm. The substrate is chosen to determine the resonance frequency and efficiency of the antenna.

Figures 18.1 and 18.2 represent the dimensions of the antenna array. The corresponding dimension values are described in Table 18.1. The array is compromised of four numbers of inverted equilateral triangular patches placed horizontally. The ground plane in the antenna array has a dimension of 2.1 mm × 5 mm. The change in the dimension of the ground plane is considered to increase the gain and front-to-back ratio. In this design, corporate arrangement of microstrip feed is used.

18.4.1 Calculation of the antenna patch

The selection of the patch shape plays a vital role in the design of an antenna. A triangular patch provides radiation characteristics similar to that of a rectangular patch with a smaller size [36]. The simplest shape of the triangular patch is the equilateral shape. The geometry of the triangular patch is given in Figure 18.3.

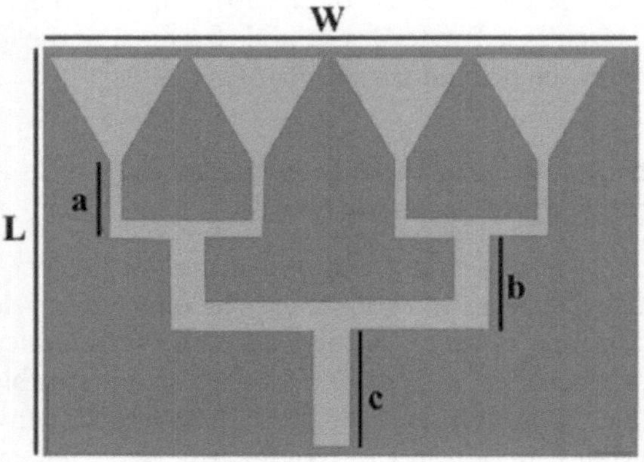

Figure 18.1 Top view array antenna.

Figure 18.2 Back view of antenna array.

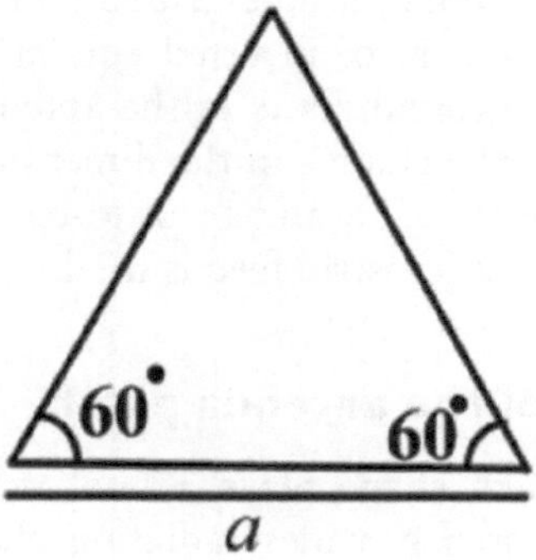

Figure 18.3 Equilateral triangle [38].

An equilateral antenna of side length a is considered in the design. The resonant frequency of the antenna is calculated using the following formula [35]:

$$f = \frac{2c}{3a\sqrt{\varepsilon_r}} \tag{18.1}$$

The smaller value of the side length of the triangle increases the resonating frequency of the antenna.

18.5 RESULT AND DISCUSSION

The return loss of the array describes the extent to which the antenna matches with the input port. The lower value of the return loss enhances the performance of the antenna array. The variation of the return loss with respect to the frequency is shown in Figure 18.4. The antenna resonates at 38.6 GHz with a return loss of –33 dB. The antenna has a bandwidth of 1.3 GHz at the resonant frequency.

The gain plot of the antenna at the resonant frequency is plotted and shown in Figure 18.5. The antenna has a gain of 10.5 dB at 38.6 GHz. The antenna exhibits a front-to-back ratio of 22 dB and a radiation efficiency of 96% at the resonating frequency. The summary of simulated results is presented in Table 18.2.

Figure 18.6 represents the surface current distribution of the patch array. The flow of current in the antenna indicates good radiation.

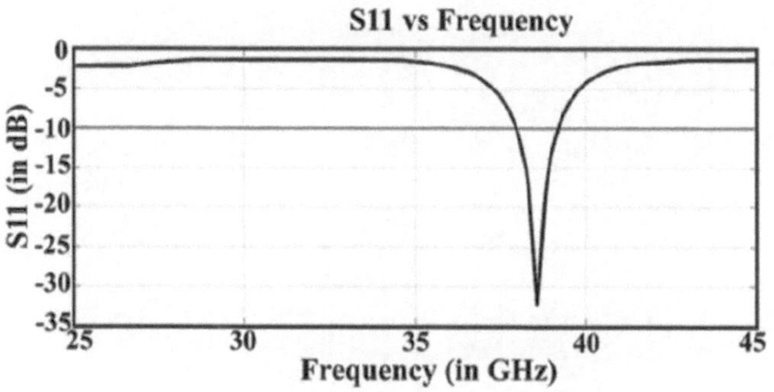

Figure 18.4 S_{11} vs frequency plot at 38.6 GHz.

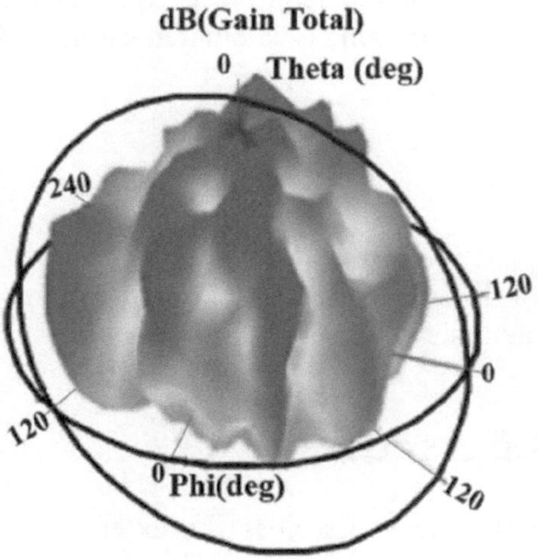

Figure 18.5 Gain plot at 38.6 GHz.

Table 18.1 Dimension of the antenna (in mm)

L	*W*	*a*	*b*	*c*	*d*
3.5	5	0.6	0.1	1	2.1

Table 18.2 Summary of simulated results

Frequency (GHz)	S_{11} *(< −10 dB)*	*Bandwidth (GHz)*	*Radiation efficiency (%)*	*Front-to-back ratio*	*Peak gain (dB)*
38.6	−33	1.3	96	22 dB	10.5

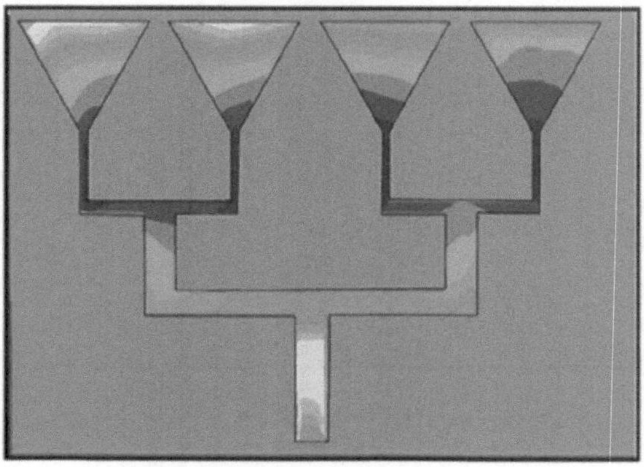

Figure 18.6 Electric field distribution at 38.6 GHz.

18.6 CONCLUSION

The suitability of an antenna for an IoT application depends on many parameters like gain, bandwidth, and radiation efficiency. The substrate, the shape of the patch, the shape of the ground plane, and the feed of the antenna affect most of the vital parameters of the antenna. In this work, an array antenna minimization design in 5G application for IoT is proposed. The antenna dimension is very compact to be used in handheld devices for IoT. The antenna resonates at 38.6 GHz with a bandwidth of 1.35 GHz and a gain of 10.5 dB considering Bakelite as the substrate with a thickness of 0.8 mm.

REFERENCES

1. S. Katoch, H. Jotwani, S. Pani, and A. Rajawat, A compact dual band antenna for IOT applications. *International Conference on Green Computing and Internet of Things*, pp. 1594–1597. IEEE. 2015.
2. A. Bekasiewicz, and S. Koziel, Compact UWB monopole antenna for internet of things applications. *Electronics Letters*, Vol. 52, pp. 492–494. 2016.
3. L. Lizzi, F. Ferrero, P. Monin, C. Danchesi, and S. Boudaud, Design of miniature antennas for IoT applications. *IEEE Sixth International Conference on Communications and Electronics (ICCE)*, pp. 234–237. IEEE. 2016.
4. N. Vikram. Design of ISM band RFID reader antenna for IoT applications. *International Conference on Wireless Communications, Signal Processing and Networking (WiSPNET)*, pp. 1818–1821. IEEE. 2016.
5. A. Satheesh, R. Chandrababu, and I. S. Rao, A compact antenna for IoT applications. *International Conference on Innovations in Information, Embedded and Communication Systems (ICIIECS)*, pp. 1–4. IEEE. 2017.
6. M. Gupta, and V. Mathur, Sierpinski fractal antenna for internet of things applications. *Materials Today: Proceedings*, Vol. 4(9), pp. 10298–10303. 2017.
7. F. A. Asadallah, J. Costantine, Y. Tawk, L. Lizzi, F. Ferrero, and, C. G. Christodoulou. A digitally tuned reconfigurable patch antenna for IoT devices. *IEEE International Symposium on Antennas and Propagation & USNC/URSI National Radio Science Meeting*, pp. 917–918. IEEE. 2017.
8. P. P. Singh, P. K. Goswami, S. K. Sharma, and G. Goswami, Frequency reconfigurable multiband antenna for IoT applications in WLAN, Wi-Max, and C-band. *Progress in Electromagnetics Research C*, Vol. 102, pp. 149–162. 2020
9. L. Santamaria, F. Ferrero, R. Staraj, and L. Lizzi, Electronically pattern reconfigurable antenna for IoT applications. *IEEE Open Journal of Antennas and Propagation*, Vol. 2, pp. 546–554. 2021.
10. P. Kumar, G. C. Ghivela, and J. Sengupta, Design and analysis of multiple bands spider web shaped circular patch antenna for IoT application. *8th IEEE India International Conference on Power Electronics (IICPE)*, pp. 1–5. IEEE. 2018.

11. A. Al-Sehemi, A. Al-Ghamdi, N. Dishovsky, G. Atanasova, and N. Atanasov, A flexible broadband antenna for IoT applications. *International Journal of Microwave and Wireless Technologies*, Vol. 12(6), pp. 531–540. 2020.
12. T. K. Saha, T. N. Knaus, A. Khosla, and P. K. Sekhar, A CPW-fed flexible UWB antenna for IoT applications. *Microsystem Technologies*, pp. 1–7. 2018.
13. M. E. de Cos Gómez, H. F. Álvarez, C. G. González, B. P. Valcarce, J. Olenick, and F. Las-Heras, Ultra-thin compact flexible antenna for IoT applications. *13th European Conference on Antennas and Propagation (EuCAP)*, pp. 1–4. IEEE. 2019.
14. D. Singh, K. R. Jha, and S. K. Sharma. Low cost flexible antenna for IoT applications. *IEEE International Symposium on Antennas and Propagation and North American Radio Science Meeting*, pp. 1929–1930. IEEE. 2020.
15. Q. Awais, H. T. Chattha, M. Jamil, Y. Jin, F. A. Tahir, and M. U. Rehman, A novel dual ultrawideband CPW-fed printed antenna for Internet of Things (IoT) applications. *Wireless Communications and Mobile Computing*, 2018.
16. S. Ashok Kumar, and T. Shanmuganantham, Design of CPW-fed inverted six shaped antenna for IOT applications. *Transactions on Electrical and Electronic Materials*, Vol. 21, pp. 524–527. 2020.
17. R. M. Bichara, F. A. Asadallah, J. Costantine, and M. Awad, A folded miniaturized antenna for IoT devices. *IEEE Conference on Antenna Measurements & Applications (CAMA)*, pp. 1–3, IEEE. 2018.
18. Y. Koga, and M. Kai, A transparent double folded loop antenna for IoT applications. *IEEE-APS Topical Conference on Antennas and Propagation in Wireless Communications (APWC)*, pp. 762–765. IEEE. 2018.
19. N. A. Eltresy, D. N. Elsheakh, E. A. Abdallah, and H. M. Elhennawy, RF energy harvesting using transparent antenna for IoT application. *International Conference on Innovative Trends in Computer Engineering (ITCE)*, pp. 287–291. IEEE. 2019.
20. L. H. Trinh, N. V. Truong, and F. Ferrero, Low cost circularly polarized antenna for IoT space applications. *Electronics*, Vol. 9, p. 1564. 2020.
21. A. A. Elijah, and M. Mokayef, Miniature microstrip antenna for IoT application. *Materials Today: Proceedings*, Vol. 29, pp. 43–47. 2020.
22. R. Ramasamy. Design and analysis of multiband bloom shaped patch antenna for IoT applications. *Turkish Journal of Computer and Mathematics Education (TURCOMAT)*, Vol. 12, pp. 4578–4585, 2021.
23. O. M. Dardeer, H. A. Elsadek, and E. A. Abdallah, 2 × 2 circularly polarized antenna array for RF energy harvesting in IoT system. *IEEE Global Conference on Internet of Things (GCIoT)*, pp. 1–6, IEEE. 2018.
24. H. Raad, A Yagi-Uda antenna array for conformal IoT and wireless charging applications. *Microwave and Optical Technology Letters*, Vol. 61, pp. 633–637. 2019.
25. H. K. Raad. An UWB antenna array for flexible IoT wireless systems. *Progress in Electromagnetics Research*, Vol. 162, pp. 109–121. 2018.
26. K. N. Olan-Nunez, R. S. Murphy-Arteaga, and E. Colin-Beltran, Miniature patch and slot microstrip arrays for IoT and ISM band applications. *IEEE Access*, Vol. 8, pp. 102846–102854. 2020.

27. T. Varum, and J. N. Matos, Compact slot antenna array for 5G communications. *IEEE International Symposium on Antennas and Propagation and USNC-URSI Radio Science Meeting*, pp. 1415–1416. IEEE. 2019.
28. L. Chi, Z. B. Weng, S. Meng, Y. Qi, J. Fan, W. Zhuang, and J. L. Drewniak, Rugged linear array for IoT applications. *IEEE Internet of Things Journal*, Vol. 7, pp. 5078–5087. 2020.
29. B. K. Tripathy, and J. Anuradha, *Internet of Things (IoT): Technologies, Applications, Challenges and Solutions*. CRC Press, 2017.
30. https://www.embeddedcomputing.com/application/networking-5g/5g/antenna-selection-for-iot-projects.
31. *Planar Antenna: Design, Fabrication, Testing, and Application*. Nova Science Publishers Inc.: New York, 2021, ISBN: 9781536198980.
32. P. Malik, J. Lu, B. T. P. Madhav, G. Kalkhambkar, S. Amit (Eds.) *Smart Antennas: Latest Trends in Design and Application*. Springer, ISBN: 9783030766368, doi: 10.1007/978-3-030-76636-8.
33. P. K. Malik, S Padmanaban, J. B. Holm-Nielsen, *Microstrip Antenna Design for Wireless Applications*. Taylor and Francis, 2021, ISBN: 9780367554385.
34. P. K. Malik, P. Kumar, S. Kumar, D. K. Singh, *Smart Antennas: Recent Trends in Design and Applications*. Bentham Science: Sharjah, Aug 2021, ISSN: 2717–5421 (Print), ISSN: 2717-543X (Online), ISBN: 9781681088600.
36. C. A. Balanis, *Antenna Theory Analysis and Design*, 2nd Edition. Wiley: India, 2007.
35. R. Garg, P. Bhartia, I. J. Bahl, and A. Ittipiboon, *Microstrip Antenna Design Handbook*. Artech House, 2001.
37. C. R. Paul, *Introduction to Electromagnetic Compatibility* (Vol. 184). John Wiley & Sons, 2006.
38. Y. S. Khraisat, and M. M. Olaimat, Comparison between rectangular and triangular patch antennas array. *19th International Conference on Telecommunications (ICT)*, pp. 1–5. IEEE. 2012.

Chapter 19

Wideband Wearable Antenna for IoT and Medical Applications

Mihir Narayan Mohanty and Shaktijeet Mahapatra
ITER, Siksha 'O' Anusandhan

Gyoo-Soo Chae
Baekseok University

CONTENTS

19.1 INTRODUCTION

With the arrival of the Internet and lowering of the data charges, the Internet-of-Things (IoT) has become ubiquitous. The need for remote monitoring of the environment and collection of data has pushed the number of active IoT devices beyond one billion worldwide. Wireless devices form the majority of the deployed devices.

IoT devices require interconnection between different devices. The antennas form the backbone of the wireless devices for transmission and reception of the signals over different communication bands. As each IoT device may need a different antenna to serve a particular band and to ensure connectivity, the design and deployment of the antenna becomes a major research topic among researchers.

DOI: 10.1201/9781003347057-19

Several microstrip patch antennas have been developed for IoT devices that are compact, multi-functional, and operate over several wireless bands. In this paper, we examine and compare various antennas that have been developed for IoT and biomedical purposes.

The relative electrical permittivity and the thickness of the substrate determine the resonant frequency and the radiation efficiency of the antenna. The rigidity or the flexibility of the substrate determines the suitability of the antenna for a particular application. The antennas that are destined to be deployed on the printed circuit boards are designed on rigid substrates for easy integration with the back-end electronics. The antennas that are designed for biomedical purposes are generally designed using flexible substrates as it allows for hassle-free deployment on the living subjects.

The feed of the antenna plays a vital role in determining its suitability for a particular application. The microstrip antennas can be fed through probes, microstrip lines, proximity coupling, apertures, and coplanar waveguides. The antennas that are designed for IoT and biomedical applications are generally excited through the microstrip lines through one of the edges. The type and position of the feed greatly affect the antenna performance in terms of gain, bandwidth, and polarization. A careful look into the nature of applications is required before deciding on the most suitable type of feed.

We also propose an antenna developed for IoT and biomedical applications that uses a rectangular antenna with slots and a modified ground for gain and bandwidth enhancement. The antenna is designed on a rigid cylindrical substrate.

This chapter is further divided into sections that discuss the popular choice of substrates, various antennas for IoT, antennas for biomedical applications, and an experimental design for verification.

19.2 SUBSTRATE CHOICES

The substrate choice for IoT or biomedical applications is governed by many factors. Researchers can choose any substrate depending on availability and thickness. On the other hand, researchers have limited choice of substrate for antennas meant for biomedical applications because body secretions like sweat can result in detuning of the antenna. Flexibility, ease of fabrication, mechanical strength, resistance to moisture and temperature, and place of final deployment of the antenna are the major governing factors.

For rigid antennas, double-side copper-clad laminates like Rogers RT duroids, FR4, or Arlon are used. The fabrication involves chemical etching only. Addition of the connectors is also easy as the laminates provide enough mechanical support. These substrates can also easily support the necessary electronics.

For flexible antennas, the substrates used are of polymer type like polyethylene terephthalate (PET), polyoxydimethyl siloxane (PDMS), polytetrafluoroethylene (PTFE or Teflon), etc. The fabrication of the antenna might involve additional steps of creating and stabilizing the substrate in the lab but the thickness of the substrate can be controlled. The metal for the patch and the ground are set into the polymers just before they are completely set to ensure adhesion. A thin sheet of acrylic, Kapton polyimide, paper, and even fabrics like felt, jeans, etc. provides a readymade base to start the fabrication, mostly by printing.

19.3 ANTENNAS FOR IoT

In this section, various antennas for IoT working at different bands have been examined. The antennas have been designed considering various applications including wireless power transfer and communication. Most of the works employed a coplanar waveguide (CPW) feeding mechanism. In CPW feeding, the antenna's microstrip feed is flanked by a coplanar ground plane. This arrangement of the feed and the ground plane acts as a waveguide. The CPW feed cuts down production costs to a large extent and can be easily integrated with the printed circuit boards. This feed results in a higher gain and wider bandwidth as compared to a simple microstrip feed or the coaxial probe feed.

19.3.1 Slotted antennas

The design of the slotted antennas starts with the basic antenna shape and then depending upon the requirements of the gain and the bandwidth, slots and slits are cut [1–10]. The slots and the slits play a major role in determination of the bandwidth. The slots in the ground plane affect the level of the gain. These antennas are simple to design. If copper-clad laminates like Rogers RT 5880, FR4, or Arlon are used, the fabrication involves chemical etching only. Addition of the connectors is also easy as the laminates provide enough mechanical support. If the substrates used are polymer type like polyethylene terephthalate (PET), polyoxydimethyl siloxane (PDMS), etc., the fabrication of the antenna might involve additional steps of creating and stabilizing the substrate in the lab but the thickness of the substrate can be controlled. The starting shape of the radiator can be a rectangular antenna [1–5], triangular [6], circular [7] monopole, or a dipole [8–10]. The foldable dipole in Ref. [8] simply uses two metallic rectangular sheets separated by a small gap. The antenna in Ref. [9] is made of small segments of rectangular patches forming a meandered line. Table 19.1 gives an overview of the slotted antennas in terms of dimension, the substrate used, and antenna geometry with the feed type. Table 19.1 compares the performance of the antennas in terms of bandwidth and gain.

Table 19.1 Overview of works related to slotted antennas

Ref. no.	*Antenna dimensions (mm^3)*	*Flexible*	*Substrate*	*Antenna type*	*Bandwidth (GHz)*	*Gain (dB or dBi)*	*Target application*
[1]	16×26×0.6	X	RT5880	CPW-fed rook-shaped	2.6/1.5	4.42	5G/IoT
[2]	25×35×1.6	X	FR4	CPW-fed rectangular antenna with slits	1.6/0.5	8.9	IoT
[3]	54×38×1	X	FR4	CPW-fed six-shaped antenna	3.4–3.69	6	IoT
[4]	18×24×0.8	X	FR4	Patch on rear with coupled-feed	1.5–2.8	3.1–4.2	IoT
[5]	138×40×1.6	X	FR4	CPW-fed slotted rectangular patch with beveled edges with tapered feed line	163%	1.2–4.2	IoT
[6]	64×42×0.135	Yes	PET	CPW-fed Bowtie-shaped slot antenna with asymmetrical flare angles	2.1–4.35	6.3	IoT
[7]	64×54×1.6	X	FR4	Microstrip-fed multi-bloom-shaped antenna with partial ground	1.6–2.45	2.5	IoT
[8]	43×29	Yes	X	Foldable dipole	1.3–2.9		GPS, ISM, 3GPP WCDMA, LTE
	36×25×0.8 41×25×0.8	X	FR4	Non-foldable dipole	1.35–2.75		GPS, WCDMA/ LTE, ISM
[9]	220×80×0.04	Yes	Acrylic sheet	Meandered line dipole antenna	0.730–0.9	2.1–2.25	LTE 700, LTE 800
[10]	Dipole antenna: 20×60×0.44; Loop antenna: 75×80×0.44	Yes	Jeans cloth	Dipole antenna/loop antenna	0.130/0.045	2.3/4.3	

19.3.2 Printed antennas

This section examines antennas that were printed using silver nanoparticle ink [11,12]. Silver nanoparticles provide superior conductivity, stability, adhesion, and resolution [12]. An inkjet printer is primarily used for printing the antennas on flexible substrates. The printed antenna can be designed to have any shape. Table 19.2 gives a comparison of some printed antennas based on dimensions, bandwidth, and gain.

19.3.3 Circularly polarized antennas

Antennas radiate with elliptical polarization that can be characterized by axial ratio, tilt, and the sense of rotation. Circular polarization can be achieved if we can make the axial ratio equal to one. The Antenna produces circularly polarized waves when it radiates two orthogonal field components with equal amplitude. To ensure the orthogonality of the field components, researchers prefer using the dual feed method as a direct method. However, in IoT applications due to limited space availability, researchers use stubs [14], Yagi antenna like structures having directors and reflectors [15] or strategically placed slots [16]. Circularly polarized antennas have found use in indoor applications, home automation applications, and Global Positioning System (GPS) applications. Table 19.3 compares some antennas with circular polarization based on 3 dB axial ratio bandwidth, impedance bandwidth, and gain.

19.3.4 Reconfigurable antennas

Antennas that can be tuned to resonate at different frequencies through external control are called reconfigurable antennas or tunable antennas. These generally have varactor capacitors [17,18] and/or PIN diodes [18]. The current flow to a particular part of the arm can be controlled by charging/discharging the capacitor [17,18], and by switching the PIN diode on/off [18], thus changing the effective length of the arms or slots. This operation leads the antenna to resonate at different frequencies. The current through the PIN diode can be controlled through the addition of automatic tuning circuits. The reconfigurable antenna in Ref. [17] uses a digitally tunable capacitor (DTC) with one arm of the IFA connected to the ground through a 7 mm short circuit. The other side is folded and printed on back of the printed circuit board. The DTC was found to consume only 1% of the entire power. Table 19.4 compares the performance of the reconfigurable antennas for IoT applications. The antenna in Ref. [18] is a quasi-square patch antenna with variable band-notch filtering. It has a partial ground plane with two circular and one rectangular slots. The activation or the deactivation of the PIN diode along with the charging and discharging of variable capacitors filters out the 3.5–6.087 GHz band.

Table 19.2 Overview of works related to printed antennas

Ref. no.	*Antenna dimensions (mm^3)*	*Flexible*	*Substrate*	*Antenna type*	*Bandwidth (GHz)*	*Gain (dB or dBi)*	*Target application*
[11]	40×5×1.254		RO4003	Printed planar inverted-F	0.013/0.029	1.7/2.3	IoT
[12]	33.1×32.7×0.254	Yes	Photo paper	CPW-fed circular monopole patch antenna	27	4.87	IoT
[13]	47×25×0.135	Yes	PET	CPW-fed circular monopole antenna	3.78–10.70/15.18–18	4	IoT

Table 19.3 Overview of works related to antennas with circular polarization

Ref. no.	*Antenna dimensions (mm^3)*	*Flexible*	*Substrate*	*Antenna type*	*3 dB axial ratio bandwidth*	*Impedance bandwidth (GHz)*	*Gain (dB or dBi)*	*Target application*
[14]	34×25×0.1	Yes	Ultralam 3850	CPW-fed scoop-like slot antenna	21%	4–11	2.7 dB	IoT
[15]	Total height: 19.6 mm	X	FR4	Three-layer rotated stack Yagi antenna	8%	6.3% around 2.45 GHz	4.25	IoT
[16]	55×55×1.6	X	FR4	CPW-fed rectangular radiator with a square ground plane with a wide slot	11%	1.12–4.72		L1, L2, L5 bands

Table 19.4 Overview of works related to reconfigurable antennas

Ref. no.	*Antenna dimensions (mm^3)*	*Flexible*	*Substrate*	*Antenna type*	*Bandwidth (GHz)*	*Target application*
[17]	60×13×7	X	FR4	Reconfigurable inverted-F antenna using a DTC	0.06	Low-power LoRa
[18]	30×30×1.6	X	FR4 and Rogers TMM4	Tunable quasi-square patch antenna with partial ground plane with two circular and one rectangular slot	3–51 with 3.5–6.087 GHz band filtered by switching on PIN diode	IoT applications

19.3.5 Antenna arrays for IoT

Antenna arrays have also been developed for IoT communications. Multiple Input, Multiple Output (MIMO) type antenna arrays [19–22] can provide diversity in terms of space and time. However, other antenna arrays [23] can provide much-needed gain without compromising the bandwidth. The MIMO antenna in Ref. [20] has four elements. Each antenna element is a sickle-shaped radiator with an L-shaped slit and a complementary split-ring resonator (CSRR) to reject the Bluetooth band (2.45 GHz), WLAN band (5.5 GHz), and downlink band of X-band satellites (7.5 GHz) from the super wide bandwidth. Each element is fed through a tapered line. Every element has a complementary slot etched in the ground plane. An octa-port MIMO antenna in Ref. [22] has eight elements. Each element is a modified circular patch on the same side as the modified ground and is fed through metallic vias from the feedline situated on the top face. Each element is placed within a semi-circular decoupling ring along with its ground structure for minimizing the mutual coupling with other elements. Three series-fed log-periodic antenna arrays are designed in Ref. [23]. Each element is an inset-fed square patch antenna. Table 19.5 examines some of the arrays for IoT applications.

19.4 ANTENNAS FOR BIOMEDICAL APPLICATIONS

Antennas that are meant for biomedical applications can be used for telemetry, diagnosis, and therapy [24,25]. The antennas that are used for telemetry should have high directivity away from the body and a low Specific Absorption Rate (SAR), while those destined for diagnosis and therapeutic applications should produce a strong SAR in the region of interest and have to be highly directional. The antennas may be placed on body or implanted inside the body, ingested, or can be placed outside of the body. However, the prime application is to connect to the Internet through a smartphone or computer for sending data of various vital information for continuous monitoring through the IoT network.

Different frequency bands widely used for biomedical applications are Medical Implant Communication Services (MICS) which works in 402–405 MHz and is used by devices that are implanted inside the body. Medical Radio Communications Service (Med Radio) which uses 401–406, 413–419, 426–432, 438–444, and 451–457 MHz bands, and Industrial, Scientific, and Medical (ISM) band comprising 433.05–434.79, 902–928, 2,400–2,483.5, and 5,725–5850 MHz. This is an unlicensed band and is used by Bluetooth and WLAN technologies. Wireless Medical Telemetry Service bands comprise 608–614, 1,395–1,400, and 1,427–1,432 MHz. This band is widely used for telemetry services. Medical Body Area Network (MBAN) uses a 2,360–2,400 MHz band. Ultra-Wideband (UWB) covers 3,100–10,600 MHz.

Table 19.5 Overview of works related to antenna arrays

Ref. no.	*Antenna dimensions (mm^3)*	*Flexible*	*Substrate*	*Antenna type*	*Bandwidth (GHz)*	*Gain (dB or dBi)*	*Isolation (dB)*	*Target application*
[19]	50.8 μm thickness	Yes	Kapton polyimide	Two element CPW-fed semi-elliptical MIMO antenna	Entire UWB range	0.2–5.5	More than −20 dB	IoT
[20]	0.5 mm thickness	X	Rogers RO4003C	Four-element sickle-shaped MIMO antenna.	1.3–40 with rejection of the Bluetooth band (2.45 GHz), WLAN band (5.5 GHz) and downlink band of X-band satellites (7.5 GHz)	7	More than −22 dB	IoT
[21]	44×44×1.6	X	FR4	Four-port MIMO antenna of compact split-ring resonator type	1.44, 2.3, 4.2	2	Max of 24.8 dB	IoT
[22]	70×70×1.6	X	FR4	Octa-port UWB modified circular MIMO antenna	3–12 GHz			IoT
[23]	Five-element: 14×25.5×0.762 Seven-element: 14×30.5×0.762 Nine-element: 14×38.5×0.762	X	Rogers RO4350B	Five-element, seven-element, nine-element series-fed log-periodic microstrip arrays	Five-element: 24.52–31.43 Seven-element: 24.48–32.48 Nine-element: 21.71–33.91	Five-element: 11 Seven-element: 11 Nine-element: 12		

All of the antennas described above can be used for biomedical applications as well. Implantable antennas have been reviewed in Refs. [26–28]. In this section, we examine some wideband antennas designed specifically for biomedical purposes.

19.4.1 UWB antennas

Antennas with wideband operation can be used for sensing applications in one part of the spectrum while being used for handling high data rate communication using other parts of the spectrum. Ultra-Wideband (UWB) antennas [29–32] are designed by strategically cutting slots from the patch and the ground plane. The antenna in Ref. [29] has meandering slots cut from the rectangular patch and the ground plane. The UWB antennas in Refs. [30,32] are designed with a simple rectangular patch antenna and with a partial and modified ground. The antenna in Ref. [31] uses slots on both patch and ground for enhancing bandwidth. Table 19.6 examines some of the UWB antennas for biomedical applications.

19.4.2 Flexible antennas

These antennas are generally designed for diagnosis and therapeutic purposes. Since the antennas are in direct contact with the human or the animal body, it becomes imperative for the researcher to adapt designs that do not cause any discomfort in the long-run use. Moreover, the antenna must not exhibit high detuning when in contact with the body or while it is bent. Kapton [33], PDMS [34], and textiles [35,36] become the first choice of substrates for these antennas. The antennas in Ref. [33] comprise four loops each. The radius of loops determines the frequency of operation. The antenna in Ref. [34] is a multilayer patch antenna which has been folded and shorted with the ground. The antenna in Ref. [35] is an inset microstrip patch with a defected ground structure using a felt and Teflon double layer. The antenna in Ref. [36] is a conformal patch antenna made entirely of textiles. The conducting parts are made of 0.08 mm thick copper-plated woven polyester fabric and a 0.32 mm thick woven polyester fabric serving as the dielectric. A padding layer of the same dielectric fabric is provided to avoid corrosion of the antenna. The antenna is used specifically for breast hyperthermia and it is found that SAR stands at more than a 100 W/kg on superficial level tissues, resulting in a temperature rise of 3.3°C (Table 19.7).

19.5 A DESIGN FOR VERIFICATION

In this section, we propose a miniaturized antenna for IoT and wearable applications. The antenna is based on our study of literature and for the

Table 19.6 Overview of works related to UWB antennas for biomedical applications

Ref. no.	Antenna dimensions (mm^3)	Substrate	Antenna type	Bandwidth (GHz)	Gain (dB or dBi)	SAR (W/kg)	Target application
[29]	40×40×1.28	FR4	Microstrip-fed rectangular patch antenna with meandering slots and a modified ground	UWB range	5.36	0.467	Biomedical applications
[30]	23×21×1.58	FR4	Microstrip-fed rectangular patch antenna with a modified ground	UWB range	0.5–4.5		Breast cancer detection
[31]	17×25×0.787	RT/duroid 5880	Slotted rectangular radiator with partial ground	1.38	6	2.5	Biomedical applications
[32]	30×22×1.6	FR4	Microstrip-fed rectangular patch with sawtooth projections on partial ground	UWB range (14.3 GHz)	4.2		

Table 19.7 Overview of works related to flexible antennas for biomedical applications

Ref. no.	Antenna dimensions (mm^3)	Substrate	Antenna type	Bandwidth (GHz)	Gain (dB or dBi)	SAR (W/kg)	Target application
[33]	75 microns thick	Kapton	3 quadruple loop antennas for GSM 900, GSM 1800, and BLE				Biotelemetry IoT applications
[34]	200×200×0.1	PDMS	Folded-shorted multilayer patch	0.013/0.180	1.37/3.10	0.14/1.06	Biomedical applications
[35]	72.54×72.54×1.6	Felt and teflon double layer	Inset-fed microstrip patch antenna with defected ground structure (DGS)		6		Healthcare monitoring
[36]	3.2 mm thickness	Woven polyester fabric	Conformal patch antenna			More than 100	Breast hyperthermia

verification of the same. The authors have also worked on multi-band antennas for wearable and IoT applications [37–40].

The antenna structure has a cylindrical profile and is simulated on a 0.8 mm thick FR4 substrate. The design starts with a basic rectangular patch of 18.5 mm×22.5 mm on a cylindrical substrate of 34 mm diameter [41–43]. This results in a small antenna with a total volume of 726 mm^3.

To bring down the frequency to the target frequency, an inner ring is added. This addition of the ring enhances the bandwidth of the antenna. The addition of the parasitic patch in the gap greatly enhances the gain of the antenna. A circular slot cut in the middle of the ground aids in matching the port to the radiating patch. Figure 19.1 shows the final design of the antenna with all the dimensions in mm.

It can be seen from Figure 19.2 that the antenna resonates at 2.45 GHz with an $S11$ of −43 dB. This can be attributed to the circular slot in the ground. The impedance bandwidth of the antenna is observed to be 500 MHz (2.23 through 2.75 GHz) covering the 2.45 GHz ISM band completely. This band is important as it covers the band for Bluetooth and WLAN. From Figure 19.3, it can be seen that the antenna has a gain of 18 dB. Hence, the front-to-back ratio is calculated to be 29 dB. The radiation efficiency is found to be 84.32%. As the magnitude of the back lobes is less than −10 dB, the antenna is a very suitable candidate for off-body communications and IoT applications. The antenna can be used for biomedical applications like a connection with a base station for high data rate communications. Figure 19.4 shows the electric field distribution of the antenna at 2.45 GHz. The presence of the slots disturbs the surface current flow on the antenna as can be seen from Figure 19.5, leading to a wider bandwidth.

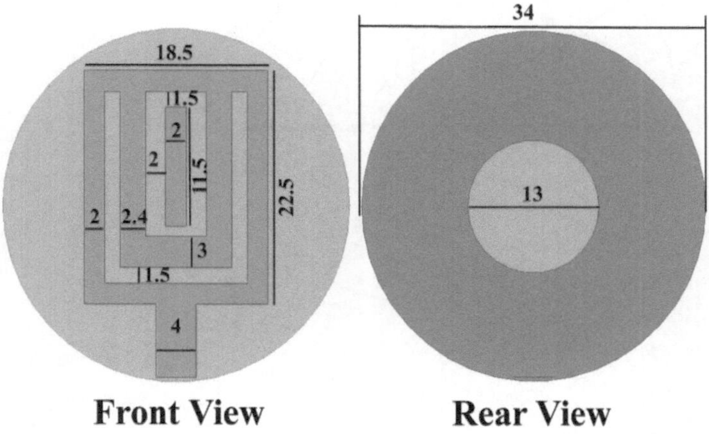

Figure 19.1 Design of proposed antenna with dimensions in mm.

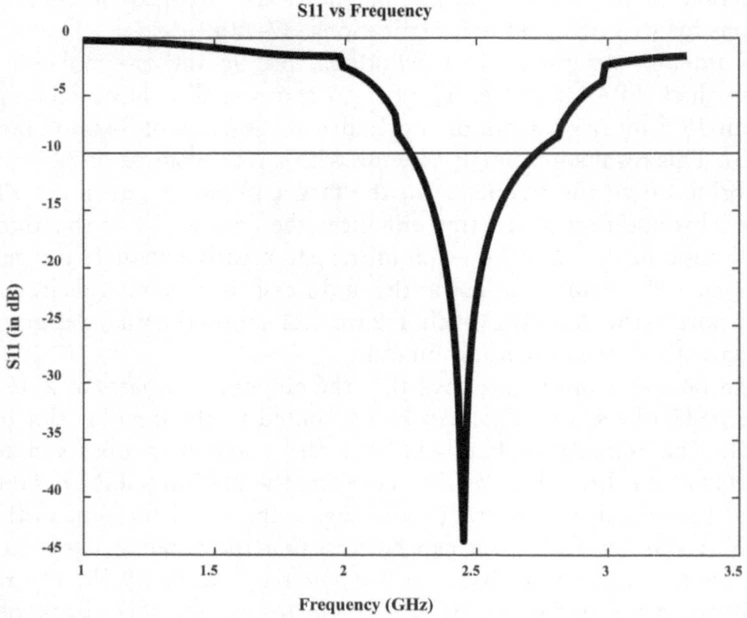

Figure 19.2 S11 vs frequency of the proposed antenna.

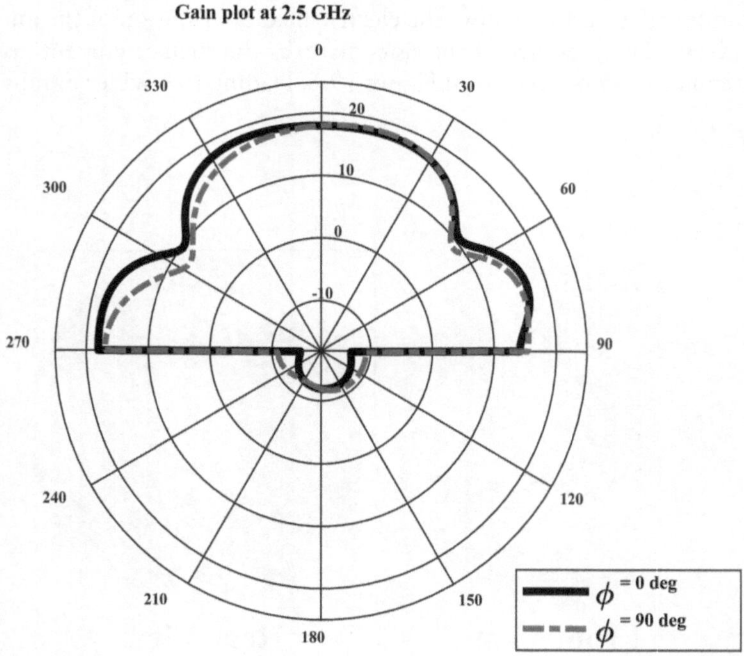

Figure 19.3 *E*-plane and *H*-plane plots of the antenna.

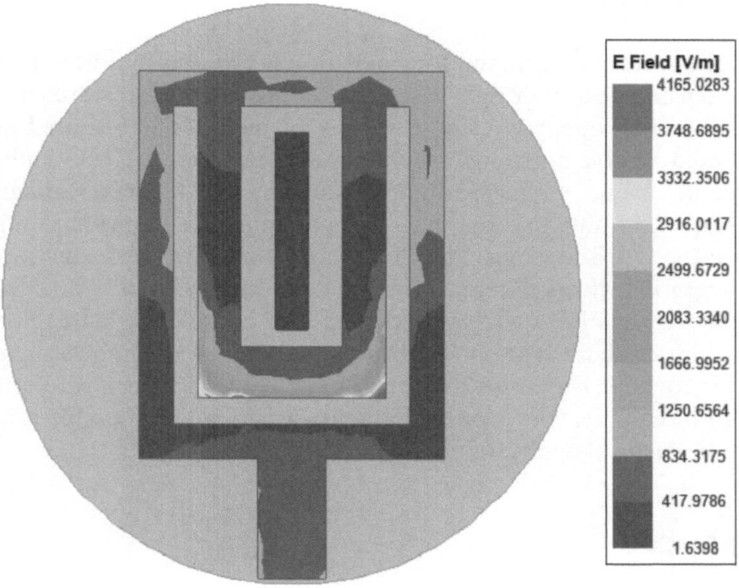

Figure 19.4 Electric field distribution of the antenna at 2.45 GHz.

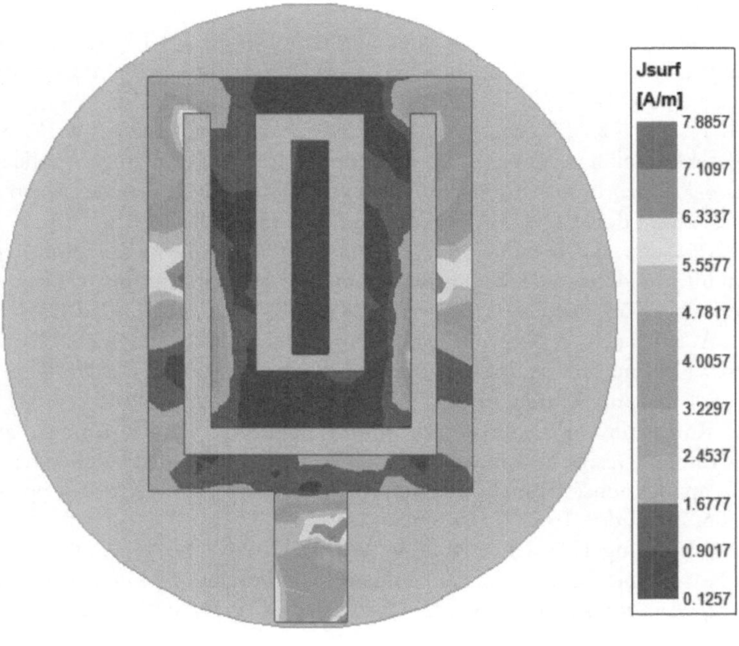

Figure 19.5 Current distribution on the antenna at 2.45 GHz.

19.6 CONCLUSIONS

Due to the increasing demand for multi-functional and multi-band antennas for sensing and high data rate connections for IoT, researchers have been developing antennas using a variety of methods like slotted antennas, printed flexible antennas, reconfigurable antennas, MIMO antennas, and antenna arrays. Biomedical applications include telemetry, diagnosis, and therapy. The antennas that have been designed for biomedical applications are the UWB antennas and flexible antennas. As the focus is on the easy integration of the antenna with the circuit boards and wider bands, the CPW feed has been widely used. A slotted rectangular patch antenna on a cylindrical base is proposed in this work. The antenna resonates at 2.45 GHz covering the entire 2.45 GHz ISM band with a peak gain of 18 dB and high directivity. The proposed antenna is suitable for a wide range of IoT and biomedical applications.

REFERENCES

1. M. Mokayef and M. A. Summakieh, An ultra-wideband antenna for IoT connectivity, *Int. J. Internet Things Web Serv.*, vol. 2, pp. 76–79, 2017.
2. Q. Awais, H. T. Chattha, M. Jamil, Y. Jin, F. A. Tahir, and M. U. Rehman, A novel dual ultrawideband CPW-fed printed antenna for Internet of Things (IoT) applications, *Wirel. Commun. Mob. Comput.*, vol. 2018, pp. 1–9, 2018, doi: 10.1155/2018/2179571.
3. S. Ashok Kumar and T. Shanmuganantham, Design of CPW-fed inverted six shaped antenna for IoT applications, *Trans. Electr. Electron. Mater.*, vol. 21, no. 5, pp. 524–527, 2020, doi: 10.1007/s42341-020-00213-z.
4. M. H. Tsoi, K. M. Wu, J. S. M. Yuen, Y. S. Choy, and S. W. Y. Mung, Wideband planar coupled-feed antenna for Internet of Things Applications, *Asia-Pacific Microw. Conf. Proceedings, APMC*, vol. 2020-Decem, pp. 460–462, 2020, doi: 10.1109/APMC47863.2020.9331528.
5. S. N. Azemi, N. K. Jiunn, M. A. Kamarudin, C. M. N. C. Isa, and A. Amir, An ultra-wideband CPW fed slot antenna for IoT applications, *J. Phys. Conf. Ser.*, vol. 1755, no. 1, 2021, doi: 10.1088/1742–6596/1755/1/012029.
6. M. A. Riheen, T. T. Nguyen, T. K. Saha, T. Karacolak, and P. K. Sekhar, CPW fed wideband bowtie slot antenna on pet substrate, *Prog. Electromagn. Res. C*, vol. 101, no. March, pp. 147–158, 2020, doi: 10.2528/PIERC20031402.
7. R. Ramasamy, V. Rajavel, M. Vasim Babu, C. Vinoth Kumar, and S. Parthiban, Design and analysis of multiband bloom shaped patch antenna for IoT applications, *Turkish J. Comput. Math. Educ.*, vol. 12, no. 3, pp. 4578–4585, 2021, doi: 10.17762/turcomat.v12i3.1848.
8. S. W. Y. Mung, C. Y. Cheung, K. M. Wu, and J. S. M. Yuen, Wideband rectangular foldable and non-foldable antenna for Internet of Things Applications, *Int. J. Antennas Propag.*, vol. 2019, 2019, doi: 10.1155/2019/2125713.

9. S. Jamwal, S. Gupta, Z. A. P. Zibran, K. R. Jha, and C. Singh, An optically transparent antenna for IoT applications, in *International Conference on Electrical and Electronics Engineering, ICE3 2020*, pp. 696–698, 2020, doi: 10.1109/ICE348803.2020.9122979.
10. S. Varma, S. Sharma, M. John, R. Bharadwaj, A. Dhawan, and S. K. Koul, Design and performance analysis of compact wearable textile antennas for IoT and body-centric communication applications, *Int. J. Antennas Propag.*, vol. 2021, pp. 1–12, 2021, doi: 10.1155/2021/7698765.
11. K. Diallo, A. Ngom, A. Diallo, J. M. Ribero, I. Dioum, and S. Ouya, Efficient dual-band PIFA antenna for the Internet of Things (IoT), in *2018 IEEE Conference Antenna Measurements and Applications CAMA 2018*, pp. 1–4, 2018, doi: 10.1109/CAMA.2018.8530584.
12. T. K. Saha, T. N. Knaus, A. Khosla, and P. K. Sekhar, A CPW-fed flexible UWB antenna for IoT applications, *Microsyst. Technol.*, no. December, 2018, doi: 10.1007/s00542-018-4260-0.
13. S. G. Kirtania, B. A. Younes, A. R. Hossain, T. Karacolak, and P. K. Sekhar, Cpw-fed flexible ultra-wideband antenna for iot applications, *Micromachines*, vol. 12, no. 4, 2021, doi: 10.3390/mi12040453.
14. S. R. Zahran, M. A. Abdalla, and Z. Hu, A flexible circular polarized wide band slot antenna for indoor IoT applications, in *2017 IEEE Antennas and Propagation Society International Symposium, Proceedings*, pp. 1163–1164, 2017, doi: 10.1109/APUSNCURSINRSM.2017.8072624.
15. J. H. Kim, C. H. Jeong, and W. S. Lee, Rotated stacked yagi antenna with circular polarization for IoT applications, *Appl. Comput. Electromagn. Soc. J.*, vol. 34, no. 8, pp. 1246–1249, 2019.
16. A. Birwal, S. Singh, B. K. Kanaujia, and S. Kumar, Broadband CPW-fed circularly polarized antenna for IoT-based navigation system, *Int. J. Microw. Wirel. Technol.*, vol. 11, no. 8, pp. 835–843, 2019, doi: 10.1017/S1759078719000461.
17. T. Houret, L. Lizzi, F. Ferrero, C. Danchesi, and S. Boudaud, Energy efficient reconfigurable antenna for ultra-low power IOT devices, in *2017 IEEE Antennas Propagation Society International Symposium*, vol. 2017, pp. 1153–1154, 2017, doi: 10.1109/APUSNCURSINRSM.2017.8072619.
18. H. R. D. Oskouei, A. R. Dastkhosh, A. Mirtaheri, and M. Naseh, A small cost-effective super ultra-wideband microstrip antenna with variable band-notch filtering and improved radiation pattern with 5g/IoT applications, *Prog. Electromagn. Res. M*, vol. 83, no. July, pp. 191–202, 2019, doi: 10.2528/PIERM19051802.
19. H. Raad, An UWB antenna array for flexible IoT wireless systems, *Prog. Electromagn. Res.*, vol. 162, no. June, pp. 109–121, 2018, doi: 10.2528/PIER18060804.
20. P. Kumar, S. Urooj, and A. Malibari, Design and implementation of quad-element super-wideband MIMO antenna for IoT applications, *IEEE Access*, vol. 8, pp. 226697–226704, 2020, doi: 10.1109/ACCESS.2020.3045534.
21. R. Nagendra and S. Swarnalatha, Design and performance of four port MIMO antenna for IOT applications, *ICT Express*, no. xxxx, pp. 4–7, 2021, doi: 10.1016/j.icte.2021.05.008.

22. P. Palanisamy and M. Subramani, Design of metallic via based octa-port UWB MIMO antenna for IoT applications, *IETE J. Res.*, no. March, 2021, doi: 10.1080/03772063.2021.1892540.
23. T. Varum, J. Caiado, and J. N. Matos, Compact ultra-wideband series-feed microstrip antenna arrays for iot communications, *Appl. Sci.*, vol. 11, no. 14, pp. 1–15, 2021, doi: 10.3390/app11146267.
24. A. Gupta, A. Kansal, and P. Chawla, A survey and classification on applications of antenna in health care domain: Data transmission, diagnosis and treatment, *Sadhana – Acad. Proc. Eng. Sci.*, vol. 46, no. 2, 2021, doi: 10.1007/s12046-021-01586-4.
25. S. Mahapatra and M. N. Mohanty, A review on state-of-art techniques of antennas for body area networks, *Int. J. Sensors, Wirel. Commun. Control*, vol. 11, no. 6, pp. 604–618, 2021, doi: 10.2174/221032791099920122815254 3.
26. C. Liu, Y. Guo, S. Xiao, and R. Tmm, A review of implantable antennas for wireless biomedical devices, *Forum Electromagn. Res. Methods Appl. Technol.*, no. 2, pp. 1–8, 2010.
27. A. Kiourti and K. S. Nikita, A review of implantable patch antennas for biomedical telemetry: Challenges and solutions, *IEEE Antennas Propag. Mag.*, vol. 54, no. 3, pp. 210–228, 2012, doi: 10.1109/MAP.2012.6293992.
28. A. W. Damaj, H. M. El Misilmani; and S. A. Chahine, Implantable antennas for biomedical applications: An overview on alternative antenna design methods and challenges, in *Proceedings on 2018 International Conference on High Performance Computing & Simulation, HPCS 2018*, pp. 31–37, 2018, doi: 10.1109/HPCS.2018.00019.
29. A. Biswas, A. J. Islam, A. Al-Faruk, and S. S. Alam, Design and performance analysis of a microstrip line-fed on-body matched flexible UWB antenna for biomedical applications, in *ECCE 2017 – International Conference on Electrical, Computer and Communication Engineering*, pp. 181–185, 2017, doi: 10.1109/ECACE.2017.7912902.
30. N. Hammouch and H. Ammor, Smart UWB antenna for early breast cancer detection, *ARPN J. Eng. Appl. Sci.*, vol. 13, no. 11, pp. 3803–3808, 2018.
31. A. Smida, A. Iqbal, A. J. Alazemi, M. I. Waly, R. Ghayoula, and S. Kim, Wideband wearable antenna for biomedical telemetry applications, *IEEE Access*, vol. 8, no. January, pp. 15687–15694, 2020, doi: 10.1109/aCCESS.2020.2967413.
32. S. Sahoo, L. P. Mishra, M. N. Mohanty, and R. K. Mishra, Design of compact UWB monopole planar antenna with modified partial ground plane, *Microw. Opt. Technol. Lett.*, vol. 60, pp. 578–583, 2018, doi: 10.1002/mop.31010.
33. H. A. Damis, N. Khalid, R. Mirzavand, H. J. Chung, and P. Mousavi, Investigation of epidermal loop antennas for biotelemetry IoT applications, *IEEE Access*, vol. 6, pp. 15806–15815, 2018, doi: 10.1109/ACCESS.2018.2814005.
34. R. Joshi et al., Analysis and design of dual-band folded-shorted patch antennas for robust wearable applications, *IEEE Open J. Antennas Propag.*, vol. 1, no. February, pp. 239–252, 2020, doi: 10.1109/ojap.2020.2991343.
35. A. B. Mustafa and T. Rajendran, An effective design of wearable antenna with double flexible substrates and defected ground structure for healthcare monitoring system, *J. Med. Syst.*, vol. 43, no. 186, pp. 1–11, 2019.

36. Y. Mukai and M. Suh, Development of a conformal woven fabric antenna for wearable breast hyperthermia, *Fash. Text.*, vol. 8, no. 1, pp. 1–12, 2021, doi: 10.1186/s40691-020-00231-8.
37. S. Mahapatra, S. Satrusallya, and M. N. Mohanty, A circular ultra-wideband antenna for wearable applications, in *Advances in Intelligent Computing and Communication*, Springer, 2020, pp. 532–536.
38. S. Mahapatra, J. Mishra, and M. Dey, A dual-band inset-fed octagonal patch antenna for wearable applications, in *Advances in Intelligent Computing and Communication*, Springer, 2021, pp. 699–706.
39. S. Mahapatra and M. N. Mohanty, An optimized feed hexagonal antenna with defective ground plane for UWB body area network application, *Instrum. Mes. Metrol.*, vol. 20, no. 5, pp. 261–267, 2021, doi: 10.18280/i2m.200503.
40. S. Satrusallya, S. Mahapatra, and M. N. Mohanty, Design of array antenna for body area network, in *Advances in Intelligent Computing and Communication*, Springer, 2020, pp. 189–193.
41. P. K. Malik, Chapter 4. Mathematical modeling and principle of wireless communication, in *Energy Harvesting Technologies for Powering WPAN and IoT Devices for Industry 4.0 Up-Gradation*, Nova Science Publishers, Inc., Hauppauge, NY, April 2020, ISBN: 9781536169430.
42. R. Roges and P. K. Malik, Planar and printed antennas for Internet of Things-enabled environment: Opportunities and challenges, *Int. J. Commun. Syst.*, vol. 34, no. 15, p. e4940, 2021, https://doi.org/10.1002/dac.4940, ISSN: 1099-1131.
43. A. Rahim and P. K. Malik, Analysis and design of fractal antenna for efficient communication network in vehicular model, in *Sustainable Computing: Informatics and Systems*, Volume 31, Elsevier, 2021, https://doi.org/10.1016/j.suscom.2021.100586, ISSN 2210-5379.

Chapter 20

A compact low-cost impedance transformer-fed wideband monopole antenna for Wi-MAX N78-band and wireless applications

A. Varshney and V. Sharma
Gurukul Kangri University

T. M. Neebha
Karunya Institute of Technology and Science

Roshan Kumar
Henan University

CONTENTS

DOI: 10.1201/9781003347057-20

20.1 INTRODUCTION

UWB technology has emerged rapidly due to its advantages of low cost, low complexity, low power consumption, high-speed data rate, and high capacity. According to the ITU-R recommendations, UWB usually refers to a signal or system that either has a large fractional bandwidth (BW) that exceeds 20% or a large absolute bandwidth of more than 500 MHz [1]. Moreover, the US Federal Communications Commission (FCC) confirmed 7.5 GHz bandwidth (ranges from 3.1 to 10.6 GHz) for UWB technology in 2002 [2]. The UWB operation is achieved by a planar monopole antenna since the monopole antenna has a wider bandwidth, small dimensions, and stable radiation properties [3]. 5G frequency bands are categorized into sub-6 GHz and high-frequency millimeter waves. Earlier, the sub-6 GHz comprises 3.3–3.4, 3.4–3.6, and 4.8–5 GHz frequency bands. Furthermore, the 3.3–3.6 GHz range is named the N78-band. Chen et al. have investigated and fabricated a Γ-shaped compact base station antenna for a 5G N78-band [4]. It is a challenging task to design an antenna that may cover all the aforesaid bands well-suited with other necessary performances since the voltage standing wave ratio (VSWR) is requisite to be lower than 1.5, i.e., $S_{11} < -4$ dB. Therefore, broadening the fractional bandwidth is imperative. Li et al. utilized an E-shaped differential feeding structure to cover 3.12–3.9 GHz for VSWR < 1.5. However, the designed configuration was complex and acted without the high-frequency part of sub-6 GHz [5]. The antennas failed to meet the requirement since they had VSWRs lower than 2, while they had satisfactory fractional bandwidths to cover 3.3–5 GHz [6,7]. In literature, many techniques have been applied to achieve a wider 5G bandwidth such as parasitic elements [8], slots on the ground plane [9], and double-side patches connected by shorting pins [10]. Communication systems usually involve small antennas to achieve the requirements of miniaturization. Nevertheless, many comparatively large-sized monopole antennas have been recently reported [11–13]. Moreover, those designs are difficult to miniaturize. Besides [13,14], the majority of the existing antennas have single or dual-band characteristics. Gunamony et al. have designed and investigated a single layer miniaturized asymmetric coplanar strip (ACS)-fed monopole antenna designed to operate in LTE band-40 and 5G mid-band frequency ranges [15]. Sahoo et al. have designed a proximity-coupled-fed microstrip circular patch antenna (CPA) for 5G applications that has a fractional bandwidth of 5.7% with 7.95 dBi gain [16].

In this chapter, an impedance transformer-fed, low-cost, miniaturized, monopole, wideband (WB) spearhead antenna using a partial ground structure for 5G N-78 band and wireless applications is proposed and investigated. Partial ground is used to achieve ultra-wideband performance around the resonating frequency.

20.2 ANTENNA DESIGN AND DEVELOPMENT

The antenna is designed on a low-cost FR-4 epoxy substrate with a relative permittivity of 4.4, a thickness of 1.6 mm, and a loss tangent of 0.02.

20.2.1 Design equations

The following fundamental microstrip circular patch antenna design equations were used for the design [17–21]:

$$a = \frac{F}{\left\{1 + \frac{2h}{\pi \varepsilon_r F\left[\ln\left(\frac{\pi F}{2h}\right) + 1.7726\right]}\right\}^{0.5}} \tag{20.1}$$

where

$$F = \frac{8.791 \times 10^9}{f_r \sqrt{\varepsilon_r}} \tag{20.2}$$

20.2.2 Optimized spearhead antenna

The geometry of the miniaturized UWB monopole antenna that comprises a 50-Ω microstrip-fed line, quarter-wave transformer, and a spearhead-shaped radiating patch is illustrated in Figure 20.1. A patch is printed on an FR4 substrate with a thickness of 1.6 mm, a relative permittivity of 4.4, and a loss tangent of 0.02. It can be easily concluded from Figure 20.1 that the proposed antenna has compact dimensions of 13 mm × 26 mm. The monopole structure consists of overlapped circular shaped patches forming the radiating element on the top side of the substrate, while on the other side of the substrate, a partial rectangular ground plane [22–24]. The optimized dimensions of the proposed antenna are labeled in Figure 20.1, and their values are listed in Table 20.1.

20.2.3 Development of spearhead radiating conductor

First, the radius of the conventional patch at 3.5 GHz design frequency is calculated by fundamental design equations 20.1 and 20.2. Then, the circular patch is converted into a rhombus with the same diameter as calculated

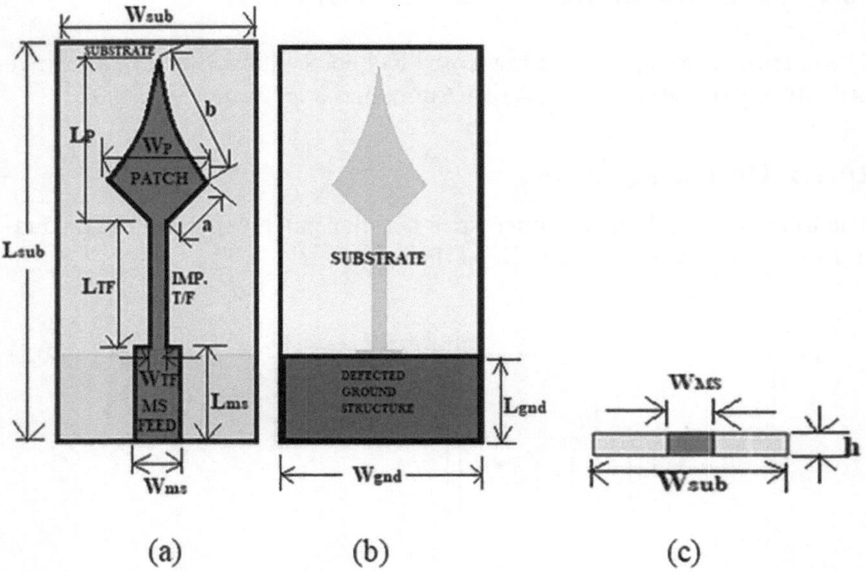

Figure 20.1 Proposed spearhead antenna structure. (a) Top view (patch), (b) rear view (ground) and (c) side view (excitation port).

Table 20.1 Optimized parameters of the proposed antenna

Element	*Designation/description*	*Value (mm)*
W_{sub}	Substrate width	13.0
L_{sub}	Substrate length	26.0
W_{gnd}	Ground width	13.0
L_{gnd}	Ground length	5.57
W_{ms}	Microstrip feed width	3.0
L_{ms}	Microstrip feed length	6.0
W_p	Radiating patch width	6.276
L_p	Radiating patch length	10.55
W_{TF}	Quarter-wave transformer width	0.849
L_{TF}	Quarter-wave transformer length	8.2
a	Slant dimension of patch	3.856
b	Arc length of patch	8.553
h	Thickness of substrate	1.6

in a conventional patch. The spearhead-shaped radiating conductor is manufactured by subtracting three circles from the rhombus as shown in Figure 20.2. The positions of the three centers (C2, C3, and C4) of the circles with respect to the rhombus' (centered at C1) three corners are displayed in Figure 20.2a. The spearhead shape of the radiating conductor is

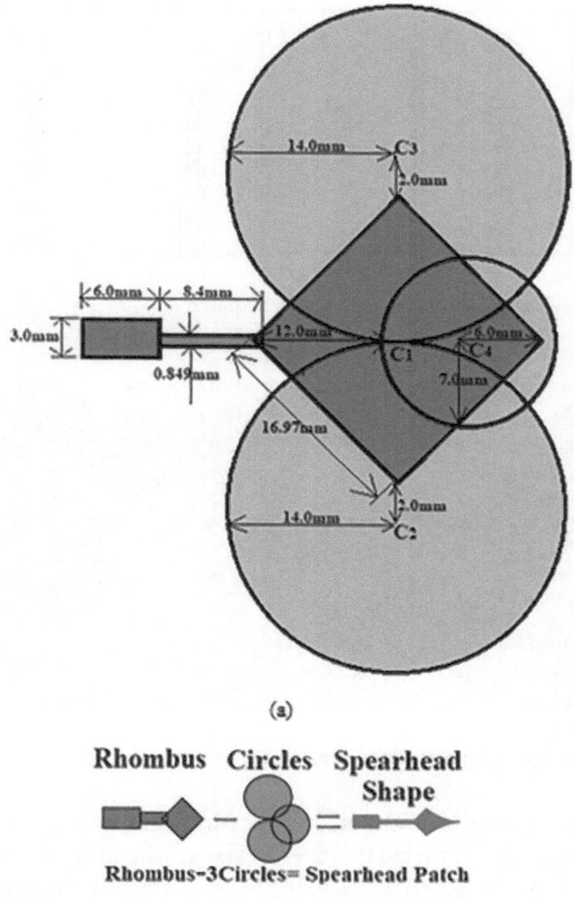

Figure 20.2 Development of spearhead patch. (a) Concept and (b) final spearhead patch.

the result of the subtraction of the three circles from the rhombus as shown in Figure 20.2b. Finally, the spearhead-like structure comes in its final shape as shown in Figure 20.2b by adding this with the 50 Ω microstrip-fed line in conjunction with the quarter-wave impedance transformer coupled line.

20.2.4 Miniaturization process of spearhead monopole antenna

The miniaturization process has been illustrated in Figure 20.3. Antenna 1 (Ant.1) through antenna 6 (Ant.2) in Figure 20.3 show the step-by-step development and miniaturization process of the spearhead microstrip monopole antenna. First, the bottom ground is compressed across the

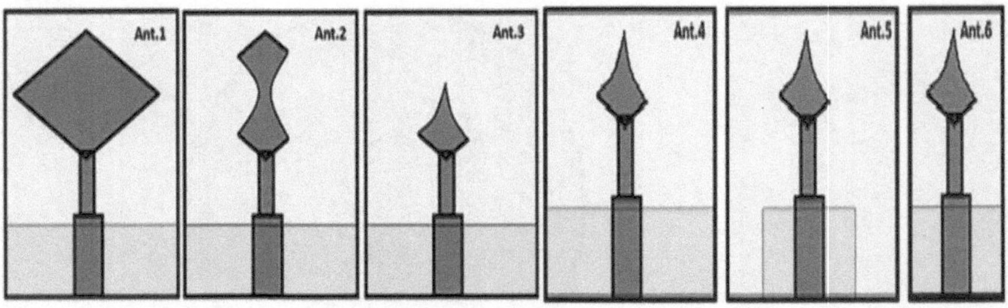

Figure 20.3 Miniaturization process of spearhead patch.

Figure 20.4 Prototype of miniaturized spearhead antenna.

length to achieve a wideband performance and the ground width is compressed (Ant.5) to miniaturize the overall size of the antenna and to set the frequency near the design frequency. In this miniaturization process, a total 85.33% reduction in antenna (Ant.6) size has been achieved with respect to the size of the initially designed conventional antenna (Ant.1).

20.2.5 Prototype of spearhead antenna

The reflection coefficient measurement and the rear and front views of the final optimized miniaturized spearhead monopole antenna are shown in Figure 20.4.

20.2.6 RLC electrical equivalent circuit of spearhead antenna

The generated RLC electrical equivalent circuit of the miniaturized spearhead monopole antenna is shown in Figure 20.5 [17,19,21,25]. Each mentioned part of the RLC equivalent circuit is clearly explained by its meaning. The 50 Ω microstrip line impedance is matched with the 243σ spearhead

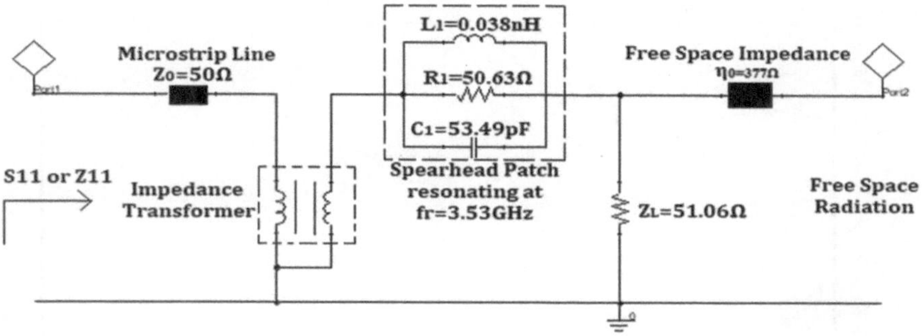

Figure 20.5 RLC electrical equivalent circuit of the spearhead antenna.

radiating conductor at a frequency of 3.5 GHz and resonance will occur. At resonance, the imaginary part of input impedance Z_{11} becomes almost zero and the circuit becomes resistive and maximum radiation will occur.

20.3 RESULTS AND DISCUSSION

The results obtained with the miniaturized monopole antenna are mentioned in Table 20.2.

20.3.1 Reflection coefficient curves

A comparative plot of simulated and measured reflection coefficients, S_{11} [dB] values with respect to frequencies in GHz, are illustrated in Figure 20.6.

Table 20.2 Simulated versus measured result comparison

Parameter	*Simulated*	*Measured*	*Comment*
Resonance frequency (GHz)	3.53	3.52	Measured resonance frequency observed much closer to designed frequency 3.5 GHz
Directivity, D (dBi)	3.04	3.21	Improves
Gain, G (dBi)	2.92	2.47	Reduced
Radiation efficiency, η (%)	95.33	84.33	Decreased
Reflection coefficient, S11 [dB]	−39.57	−25.98	Decreased because of fabrication and connector soldering losses
VSWR (mag.)	1.017	1.107	Decreased because of losses
Impedance, Z_0 (Ω)	50.23	45.88	Little amount of matching loss because of fabrication and connector soldering
−10 dB fractional bandwidth (FBW) (GHz)	3.28–3.83 (15.58%)	3.18–4.06 (25%)	Improves

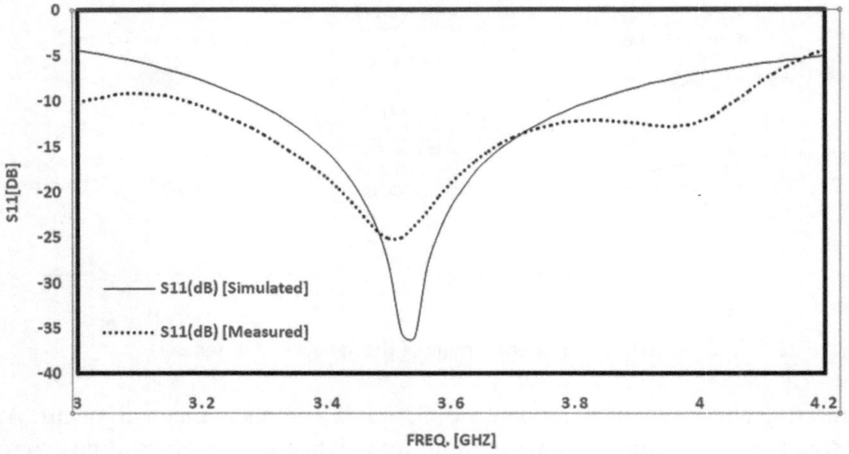

Figure 20.6 Simulated vs. measured reflection coefficient.

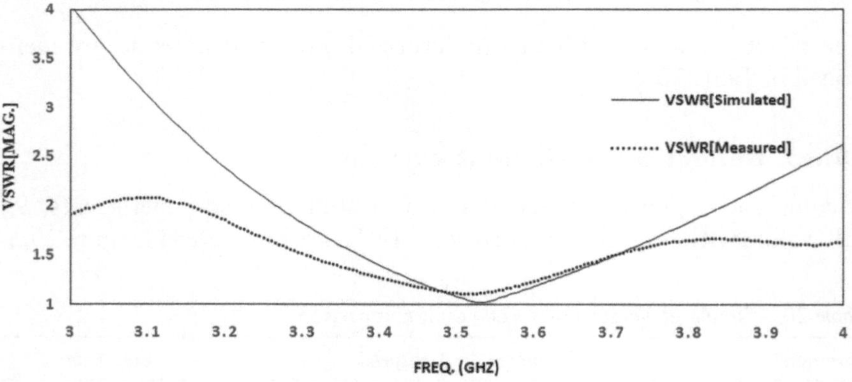

Figure 20.7 Simulated vs. measured VSWR curves.

By comparison between the two S_{11} values, it is concluded that the –10 dB values for the measured S_{11} curve has higher bandwidth than the simulated reflection coefficient values. The resonance frequencies for both the measured and simulated curves are in proximity. Therefore, a good agreement between the simulated and measured values has been observed.

20.3.2 VSWR curves

The measured voltage standing wave ratio (VSWR) is plotted together on the curve of simulated VSWR as depicted in Figure 20.7. By the curve

VSWR value 1.107 is observed. It is noticed that the measured VSWR curve shows that an excellent impedance matching has been achieved as the whole VSWR value lies below 2.0 for the entire bandwidth of interest.

20.3.3 Radiation pattern

The E-plane (φ=0°) and H-plane (φ=90°) radiations are visualized and are represented by the dotted and solid lines, respectively, as shown in Figure 20.8. The H-plane pattern has the shape of eight while the E-plane pattern is like a circle. The three-dimensional radiation pattern of gain corresponding to all azimuth and elevation angles has been shown in Figure 20.9. It is observed that an omnidirectional radiation pattern has been obtained with a gain value of 0.77 dB or 2.92 dBi corresponding to the designed frequency of 3.5 GHz.

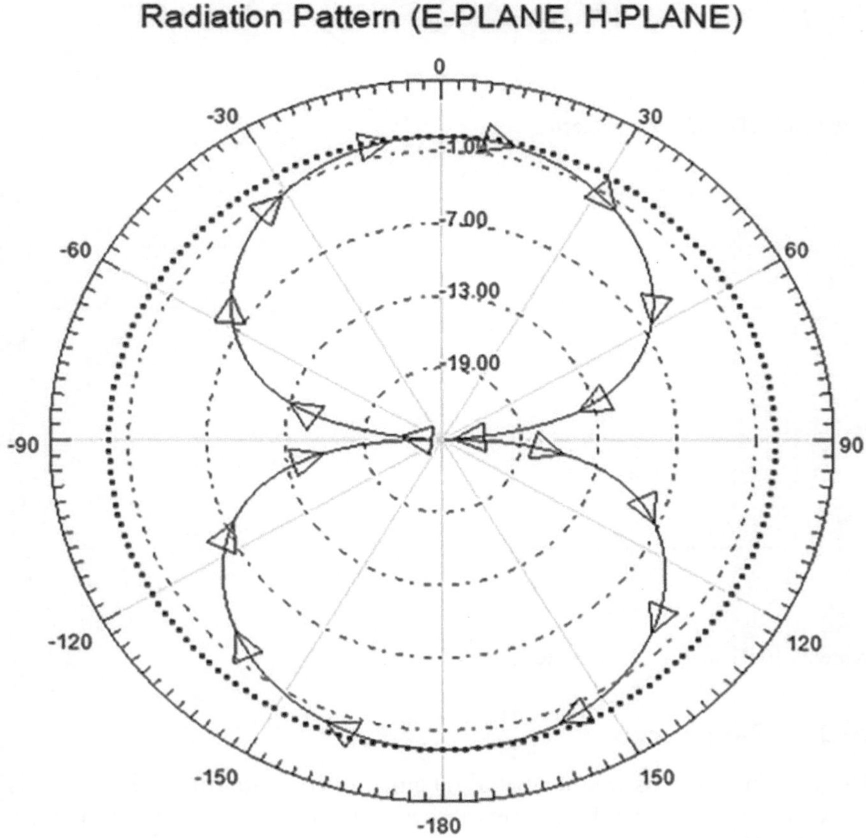

Figure 20.8 E-plane (φ=0°) and H-plane (φ=90°) radiation pattern at 3.5 GHz.

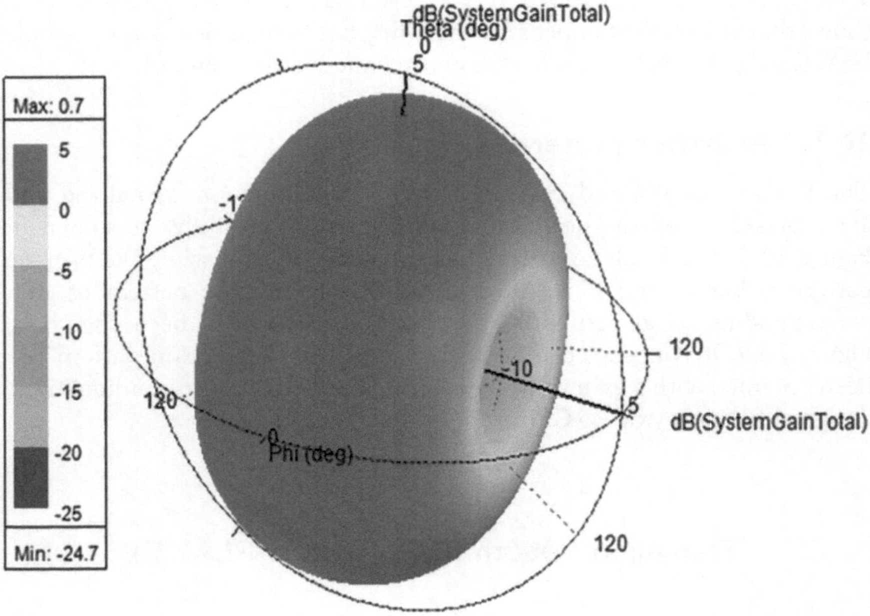

Figure 20.9 3D-radiation pattern.

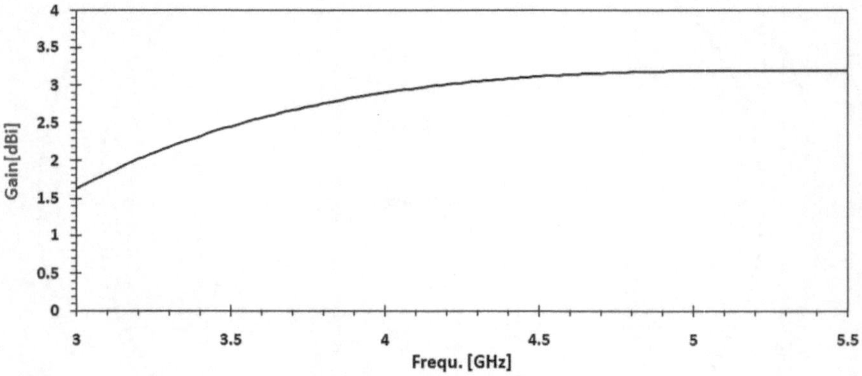

Figure 20.10 Gain vs frequency plot.

20.3.4 Gain curve

The gain (dBi) versus frequency (GHz) curve of the miniaturized antenna is shown in Figure 20.10. It is observed from the curve that gain has an average 3.20 dBi value corresponding to frequencies higher than 4.2 GHz. The gain of 2.92 dBi is noticed at resonance frequency.

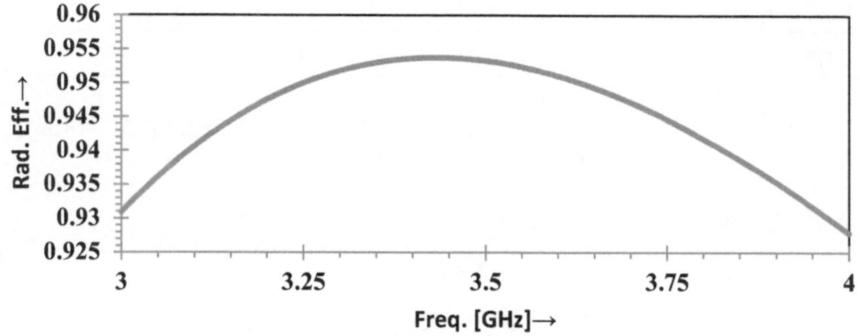

Figure 20.11 Radiation efficiency vs frequency.

20.3.5 Radiation efficiency curve

The radiation efficiency versus frequency (GHz) curve of the miniaturized antenna is shown in Figure 20.11. The highest radiation efficiency of 95.37% occurs at a frequency of 3.44 GHz.

20.3.6 Comparison with recently published work

The performance comparison between the proposed spearhead antennas with the recently published similar antennas is shown in Table 20.3. The spearhead antenna possesses an overall efficiency of more than 92.78% throughout the entire band of interest 3–4 GHz.

20.4 CONCLUSION

A miniaturized spearhead monopole antenna having an omnidirectional radiation pattern with a gain of 2.92 dBi has been developed. The antenna shows an excellent impedance matching (VSWR 1.107) at a resonant frequency of 3.5 GHz. The fractional bandwidth corresponding to a reflectance value less than –10 dB ranges from 3.18 to 4.06 GHz (25%), and thus, wideband performance is achieved. The presented antenna is miniaturized 85.33% in size with respect to that of a conventional circular microstrip antenna while maintaining excellent performance in terms of the reflection coefficient, bandwidth, and positive gain. It makes the proposed antenna a better candidate to be used in applications where the area is a constraint. In the future, the proposed antenna could be used as an element for MIMO or massive MIMO systems due to its compact size and simple geometry. The antenna is particularly suitable for the Wi-MAX and N78-band 5G wireless applications.

Table 20.3 Proposed antenna with recently published similar antennas

Refs.	Size (mm^2)	Features
[4]	75×80	Resonant frequency, f_r=3.42 GHz Fractional bandwidth FBW=3.15–3.67 GHz: 15.2% (−14 dB) & 2.90–3.80 GHz: 26.3% (−10 dB) Gain=8.2 dBi
[5]	65×65	f_r=3.50 GHz FBW=3.12–3.90 GHz: 22.2% (VSWR<1.5) Gain=8.1+0.3 dBi
[26]	78×44.6	f_r=8.06 GHz FBW=2.63–10.86 GHz: 122.02% (−10 dB) Gain=2.75 dBi
[17]	66.4×66.4	f_r=2.50 GHz FBW=1.81–3.0 GHz: 48.98% (−10 dB) Gain=7.16 dBi
[15]	19.25×10.5	f_{r1}=2.35, f_{r2}=3.6 FBW=2.23–2.35 GHz (LTE): 5.1%, & 3.3–3.6 GHz (5G): 8.33% Gain=1.05 dBi (LTE), 0.63 dBi (5G)
[16]	30×45	f_r=3.5 GHz FBW=3.40–3.60 GHz (5G): 5.7% Gain=7.95 dBi
This work	13×26	f_r=3.52 GHz FBW=3.18–4.06 (−10 dB): 25% (−10 dB) FBW=3.32–3.70 GHz: 10.79% (−14 dB) Gain=2.92 dBi

REFERENCES

1. R. S. Kshetrimayum, An introduction to UWB communication systems, *IEEE Potentials*, vol. 28, no. 2, pp. 9–13, Mar–Apr 2009, doi: 10.1109/MPOT.2009.931847.
2. Fedral Communication Commission, First order and report: Revision of part 15 of the Commission's rules regarding UWB transmission systems, 14 Feb 2002.
3. M. J. A. and Z. N. Chen, A wide-band shorted planar monopole with bevel, *IEEE Transactions on Antennas Propagation*, vol. 51, pp. 901–903, 2003.
4. Y. Chen, Q. Liu, Y. Zhang, H. Li and W. Zong, Design of a compact base station antenna for 5G N78-band application, *IEEE 3rd International Conference on Electronic Information and Communication Technology (ICEICT)*, pp. 229–231, 2020, doi: 10.1109/ICEICT51264.2020.9334286.
5. Y. Li, Z. Zhao, Z. Tang and Y. Yin, Differentially-fed, wideband dual-polarized filtering antenna with novel feeding structure for 5G sub-6 GHz base station applications, *IEEE Access*, vol. 7, pp. 184718–184725, 2019, doi: 10.1109/ACCESS.2019.2960885.
6. Q. Hua et al., A novel compact quadruple-band indoor base station antenna for 2G/3G/4G/5G systems, *IEEE Access*, vol. 7, pp. 151350–151358, 2019, DOI:10.1109/ACCESS.2019.2947778.

7. Y. Zhu, Y. Chen and S. Yang, Integration of 5G rectangular MIMO antenna array and GSM antenna for dual-band base station applications, *IEEE Access*, vol. 8, pp. 63175–63187, 2020, doi: 10.1109/ACCESS.2020.2984246.
8. S. Wen and Y. Dong, A low-profile wideband antenna with monopole-like radiation characteristics for 4G/5G indoor micro base station application, *IEEE Antennas and Wireless Propagation Letters*, vol. 19, no. 12, pp. 2305–2309, Dec 2020, doi: 10.1109/LAWP.2020.3030968.
9. M. R. Hasan, M. A. Riheen, P. Sekhar and T. Karacolak, Compact CPW-fed circular patch flexible antenna for super-wideband applications, *IET Microwaves, Antennas & Propagation*, vol. 14, pp. 1069–1073, 2020.
10. S. X. Ta, D. M. Nguyen, K. K. Nguyen, C. N. Dao and N. NguyenTrong, Wideband differentially-fed dual-polarized antenna for existing and sub-6 GHz 5G communications, *IEEE Antennas and Wireless Propagation Letters*, in press.
11. G. K. S. Pandey, H. S. Bharti, P. K. Meshram and M. Kumar, Design and analysis of Ψ-shaped UWB antenna with dual band notched characteristics, *Wireless Personal Communications*, vol. 89, pp. 79–92, 2016.
12. G. M. and S. Sahu, Compact circular patch UWB antenna with wlan band notch characteristics, *Microwave and Optical Technology Letters*, vol. 58, pp. 1068–1073, 2016.
13. G. K. Pandey, H. S. Singh, P. K. Bharti and M. K. Meshram, Design and analysis of multiband notched pitcher-shaped UWB antenna, *International Journal of RF and Microwave Computer-Aided Engineering*, vol. 25, no. 7, pp. 601–609, Sept 2015.
14. W. Xiao, T. Mei, Y. Lan, Y. Wu, R. Xu and Y. Xu, Triple band notched UWB monopole antenna on ultra-thin liquid crystal polymer based on ESCSRR, *Electronics Letters*, vol. 53, pp. 57–58, 2017.
15. S. L. Gunamony, J. B. Gnanadhas and D. E. Lawrence, Design and investigation of a miniaturized single-layer ACS-fed dual band antenna for LTE and 5G applications, *Journal of Electromagnetic Engineering and Science*, vol. 20, no. 3, pp. 213–220, July 2020, doi: 10.26866/jees.2020.20.3.213.
16. A. B. Sahoo, N. Patnaik, A. Ravi, S. Behera and B. B. Mangaraj, Design of a miniaturized circular microstrip patch antenna for 5G applications, *International Conference on Emerging Trends in Information Technology and Engineering (ic-ETITE)*, pp. 1–4, 2020, doi: 10.1109/ic-ETITE47903.2020.374.
17. A. Varshney, N. Cholake and V. Sharma, Low-cost ELC-UWB fan-shaped antenna using parasitic SRR triplet for ISM band and PCS applications, *International Journal of Electronics Letters(Online)*, 10 Aug 2021, doi: 10.1080/21681724.2021.1966655.
18. C. A. Balanis, *Antenna Theory Analysis and Design*, John Wiley & Sons, India, 2nd edition, reprints, 2009.
19. C. Li, K. Zhu, L. Li, Y.-M. Cai and C.-H. Liang, Design of electrically small metamaterial antenna with ELC and EBG loading, *IEEE Antennas and Wireless Propagation Letters*, vol. 12, pp. 678–681, May 2013, doi: 10.1109/LAWP.2013.2264099.
20. A. Varshney and V. Sharma, Tri-blade table fan shaped ultra-wideband microstrip antenna using parasitic SRR triplet, Australia Patent 2021101898, 19 May 2021.

21. A. Varshney, V. sharma and N. Sharma, A low cost UWB windmill-shaped antenna using CSRR for industrial and society applications, Australia Patent 2021103794, 11 Aug 2021.
22. N. Shaik and P. K. Malik, A comprehensive survey 5G wireless communication systems: Open issues, research challenges, channel estimation, multi carrier modulation and 5G applications, *Multimedia Tools and Applications*, 2021, https://doi.org/10.1007/s11042-021-11128–z.
23. P. Tiwari and P. K. Malik, Wide band micro-strip antenna design for higher "X" band, *International Journal of e-Collaboration (IJeC)*, vol. 17, no. 4, pp. 60–74, 2021, http://doi.org/10.4018/IJeC.2021100105, ISSN: 1548-3673.
24. D. S. Wadhwa, P. K. Malik and J. S. Khinda, High gain antenna for n260- & n261-bands and augmentation in bandwidth for mm-wave range by patch current diversions, *World Journal of Engineering*, 2021, https://doi.org/10.1108/WJE-03-2021-0133, ISSN: 1708-5284.
25. A. Varshney, V. Sharma and A. Srivastava, A novel simplified equivalent modeling method for microstrip line interconnects, India Patent 202111019468, 7 May 2021.
26. L.-C. Tsai, A ultra-wideband antenna with dual-band bandnotch filters, *Microwave and Optical Technology Letters*, vol. 59, no. 8, pp. 1861–1866, 27 May 2017, doi: 10.1002/mop.30639.

Chapter 21

Printed SIW cavity-backed slot antenna

T. Shanmuganantham
Pondicherry University

Nanda Kumar M.
Sreenidhi Institute of Science and Technology

CONTENTS

21.1 INTRODUCTION

The millimeter-wave [1,2] frequency ranges from 30 to 300 GHz and is needed to meet the needs of the current day-to-day situations. During the past decade, millimeter waves (MMWs) saw a tremendous rise in popularity due to the rise in academic, industrial, and personal use. In the millimeter-wave frequency (MMW), the millimeter communication application is an unlicensed band and frequency is in the range of 57–64 GHz [2–5].

Since the antennas function in a line of sight, a larger number of people can use this frequency at the same time, even if interference is minimal. When compared to other frequencies, the frequency reuse of 60 GHz is higher, as shown in Figure 21.1. In 60 GHz, radio terminals operate on the same frequency in a favorable configuration due to the narrow beam and frequency reuse. This should reduce the possibility of receiver interference and this frequency is suitable for short-range use between unlicensed devices [6,7].

The oxygen absorption per kilometer is represented in Figure 21.2. At 60 GHz, 98% of the energy is absorbed by oxygen and the frequency spectrum would be fit for short-range, high-data rates (multi-gigabits per second), and broadband applications such as a computer-to-computer

DOI: 10.1201/9781003347057-21

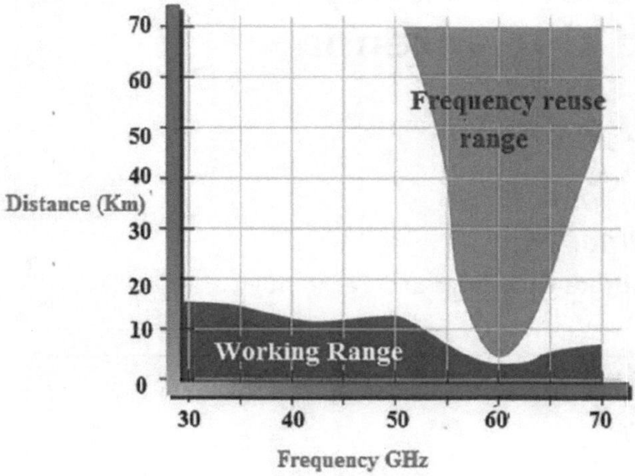

Figure 21.1 Frequency reuse.

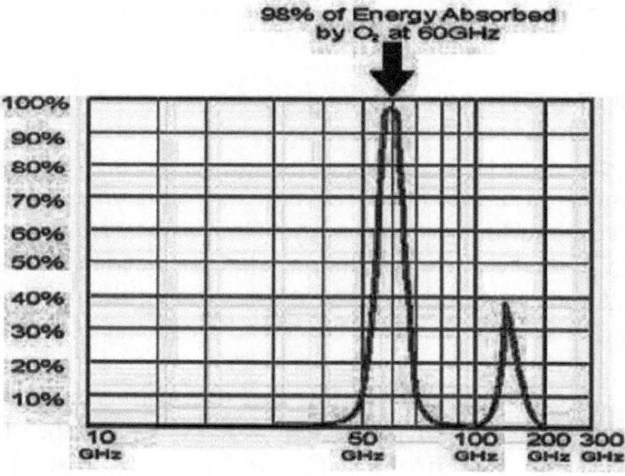

Figure 21.2 Atmospheric absorption per kilometer.

applications. The user doesn't need permission from the FCC to operate in this particular spectrum and different counties follow a different range of frequencies, as shown in Figure 21.3 and it ranges from 3 to 9 GHz. Due to high bandwidth and high-data rates, this band can be used for WLAN, GiFi, WPAN, WiFi, automotive applications like opening a car door, closing and opening doors, etc. and the minimum required bandwidth to fulfill all applications mentioned above is 3 GHz [8,9].

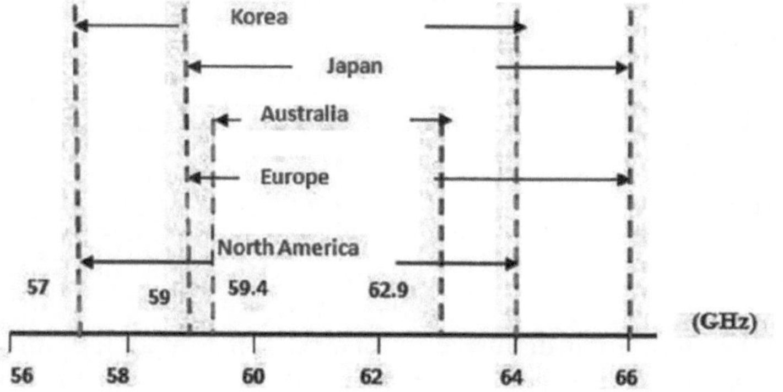

Figure 21.3 World wide spectrum available for 60 GHz.

The 60 GHz band is of particular interest to academics and academicians because of its enormous capacity and because the wireless system plays a major role in gigabit fidelity (GiFi) [10–12]. Short-range applications benefit from the 60 GHz communication, but its range is limited by high levels of oxygen absorption and rain attenuation in 60 GHz propagation characteristics [13,14]. The short-range operation makes this a very secure and noise-free method of communication [15–19].

The use of high frequencies in one of the substrate-integrated circuits (SICs) is the laminated waveguide (also known as a post-wall waveguide), two rows of vias etched through the substrate, which can be used for active or passive components and antennas. Lens and horn antennas are commonly employed for MMW applications, however, their enormous size and high cost make them unsuitable for lower-cost applications. For mmW applications, slotted antennas have emerged as a viable alternative because of their smaller size, moderate gain, and wide bandwidth [20–22].

The spacing and diameter are extremely important parameters in SIW design and choose as $d \leq \frac{\lambda_g}{5}$ and $s \leq 2d$ and SIW width is stated in below [23–27]:

$$a_R = a_S - \frac{d^2}{0.95\,s} \tag{21.1}$$

For whatever reason, the equation above doesn't include the diameter (d) to a width (a_R) ratio, which can result in an incorrect answer.

$$a_{\text{siw}} = a_R + 1.08 * \frac{d^2}{s} - 0.1 * \frac{d^2}{a_{\text{SIW}}} \tag{21.2}$$

Generalized equation of the slot as given below:

Width (W):

$$\frac{\lambda}{4} \le W \le \frac{\lambda}{2} \tag{21.3}$$

Length (L):

$$L \le W \tag{21.4}$$

Some of the literature is discussed as follows. Tomas et al. [20] introduced an array-based microstrip patch antenna fed by microstrip for high-gain applications with a 6×8 array and an impedance bandwidth of 1.1 GHz and gain of 21.6 dBi. P. Srivastava et al. [21] proposed an SIW feed antipodal linear-tapered slot antenna (ALTSA) for millimeter wireless application with a resonant frequency of 60 GHz having a bandwidth of 1.5 GHz, a gain of 16.3 dBi, and a slot loaded with different dielectric shapes for gain improvement and those are rectangular, triangular, and exponential.

It was proposed by K gong and his team [22] for millimeter wireless applications. They looked at a wide width slot to meet the operating band of millimeter wireless applications. The impedance bandwidth is 3 GHz and the gain is 6 dBi [23] making an exponentially tapered slot antenna for 60 GHz applications that has an impedance bandwidth of 0.8 GHz and it can gain 11.2 dBi. It has a frequency range of 60 GHz, and it has a gain of 10 dBi [23] made an antenna for V-band applications that has an impedance bandwidth of 1.33 GHz, and it can gain 11.2 dBi, too.

MNK [24] proposed an SIW-based slot antenna for 60 GHz applications. The shape of the antenna is an h-shaped slot and its size is 14×8×0.381. The bandwidth of the antenna is 4.1 GHz and the gain at 60 GHz is 6.2 dBi. Shanmuganantham et al. [25] introduced an antenna for 60 GHz applications with a size of 14×8.4×0.381 mm^3 and an occupied bandwidth of 3.5 GHz. Kumar et al. [26–29] introduced a cavity-backed antenna for 60 GHz application with a bandwidth of 4.1 GHz and a gain of 6 dBi at a 60 GHz frequency band [30–32].

In this article, the geometry of the antenna and its design flow is discussed in Section 21.2. Section 21.3 gives a clear overview of the results described; these are simulated, measured results, and finally, Section 21.4 is the conclusion of the chapter.

21.2 PROPOSED ANTENNA STRUCTURE

Although SIW's TE10 distribution mode is the most common, TE110 is the most common distribution mode for cavity-backed SIW. Equation 21.5 shows the distribution of the normalized electric field.

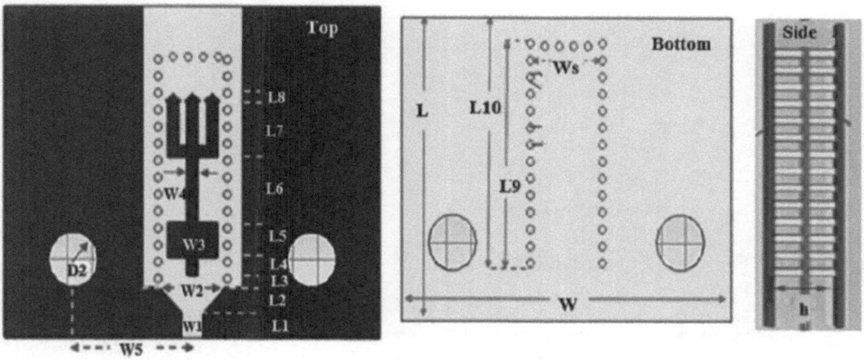

Figure 21.4 Evolution of the antenna projected.

Table 21.1 Optimized parameters

Parameters	*Dimension (mm)*	*Parameters*	*Dimension (mm)*
W1/W2/W3/W4	0.85/2.8/2/0.35	L1 = L2	0.85
W	14	L3	0.4
W5	4.55	L4	0.65
D2/D/S	1/0.3/0.55	L5	1.25
W-S/W-g	2.85/3.98	L6/L7/L8	2.3/2/0.2
L9/L10	8.2/10	W4	0.38

$$E_{zn} = \sin\left(\frac{\pi x}{L}\right) * \sin\left(\frac{\pi y}{W}\right) \tag{21.5}$$

The proposed design side, top and bottom views are represented in Figure 21.4. Microstrip feed with input impedance 50 ohms is used and implemented for 60 GHz applications. The proposed design integrates with Rogers substrate material. The permeability is 2.2, and the thickness chosen is 0.381 mm. The slot is etched on the top side of a substrate, and the shape chosen first is a rectangular slot. Their width and height equations are represented in equations 21.3 and 21.4 and finally converted into Neptune shape for improving the bandwidth.

The standard equations used in microstrip and tapering of microstrip are introduced to match the impedance bandwidth of the antenna. The parameters used in this design are mentioned in Table 21.1, and the side view states that holes are interlinked with two ground planes via the substrate. The top and bottom views of the fabricated prototype are revealed in Figure 21.5; 1.85 mm diameter female connector is used for testing, and a flow chart is used to implement the proposed design as shown in Figure 21.6. The diameter of the holes is 0.3 mm for this design and slight fluctuations occur while

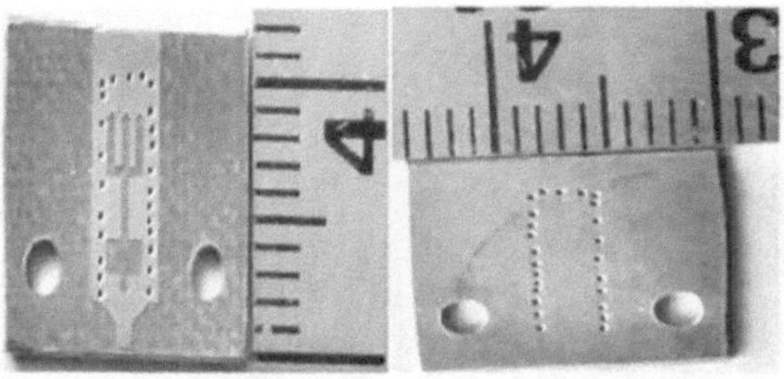

Figure 21.5 Fabricated prototype of an antenna.

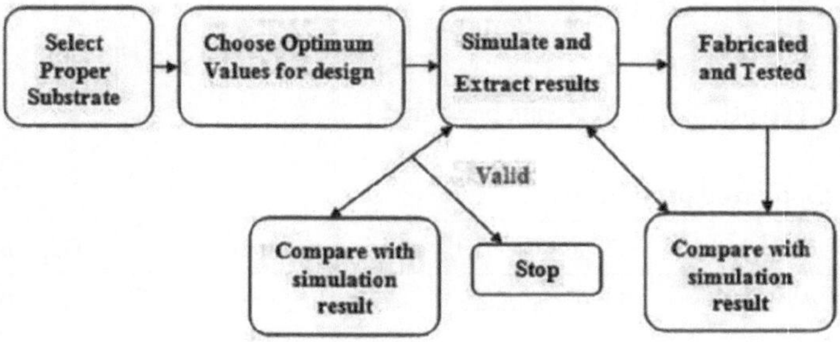

Figure 21.6 Flowchart of proposed design.

inserting the holes of the fabricated prototype and they may influence the slight variations in the bandwidth and reflection coefficient.

Steps to follow and implement the proposed antenna are as follows:

Step 1: Introduce the SIW cavity-backed antenna by microstrip followed by microstrip tapering
Step 2: Introduce two holes in outside the cavity to hold connector
Step 3: Rectangular slot is introduced on the top of the SIW cavity
Step 4: Rectangular slot is converted to Neptune-based slot antenna for bandwidth improvement

21.2.1 Microstrip design equations

The width (W) and height (h) of the microstrip is [24–29]

$$\frac{W}{h} = \begin{cases} \dfrac{2}{\pi}\left\{ \begin{array}{c} a - 1 - \ln(2a-1) + \cdots \\ \dfrac{\varepsilon_r - 1}{2\varepsilon_r}\left[\ln(a - 1 + 0.39) - \dfrac{0.1}{\varepsilon_r}\right] \end{array} \right\} & \dfrac{W}{h} > 2 \\ \dfrac{8e^a}{e^a - 2}, & W/h < 2 \end{cases} \tag{21.6}$$

where

$$a = \frac{377\pi}{2Z_o \sqrt[2]{\varepsilon_r}} \tag{21.7}$$

$$b = \frac{Z_o}{60\sqrt[2]{\varepsilon_r}} \sqrt[2]{\frac{\sqrt[2]{\varepsilon_r}+1}{2} + \frac{\varepsilon_r - 1}{\varepsilon_r + 1}\left(0.23 + \frac{0.11}{\varepsilon_r}\right)} \tag{21.8}$$

The length of the microstrip is [8,9,26–29,33]

$$L_m = n^* \lambda_g; \qquad n = 1,3,5,7\ldots \tag{21.9}$$

21.3 RESULTS AND DISCUSSION

The simulation results of $S11$ and VSWR are represented in Figure 21.7 and pragmatic in the frequency range of 57–64 GHz applications. Figure 21.7a indicates the reflection coefficient versus frequency of the proposed design and resonates two frequencies; those values are 57.53 and 60 GHz. The bandwidth is 4.35 GHz with reference of S_{11}=–10 dB and ranging from 57.05 to 61.4 GHz. Figure 21.7b represents the VSWR of antenna matched a impedance bandwidth with VSWR=2 reference line and VSWR, S_{11} values at resonant frequencies is –19.1 dB, 1.25 at 57.55 GHz and –25.869 dB, 1.11 at 60 GHz.

Figure 21.8 shows the S_{11} analysis for various W_4 values, and four different W4 values are taken into account while assessing the S_{11} results: 0.35, 0.36, 0.38, and 0.4 mm. The resonant frequency, bandwidth and reflection coefficient will be affected by the change in W_4.

The surface current at two frequencies is represented in Figure 21.9 and it is observed that the starting of the slot and the top section of the slot has more current flow at 60 GHz. The maximum current distribution is indicated with red color. The comparison of the proposed design with existing literature is shown in Table 21.2. The parameters considered for comparison are size, gain, and bandwidth. The substrate used for design is compared to previous literature. The proposed design has an average bandwidth that is 1.5 times greater while also being smaller in size.

Figure 10a and b represents the comparison of simulated, measured with reflection coefficient, and VSWR results. A small deviation is observed with the simulation and measurement results due to a small deviation in creating holes in the fabricating prototype. The measurement result as impedance bandwidth of 4.15 GHz with respect to S_{11} = –10 dB and VSWR=2 and ranges from 57.05 to 62 GHz, produces a two resonant frequencies those are 58.2, 59.8 GHz. The clear comparison of S_{11} and the VSWR results is described in Table 21.2.

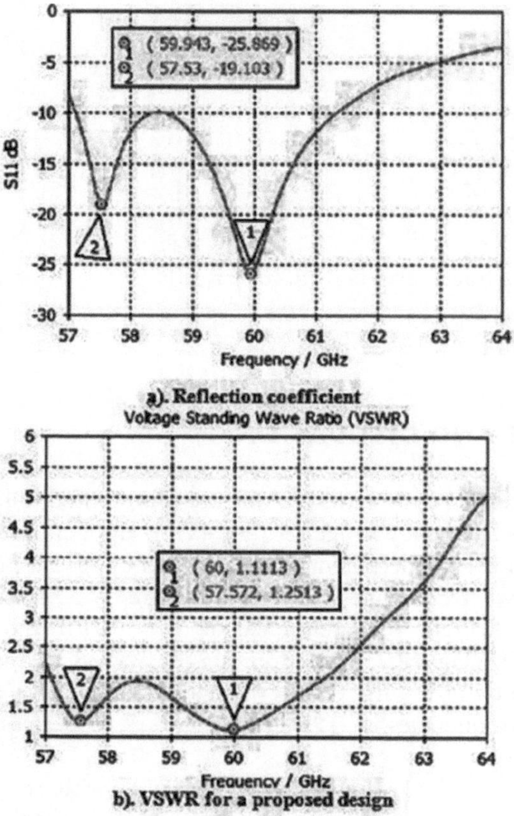

Figure 21.7 Proposed antenna results.

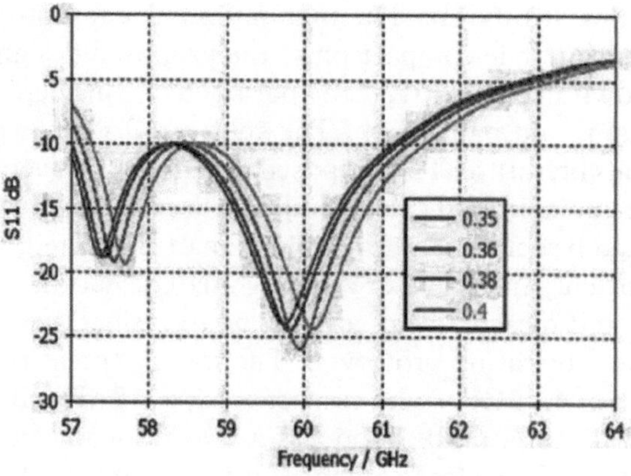

Figure 21.8 S_{11} for different values of W_4.

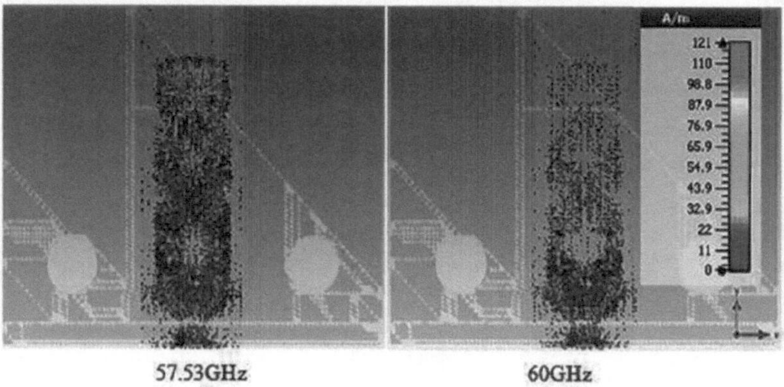

Figure 21.9 Surface current distributions of the proposed design.

Table 21.2 Comparison with existing literature

S. no.	*Ref.*	*Antenna size (mm³)*	*Substrate used in the design*	*Gain (dB)*	*Bandwidth (GHz)*
1	[20]	34×24.75×0.85	Rogers substrate $\varepsilon_r = 2.2$	21.6	1.1
2	[21]	44.61×9.93×0.381	Rogers substrate $\varepsilon_r = 2.2$	13.7	3
3	[22]	25×16×0.635	Rogers substrate $\varepsilon_r = 2.2$	6	3
4	[23]	33.5×8×0.787	Rogers substrate $\varepsilon_r = 2.2$	10	0.8
5	[26]	14×8.4×0.381	Rogers substrate $\varepsilon_r = 2.2$	6.65	3.5
6	[24]	14×8×0.381	Rogers substrate $\varepsilon_r = 2.2$	6	4.1
7	[25]	14×8.4×0.381	Rogers substrate $\varepsilon_r = 2.2$	6.2	3.64
8	Proposed	15×10.7×0.381	Rogers substrate $\varepsilon_r = 2.2$	8	4.35

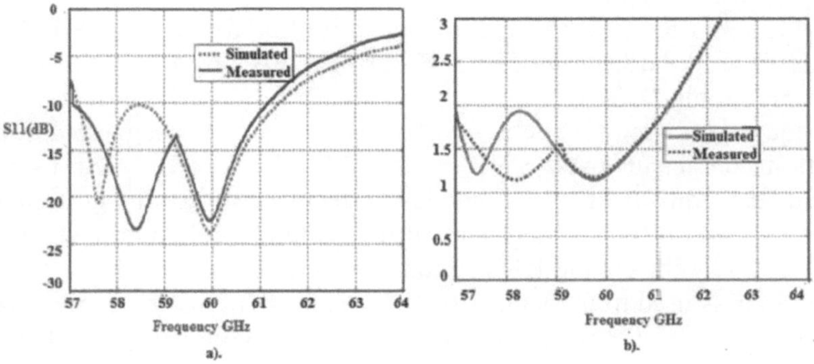

Figure 21.10 Simulated and measured S_{11}/VSWR.

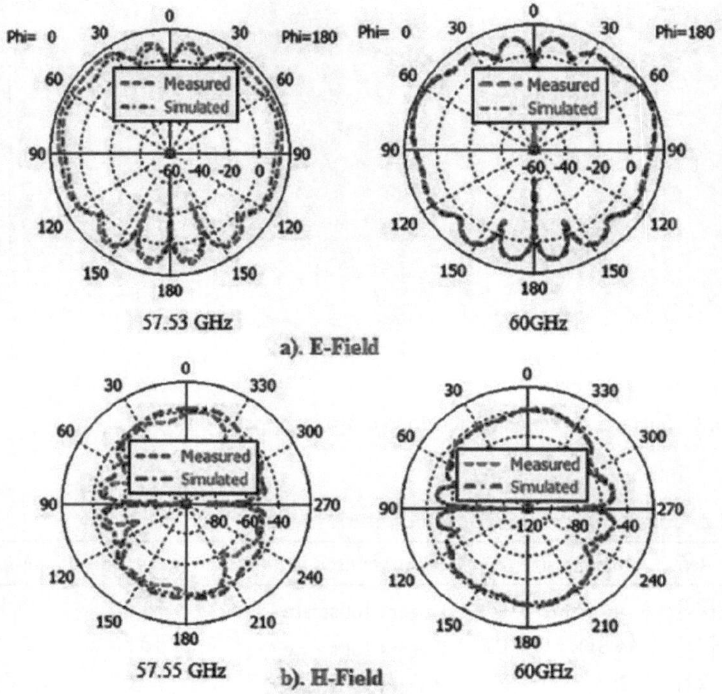

Figure 21.11 Reflection coefficient of a proposed antenna.

The far-field emission patterns of the proposed antenna at resonant frequencies are represented in Figure 21.11. Here, the red and green colors indicate the result of the measured, simulation results and bi-directional radiation patterns at the E-plane as well as H-plane (two resonant frequencies) it are observed. This is well suited for wireless applications. The simulation and measurement results of the proposed antenna gain values are represented in Figure 21.12. The measured and simulation gain values are 5.48, 5.45 dBi at 57.54 GHz, 6.95, 7 dBi at 57.5 GHz, 7.8, 8.1 dBi at 58 GHz, 8.55, 8.7 dBi at 59 GHz, 8.05, 8 dBi at 60 GHz, and 7.768, 7.6 dBi at 61 GHz (Table 21.3).

An equivalent circuit model is constructed to match the suggested design reflection coefficient and VSWR results. MATLAB software is used to evaluate the equivalent circuit model's effectiveness. Resonant frequencies of 57.5 and 60 GHz are generated by the first RLC and are shown in Figure 21.13 by the second RLC circuit, respectively.

MATLAB is used to implement the RLC equivalent model for the analysis of S_{11} and VSWR. Listed below is the conventional formula for R. L, C. Simulation results are shown in Figures 21.14 and 21.15 and are compared to the similar model with reflection and VSWR.

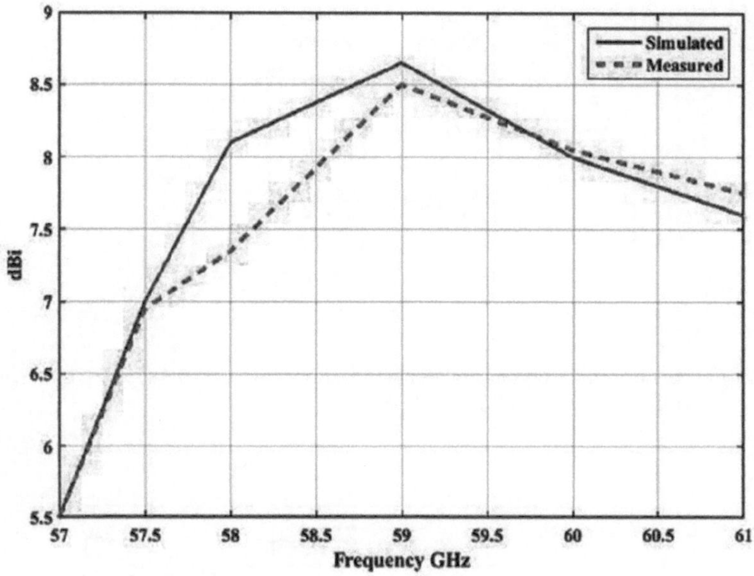

Figure 21.12 Far-field gain of the proposed design.

Table 21.3 Comparison of simulated, measured results

S. no.	*Method*	*Resonant frequency (GHz)*	*Reflection coefficient (dB)*	*VSWR*
1	SR	57.523	−19.1	1.2513
	MR	58.19	−23.52	1.143
	ER	57.52	−19.1	1.2513
2	SR	60	−25.869	1.113
	MR	59.78	−22.4	1.53
	ER	60	−25.869	1.113

$$R = 2Z_o\left[\frac{1}{|S_{11}|^2} - 1\right]\Omega \tag{21.10}$$

$$C = \frac{0.25 f_c}{\pi\left(f_o^2 - f_c^2\right)}\,\text{pF} \tag{21.11}$$

$$L = \frac{1}{4\pi^2 f_c^2 C}\,nH \tag{2.12}$$

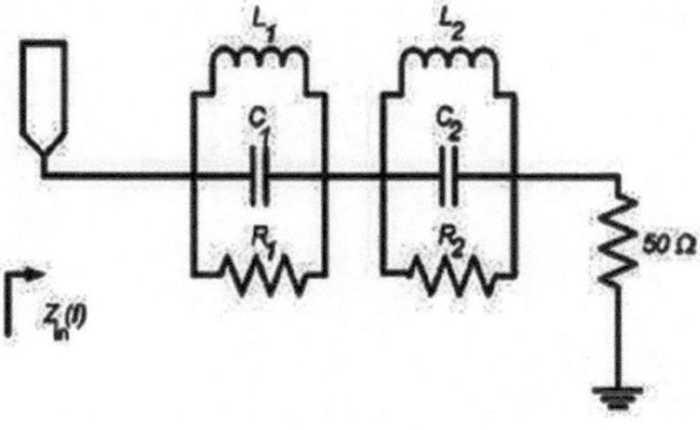

Figure 21.13 Equivalent circuit model (R1 = 18 Ω, R2 = 28Ω, L1 = 0.6455 ρH, L2 = 3.14 ρH, C1 = 11.81 ρF and C2 = 2.42 ρF).

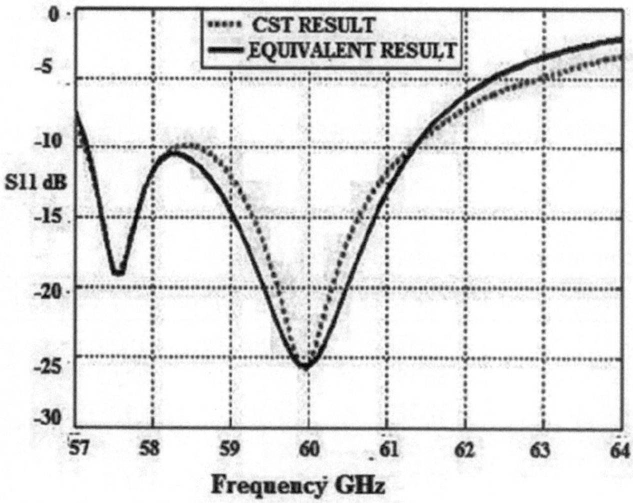

Figure 21.14 Reflection coefficient of the proposed design.

21.4 CONCLUSION

In this chapter, a SIW cavity-backed Neptune-shaped slot has been introduced for millimeter-wave wireless applications, that is 60 GHz applications. VSWR and far-field patterns, as well as reflection coefficient and gain, are examined in relation to antennas. In this study, a planned antenna was built, tested, and was found to be nearly identical to the actual results. Analyzing RLC parallel equivalent model results in confirmation of the design's S11 and VSWR. The impedance bandwidth of the simulated,

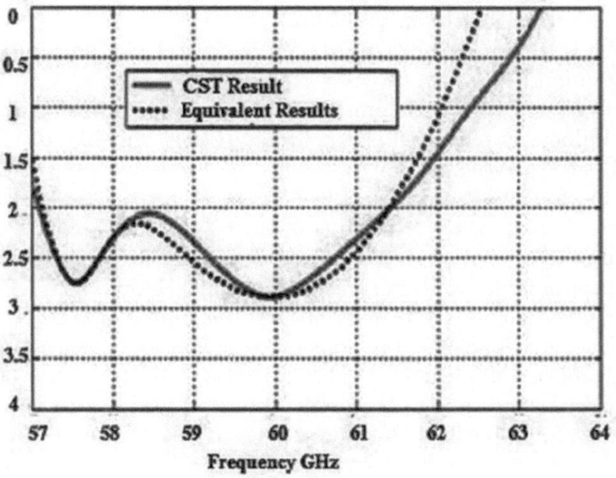

Figure 21.15 Voltage standing wave ratio of the proposed design.

measured, and equivalent model is 4.35, 4.15, 4.35 GHz, and the fabricated antenna resonates at two frequencies are 58.19 and 59.9 GHz, respectively. This antenna is suitable for WLAN, WPAN, GIFI, and short-range applications.

REFERENCES

1. H H Meinel, Millimeter wave applications and technology trends, *Annales des Telecommunication*, vol. 47, no. 11–12, pp. 456–468, Nov. 1992.
2. A Bakhtafrooz and A Borji, Novel two-layer MMW slot-array antennas based on substrate-integrated-waveguides, *Progress in PIER*, vol. 109, pp. 475–491, 2010.
3. A Lamminen, J Saily and A R Vimpari, 60 GHz Patch Ante's & arrays on LTCC with embedded cavity sub's, *IEEE Transactions on Antennas & Propgation*, vol. 56, no. 9, pp. 2865–2874, Sep. 2008.
4. L Wang, Y X Guo and W X Sheng, Wide-band high gain 60 GHz LTCC L probe patch ante. Array with a soft surface, *IEEE Transactions on Antennas & Propagation*, vol. 61, no. 4, pp. 1802–1809, Apr. 2013.
5. Y Miura, J Hirokawa, Y Shibuya, M Ando and G Yoshida, Double layer full corporate feed hollow WG slot array ante. in the 60 GHz band, *IEEE Transactions on Antennas & Propagation*, vol. 59, no. 8, pp. 2844–2854, Aug. 2011.
6. G Rahul and R Kumaralingam, The 60 GHz wireless network infrastructure, white paper, HCL Technologies, 2011.
7. K C Huang and Z Wang, *Mmw Communication Systems*, John-Wiley & Sons, New Jersey, vol. 29, 2011.

8. C Koh, The benefits of 60 GHz un-licensed wireless comm's, Whitepaper, YDI Wireless Falls Church, VA, 2002.
9. https://www.rfglobalnet.com/doc/fixed-wireless-communications-at-60GHz-unique–0001
10. D D Grieg and H F Engelmann, Micro-strip A new transmission technique for the klilomegacycle-range, *Proceedings of IRE*, vol. 40, no. 12, pp. 1644–1650, 1952.
11. F Assadourian and Rimai, Simplified theory of micro-strip-transmission systems, *Proceedings of the IRE*, vol. 40, no. 12, pp. 1651–1657, 1952
12. M Arditi, Characteristics & applications of micro-strip for microwaves, *IEEE Transactions on Microwave Circuits*, pp. 31–56, 1955.
13. S K Yong, P Xia and A V Garcia, *60 GHz Technology for Gbps WLAN and WPAN*, 1st ed., John Wiley & Sons Ltd, Chichester, UK, 2011.
14. T S Rappaport, J N Murdock and F, Gutierrez, State of the art in 60-GHz integrated circuits and systems for wireless communications, *IEEE Proceedings*, vol. 99, no. 8, pp. 1390–1436, Aug. 2011.
15. P Smulders, Exploiting the 60 GHz band for local wireless multimedia access: Prospects and future directions, *IEEE Communication Magazine*, vol. 2, no. 1, pp. 140–147, Jan. 2002.
16. K C Huang and D J Edwards, *Millimetre Wave Antennas for Gigabit Wireless Communications*, 1st ed., John Wiley, Chichester, UK, 2008.
17. M Tomas, J Lacik, J Puskely and Z Raida, Design of aperture coupled microstrip patch antenna array fed by SIW for 60 GHz band, *IET Microwaves Antennas and Propagation*, vol. 10, no. 3, pp. 288–292, 2016.
18. P Shrivastava and T Rama Rao, 60 GHz radio link characteristic studies in hallway environment using antipodal linear tapered slot antenna, *IET Microwaves Antennas and Propagation*, vol. 9, no. 15, pp. 1793–1802, 2015.
19. K Gong, Z N Chen, X Qing, P Chen and W Hong, Substrate integrated waveguide cavity-backed wide slot antenna for 60-GHz bands, *IEEE Transactions on Antennas and Propagation*, vol. 60, no. 12, pp. 6023–6026, 2012.
20. S Ramesh and T Rama Rao, *Planar High Gain Dielectric Loaded Exponentially TSA for Millimeter Wave Wireless Communications*, Wireless Press Communication (Springer), pp. 3179–3192, June 2015.
21. Z Zhang, X Cao, J Gao, S Li and J Han, Broadband SIW cavity-backed slot antenna for endfire applications, *IEEE Transactions on Antennas and Propagation*, vol. 17, no. 8, pp. 1271–1275, 2018.
22. M Farashahi, E Z Jahromi and R Basiri, A compact wideband circularly polarized SIW horn antenna for K band applications, *International Journal of Electronics and Communications*, vol. 99, pp. 376–383, 2019.
23. Z Xu, J Liu and Y Li, Gain enhanced SIW cavity backed slot antenna by using TE_{410} mode resonance, *International Journal of Electronics and Communications*, vol. 98, pp. 68–73, 2019.
24. M Nanda Kumar and T Shanmuganantham, Broad-band H-spaced head shaped slot with SIW based antenna for 60 GHz wireless communication applications, *Microwave and Optical Technology letters (MOTL-Wiley)*, vol. 61, no. 8, pp. 1911–1916, 2019.

25. M Nanda Kumar and T Shanmuganantham, Substrate integrated waveguide based slot antenna for 60 GHz wireless applications, *Microwave and Optical Technology Letters (MOTL-Wiley)*, vol. 61, no. 8, pp. 1945–1951, 2019.
26. M Nanda Kumar and T Shanmuganantham, Broad band I shaped SIW slot antenna for V-band Applications, *Applied Computational Electromagnetic Society (ACES)*, vol. 34, no. 11, Nov. 2019,
27. T Shanmugnantham and M Nanda Kumar, V-band Substrate integrated waveguide cavity backed slot antenna for millimeter-wave wireless applications, *Indian Conference on Antennas and Propagation (InCAP 2019)*, Dec. 2019.
28. M Nanda Kumar and T Shanmuganantham, Division shaped SIW slot antenna for millimeter wirelesss/automotive radar applications, *Journal of Computers and Electrical Engineering*, vol. 71, pp. 667–675, 2018.
29. M Nanda Kumar and T Shanmugannatham, Back to back Pi-shaped slot with SIW cavity-backed antenna for 60 GHz applications, *International Journal of Microwave and Optical Technology*, vol. 14, no. 6, pp. 371–380, Nov. 2019.
30. A Kaur and P K Malik, Multiband elliptical patch fractal and defected ground structures microstrip patch antenna for wireless applications, *Progress in Electromagnetics Research B*, vol. 91, pp. 157–173, 2021. doi: 10.2528/PIERB20102704, ISSN: 1937-6472.
31. N Shaik and P K Malik, A retrospection of channel estimation techniques for 5G wireless communications: opportunities and challenges, *International Journal of Advanced Science and Technology*, vol. 29, no. 05, pp. 8469–8479, June 2020, ISSN: 2005-4238.
32. P K Malik and M Singh, Multiple bandwidth design of micro strip antenna for future wireless communication, *International Journal of Recent Technology and Engineering*, vol 8, no. 2, pp. 5135–5138, July 2019. doi: 10.35940/ijrte.B2871.078219, ISSN: 2277-3878.
33. D Lockiev and D Peck, High-data-rate MMw radio's, *IEEE Microwave Magazine*, vol. 10, no. 5, pp. 75–88, 2009.

Index